Biology of
Oral Cancer
Key Apoptotic Regulators

T0225394

Biology of
Oral Cancer
Key Apoptotic Regulators

Prakash S. Bisen
Zakir Khan
Saurabh Bundela

CRC Press
Taylor & Francis Group
Boca Raton London New York

CRC Press is an imprint of the
Taylor & Francis Group, an **informa** business

CRC Press
Taylor & Francis Group
6000 Broken Sound Parkway NW, Suite 300
Boca Raton, FL 33487-2742

First issued in paperback 2017

© 2014 by Taylor & Francis Group, LLC
CRC Press is an imprint of Taylor & Francis Group, an Informa business

No claim to original U.S. Government works
Version Date: 20130520

ISBN 13: 978-1-138-07676-1 (pbk)
ISBN 13: 978-1-4665-7558-5 (hbk)

Library of Congress Cataloging-in-Publication Data

Bisen, Prakash S., author.
    Biology of oral cancer : key apoptotic regulators / Prakash S. Bisen, Zakir Khan, Saurabh Bundela.
        p. ; cm.
    Includes bibliographical references and index.
    ISBN 978-1-4665-7558-5 (hardcover : alk. paper)
    I. Khan, Zakir, 1979- author. II. Bundela, Saurabh, 1977- author. III. Title.
    [DNLM: 1. Mouth Neoplasms. 2. Apoptosis--drug effects. 3. Apoptosis Regulatory Proteins. 4. Cell Proliferation--drug effects. 5. Cell Transformation, Neoplastic. WU 280]

    RC280.M6
    616.99′431--dc23   2013019703

Visit the Taylor & Francis Web site at
http://www.taylorandfrancis.com

and the CRC Press Web site at
http://www.crcpress.com

Dedication

We dedicate this work to the late Mr. Shitla Sahai, the son of the soil and great philanthropist, founder trustee of the Cancer Hospital and Research Institute, Gwalior, Madhya Pradesh, India. He realized various socioeconomic problems of the less privileged and the poor man of this region faced in the absence of proper cancer care treatment and vowed to start a cancer hospital in this underdeveloped region. He is the pioneer of the cancer institute started in May 1971, with his limited personal investments and modest resources. The institute now has full facilities of diagnosis and management of the disease under one roof with financial support from the state central government and enjoys the status of regional cancer research and treatment center for cancer treatment and research in India.

Contents

Preface

Human civilization has made phenomenal progress during the past couple of centuries. The sequencing of the human genome was a much celebrated event and was believed to be a panacea for all human disease. The availability of wealth of knowledge about molecular events related with any disease can be surely directly or indirectly attributed to the human genome sequencing; however, we have failed to translate this knowledge effectively to alleviate miseries caused by age-old diseases like cancer, stroke, and diabetes. The change in lifestyle and environmental factors have placed humans in a vulnerable position. Cancer is caused by the complex interaction of various internal (genetics) and external (carcinogens, radiation, and smoking) factors, and therefore the key to curing cancer lies in effective identification of these causal factors and their modulation to restore the normal state. The failure of existing treatment modalities for cancer is due to a lack of precise understanding of the interplay of these factors and related molecular events leading to growth and proliferation of cancer cells. Through this book, we have attempted to communicate our understanding about oral cancer, which has consistently ranked among the top 10 causes of cancer-related mortalities worldwide. The incidence and mortality rate for oral cancer is relatively greater in developing and underdeveloped nations than in developed nations, and one of the main reasons is the lack of awareness and medical infrastructure in these countries.

The book has been written as an attempt to spread awareness about cancer-causing factors, along with state-of-the-art medical options available for management and treatment of cancer, and fill the gap between the basic learner and advanced learner who want to learn cancer biology. The informal style of writing has been adopted with a purpose of reaching out to a wider audience whose life is directly or indirectly impacted by various types of cancers, including those of the oral cavity. This book would be indispensable for research/graduate students who want to understand molecular events like the role of apoptosis in causing oral cancer. Our thoughts about cancer have been captured in nine chapters of this book. Each chapter starts with a prologue to the concept, followed by a detailed discussion of the concept, and ends with a vision for the future approach and challenges. The content of every chapter is supported by illustrations for a better understanding of the concept discussed. A large compilation of references has been added at the end of each chapter.

The first two chapters, about cancer in general and oral cancer in particular, were written keeping in mind a broad audience. Chapter 1, "Cancer: A Worldwide Menace," is recommended to all inquisitive readers irrespective of academic or professional background; for instance, it would be resourceful for a lady with a nonscientific background who wants to know about risk factors, detection, and treatment options for various cancers, and thereby take control of her health along with that of others in her family. This chapter is also recommended for graduate students who would like to get an overview of cancer, before diving deep into specific problem areas in cancer research. Chapter 2, "Oral Cancer," throws light on various aspects of oral cancer, and was written with the intention of providing a detailed appraisal of aspects around the genesis and manifestation of oral cancer.

Programmed cell death or apoptosis plays a very important role in the maintenance of cellular homeostasis. Cancer cells are known to adapt various mechanisms by which they survive and thrive, irrespective of the presence of signals that are responsible for checking the growth of normal cells and maintaining cellular integrity. Evasion of cell death by the apoptotic process is considered one of the most elaborate survival mechanisms present in cancer cells. We dedicated the next six chapters (Chapters 3–8) to explain survival strategies adapted by cancer cells. Chapter 3, "Proliferative and Apoptotic Signaling in Oral Cancer," deals with various factors involved in proliferative and apoptotic regulation in oral cancer. p53 plays a protective role in normal cells and is known to regulate a host of genes/proteins involved in key cellular processes, including apoptosis, cell division, and replication. Chapter 4, "Apoptotic Regulations," explains apoptosis in detail, along with various regulatory modules of the apoptotic pathway. In Chapter 5, "Dynamics of p53 in Oral Cancer," we highlight various roles played by p53, and how its dysregulation is implicated in cancer growth. The therapeutic and diagnostic application of p53 and other molecules from the apoptotic pathway is explained in detail in Chapter 6, "Diagnostic and Therapeutic Potential of Apoptotic Marker." The survivin, a key anti-apoptotic protein, belongs to the IAP family, which is almost exclusively expressed in tumor tissues. The anti-apoptotic role of the survivin protein, and its regulation by various molecules, has been explained in Chapter 7, "Expression and Regulation of Survivin." In Chapter 8, "Therapeutics of Survivin," we discuss various therapeutic approaches designed to control or kill cancer cells by modulation of the survivin protein.

Most of the mortalities due to oral cancer can be attributed to detection of cancer during advanced stages, which essentially reduces the chances of survival by many folds. The effective and accurate detection of oral cancer in its early stage is much desired to reduce mortalities due to it. In the past couple of decades we have witnessed a phenomenal amount of work done in the field of cancer biology. Every individual work has proved as a dot, which taken together have helped us in creating a near-complete picture of cancer at a molecular level. In Chapter 9, "Molecular Diagnosis of Oral Cancer," we have briefly discussed the key molecular events contributing to oral cancer development and molecular techniques that can be used for the effective diagnosis of oral cancer.

We hope that this book will provide a solid foundation to students who wish to pursue research in the area of oral cancer. We have taken utmost care to include all relevant information about concepts discussed in this book; however, there is a chance we might have inadvertently missed some important information, for which we encourage and request all readers to send us their comments and suggestions for improvements in this book.

ABOUT THE BOOK

Cancer has consistently maintained its status as one of the top killers since time immemorial. Developed countries like the United States have observed a decline in cancer-related death during the past couple of years, which is an encouraging result of dedicated commitment from various stakeholders, including the research community, medical society, support groups, and policy makers. However, cancer statistics from the rest of the world are far from satisfactory, and millions of people are projected to acquire a form of cancer, and more than half of them are expected to die within 5 years from the time of first detection of cancer. According to the latest cancer statistics, oral cancer has become the topmost cause of death in males in Southeast Asia. Oral cancer, like its siblings, is caused by the complex interaction of multiple factors. This book is written with the objective to spread awareness among readers by highlighting factors responsible for causing cancers, including oral cancer. There are various molecular events that lead to the genesis and growth of oral cancer. The oral cavity is one of the most accessible sites of physical examination, which should ideally negate any chance of development of oral carcinogenesis; despite this, there is a large incidence rate of oral cancer, which points to the development of effective techniques for the detection of oral cancer in its nascent stages. This book discusses various detection techniques that leverage molecular events associated with oral carcinogenesis to effectively detect oral cancer. It is written with the hope that it will find its place as a reference book on oral cancer for students, teachers, researchers, and anyone who wants to understand oral cancer.

Acknowledgments

We express our gratitude to Dr. B.R. Shrivastav, director and management trustee, Cancer Hospital and Research Institute, Gwalior, and Mr. Basudev Dalmia, management trustee, Birla Hospital and Research Center, Gwalior, for their valuable guidance, encouragement, and extending all necessary facilities to complete the task smoothly.

We gratefully acknowledge Mr. Rakesh Singh Rathore, CEO, Vikrant Group of Institutions, Gwalior, for extending financial support for this project and help in various ways for completing this work without any hindrances.

We express our sincere thanks to Prof. G.B.K.S. Prasad, chairman, School of Studies in Biochemistry, Jiwaji University, Gwalior; Dr. Ram P. Tiwari and Dr. Anubhav Jain, Diagnostic Division, RFCL Ltd. (formerly Ranbaxy Fine Chemicals Ltd.); Avantor Performance Materials, New Delhi, India; and Dr. Ruchika Raghuvanshi, Mr. Bhagwan S. Sanodiya, Mr. Gulab S. Thakur, Mr. Rakesh Baghel, and Mr. Rohit Sharma, Tropilite Foods Pvt. Ltd., Gwalior, for their valuable assistance in preparation of this book.

Our sincere thanks to Mr. Devendra Singh, Mr. Avinash Dubey, and Mr. Rahul Jha for all computational work for preparing the book in a presentable form. The cover page of the book was designed courtesy of Mr. Avinash Dubey.

Thanks are also due to Mr. Michael Slaughter, Ms. Kari Budyk, and Michele Smith, CRC Press–Taylor & Francis Group, Novato, California, for their full support in publishing this book on time with patience and interest. Finally, we thank our families for their constant support, cooperation, and understanding. We are grateful to the Council of Scientific and Industrial Research (CSIR), New Delhi, for the award of emeritus scientist to Professor P.S. Bisen.

1 Cancer
A Worldwide Menace

KEY WORDS

Cancer
Carcinogenesis
Chemotherapy
Metastasis
Oncogenes

1.1 INTRODUCTION

There are more than a hundred different types of cancers. The names for cancers are derived from the organ or type of cell in which it starts; for example, cancer that begins in the colon is called colon cancer; cancer that begins in basal cells of the skin is called basal cell carcinoma. The Greek physician Hippocrates (460–370 B.C.), who is considered the father of medicine, is credited with the origin of the word *cancer*. Hippocrates used the terms *carcinos* and *carcinoma* to describe non-ulcer-forming and ulcer-forming tumors. In Greek, these words refer to a crab, most likely applied to the disease because the finger-like spreading projections from cancer are structurally similar to the shape of a crab. After cardiovascular diseases, cancer is the second biggest cause of human death worldwide. Cardiovascular diseases and cancer together are responsible for over 80% of all deaths in industrialized countries. Global cancer incidence is an ever-increasing trend. Conventional therapies control cancer by acting upon effects (like proliferation, cell growth, etc.) rather than acting on the root cause of carcinogenesis. Because of this, there is no effective cancer therapy available, and cancer-related malignancies and deaths are increasing.

Cancer indiscriminately affects people at all ages, with a propensity toward older people. Cancer occurs predominantly in older people, with three-quarters of cases diagnosed in people aged 60 and over, and more than a third (36%) of cases in people aged 75 and over. Less than 1% of all cases occur in children (0 to 14 years); 1,367 cases of cancer were diagnosed in children in 2007, with a slightly higher incidence in boys than girls. In spite of tremendous progress in the field of research and development, in the area of oncology, cancer is still a major killer across the world. According to recent statistics, cancer accounts for about 23% of the total deaths in the United States (Jamel et al., 2007). The world population is expected to reach 7.5 billion by 2020, and approximately 15 million new cancer cases will be diagnosed, and 12 million cancer patients will die (Brayand and Moller, 2006).

In India cancer has become one of the leading causes of death. It is estimated that there are nearly 2 million cancer cases at any given point in time. More than 0.7 million new cases are reported, and 0.3 million deaths occur annually due to cancer. There is a tremendous need to upscale diagnostic, treatment, and follow-up facilities with a capability to manage over 1.5 million cases at any given time. Data from population-based registries under the National Cancer Registry Program indicate that the leading sites of cancer are the oral cavity, lungs, esophagus, and stomach among men, and the cervix, breast, and oral cavity among women. Cancers of the oral cavity and lungs in males, and cervix and breast in females account for over 50% of all cancer deaths in India. The World Health Organization (WHO) has estimated that 91% of oral cancers in South and Southeast Asia are directly attributable to the use of tobacco. Developing countries share a major global burden of deaths due to oral cancer; countries like India contribute ~26% of the global oral cancer incidence. Oral cancer is the second most common form of cancer among Indian males (Bisen, 2012a, 2012b; Dikshit et al., 2012). Oral cancer is a multifactorial disease that has implicating attributes like genetics, environment, lifestyle, and behavior. Around half of the patients detected for oral cancer will die within 5 years of initial diagnosis. The 5-year survival rate has not improved in spite of a better understanding of cancer at a molecular level and with the advent of rationally targeted drugs (Bisen et al., 2012a, 2012b).

Cancer originates in cells, which are the building blocks of the body. Normally, the body forms new cells to replace old cells, which die. Sometimes this process goes awry, and new cells grow even when the system doesn't need them, and old cells don't die when they should. These accumulating extra cells can form a mass called a tumor. Tumors can be classified as *benign* or *malignant*. Benign tumors are not cancerous, while malignant ones are. Some benign tumors are precancerous and may progress to cancer if left untreated. Other benign tumors do not develop into cancer. Cells from malignant tumors can invade nearby tissues, making their management more challenging than normal. They can also break away and spread to other parts of the body (Figure 1.1). Cancer is caused by genetic alterations in oncogenes, tumor suppressor genes, and other genes responsible for carrying on routine tasks to maintain cellular homeostasis. These alterations are usually somatic events, although germ-line mutations can predispose a person to heritable or familial cancer.

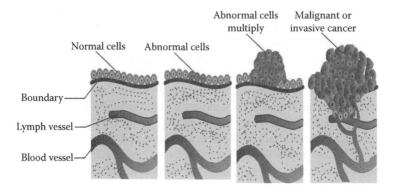

FIGURE 1.1   Presentation of multiplication of abnormal cells.

The development of a cancerous tumor often requires multiple concomitant genetic changes in oncogenes and tumor suppressor genes. The exact sequence of this genetic insult is still a mystery, even after decades of cancer research. The resulting tumors for any single cancer type can be cytogenetically quite diverse, because of myriad paths of genetic transformation possible for tumorigenesis. This heterogeneity contributes to differences in clinical behavior and responses to treatment of tumors of the same diagnostic type. Apart from the initial clone and subclones, tumors can also contain progenitor cancer cells, all of which constitute a spectrum of cells with different genetic alterations and states of differentiation. These tumor subpopulations can differ in sensitivity to chemotherapy, radiotherapy, and other treatments, making clinical management difficult. This is the primary reason why traditional treatment modalities fail, making it compelling to treat the tumor at an individual level after thorough evaluation of the underlying genetic profile.

The growth of cells in the body is a well-regulated phenomenon, which constitutes temporal and spatial regulation of genes, leading to the development and maintenance of various tissues and organ systems. Barring few instances such as wound healing, the rate of cell growth is also well regulated. In this situation, the growth of a localized group of cells is accelerated (often termed hyperplasias) to reconstitute the tissue to its previous state of normal structure and function, following which tightly regulated growth resumes. In general, hyperplasias arise to meet special needs of the body and subside once these needs are met. Hyperplasias are the result of the sustained impact over time of stimulatory influences together with a loss of growth-inhibitory factors that are normally found within or around cells. As long as the loss of inhibition of cell growth is temporary, the capacity for enhanced cell proliferation when necessary has obvious advantages. However, if cells permanently lose their ability to respond to growth-inhibitory factors, their growth becomes irrepressible, which may lead to cancerous growth.

Tumors can be broadly classified as benign and malignant; benign tumors are nonfatal and localized in nature, whereas malignant tumors are fatal and have a tendency to spread across the body through the process of invasion and metastasis. These two types of tumors are collectively referred to as neoplasms (new growths), and their study is known as oncology. Tumors are referred to as malignant or benign based on the structural and functional properties of their component cells and their biological behavior. The cells and tissues of malignant tumors differ from the tissues from which they arise. They exhibit more rapid growth and altered structure and function, including stimulation of new blood vessel growth (angiogenesis) and a capacity to invade adjacent normal tissues, enter the blood vascular system, and spread (metastasize) to distant sites. The properties of malignant tumor cells serve to enhance and support their proliferation and extension throughout the body tissues and organs, eventually leading to death of the host (Figure 1.2).

Normal Cell Structure	Cancerous Cell Structure
Large cytoplasm	Small cytoplasm
Single nucleus	Multiple nuclei
Single nucleolus	Multiple and large nucleoli
Fine chromatin	Coarse chromatin

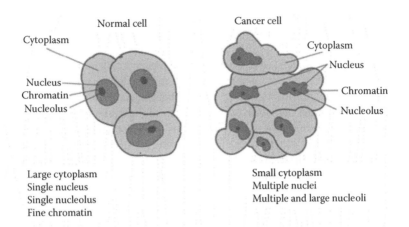

FIGURE 1.2   Differences between a normal and a cancerous cell structure.

In contrast, the cells and tissues of benign tumors tend to grow more slowly and in general closely resemble their normal tissues of origin. When the structure and function of benign tumor cells are morphologically and functionally indistinguishable from those of normal cells, their growth as a tumor mass is the sole feature indicative of their neoplastic nature. We can expect that a greater understanding of malignant cell growth and behavior will lead to the development of novel cancer therapies based on tumor cell biology that will complement or replace the current treatment options consisting of surgical excision, chemotherapy, and radiation. Populations around the world vary from each other as far as incidences of cancer type are concerned, which could very well attributed to differences in environmental/geological factors. The Scandinavian countries have less incidence and death rates for skin cancer than Australia and New Zealand, which could be due to the marked differences between these two regions in total annual hours of exposure to sunlight. The importance of environmental factors can also be ascertained by analysis of immigrants. The prostate and colon cancer rates are more common in the United States than in countries like Japan. The rate of prostate and colon cancer is much higher in U.S.-migrated Japanese immigrants than resident Japanese. However, the rates for each type of tumor among first-generation Japanese immigrants are intermediate between the rates in Japan and those in California, suggesting that environmental and cultural factors may play a more important role than genetic ones.

The irreversibility of the structural and behavioral changes of cancer cells has long been recognized and has favored the postulate that they are probably due to permanent genetic alterations. The role of genetic alterations responsible for cancer was more or less speculative until the discovery in 1979 that oncogenes (cancer-causing genes) are derived from proto-oncogenes (normal growth-regulatory cellular genes). When proto-oncogenes become mutated or deregulated, they are converted to oncogenes, which are capable of causing the malignant transformation of cells, including those of humans (Figure 1.3). Cellular proto-oncogenes code for proteins involved in cell regulation, such as growth factors, their receptors, and transmembrane signal

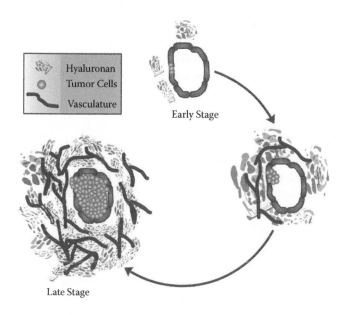

Hyaluronan
Tumor Cells
Vasculature

Early Stage

Late Stage

FIGURE 1.3   Stages of cancerous cell division in a tissue.

transducers. Thus, changes in the structure of proto-oncogenes and their conversion to oncogenes result in the synthesis of abnormal proteins that are incapable of carrying out their usual growth-regulatory functions. In identifying the genes involved in the development of cancer, researchers discovered a group of cellular genes—tumor suppressor or suppressor genes—whose protein products normally negatively regulate cell growth by suppressing cell proliferation, thus counterbalancing the growth-stimulatory effects of proteins synthesized by proto-oncogenes. Genetic analyses of various animal and human cancers have demonstrated that, in the majority, alterations of oncogenes and suppressor genes were often simultaneously present.

These analyses suggest that multiple genetic alterations involving growth-stimulatory and growth-inhibitory genes are required for the induction of malignancy. Such discoveries have ushered in a new era in cancer biology and may well lead to the eventual control, cure, and prevention of malignant diseases.

Cancer is caused by the complex interaction of intrinsic (heredity) and extrinsic (environmental and lifestyle) factors; intrinsic factors make the person more susceptible to deleterious effects of extrinsic factors. Hereditary causes of cancer are less common and are due to the inheritance of a single mutant gene that greatly increases the risk of developing a malignant tumor. Such cancers include a childhood tumor of the eye, *retinoblastoma*, and a bone tumor, *osteosarcoma*, both of which involve the loss of a tumor suppressor gene, and familial adenomatous polyposis, in which all patients develop colon cancer by age 50. The most common types of cancer that occur sporadically, such as cancers of the breast, ovary, colon, and pancreas, also have been documented to occur in familial forms. The children in such families appear to have a two- to threefold increased risk of developing a particular tumor, but the transmission pattern is unclear. A still rare hereditary cause of cancer is an inherited deficiency in the ability to repair DNA. Patients with this defect (known as *xeroderma pigmentosum*) are particularly sensitive to sunlight and develop skin cancer during early adolescence because of unrepaired mutations induced by ultraviolet (UV) light.

1.2   HISTORY

The existence of cancer can be tracked throughout the history of mankind. People have written about cancer since the dawn of history. The existence of cancer in earlier times can be found in fossilized bone tumors, human mummies in ancient Egypt, and ancient manuscripts. Evidence of bone cancer and cancer of the head and neck was revealed among Egyptian mummies. Literature references of cancer date back to about 1600 B.C. An Egyptian work called the *Edwin Smith Papyrus*, believed to be a copy of part of an ancient textbook on trauma surgery, describes eight cases of tumors or ulcers of the breast that were treated by cauterization, with a tool called the fire drill. It refers to cancer as a disease without any treatment.

During the Renaissance, beginning in the 15th century, scientists in Italy developed a greater understanding of the human body. Scientists like Galileo and Newton began to use the scientific method, which later was used to study disease. In 1761, Giovanni Morgagni of Padua was the first to do something that has become routine today—he performed autopsies to relate the patient's illness to the pathologic findings after death. This laid the foundation for scientific oncology, the study of cancer. The famous Scottish surgeon John Hunter (1728–1793) suggested that some cancers might be cured by surgery and described how the surgeon might decide which cancers to operate on. If the tumor had not invaded nearby tissue and was "moveable," he said, "there is no impropriety in removing it." A century later development of anesthesia allowed surgery to flourish, and classic cancer operations such as radical mastectomy were developed.

The advent of the modern microscope to study diseased tissues during the 19th century can be regarded as the birth of scientific oncology. Rudolf Virchow, often regarded as the founder of cellular pathology, formed the scientific basis for the modern pathologic study of cancer. The benefits of the pathological study of cancer tissue were twofold, as it not only helped in better understanding the damage cancer had done, but also formed the foundation for the development of cancer surgery. Pathological study allows more accurate diagnosis of patients and is also used routinely to confirm the extent of success among patients who have undergone therapy or surgery.

1.2.1   THEORIES

The causes of cancer have always been a bone of contention for physicians. Some of the beliefs that existed in earlier times for the causes of cancer follow.

1.2.1.1   Humoral Theory

Hippocrates believed that our body systems consist of four humors (body fluids): blood, phlegm, yellow bile, and black bile. He postulated that diseases and heath-related malignancies are caused by a disturbance of the innate balance of these humors. He suggested that cancer is caused by the excess of black bile in the body. This theory was passed on among generations for over 1,300 years. The acceptance of this theory for such a long time was also due to other factors, like prohibition of the study of the body and autopsies due to religious practices of those times.

1.2.1.2   Lymph Theory

The humoral theory of cancer was replaced by the lymph theory, which implicated the lymph as a cause of cancer. Stahl and Hoffman postulated that cancer was composed of fermenting and degenerating lymph varying in density, acidity, and alkalinity (Manfred Kuen and Heiner, 2010). This theory got rapid support and was endorsed by eminent surgeon John Hunter (1723–1792).

1.2.1.3   Blastema Theory

The lymph theory was categorically negated by German pathologist Johannes Muller in 1838, who demonstrated that cancer is made up of cells and not lymph. He believed that cancer cells arise from budding elements (blastema) between normal tissues, but not from normal cells. However, his student Rudolph Virchow (1821–1902), the famous German pathologist, demonstrated that all cells, including cancer cells, are derived from other cells.

1.2.1.4   Chronic Irritation Theory

Virchow proposed that chronic irritation was the cause of cancer, but he falsely believed that cancers "spread like a liquid." A German surgeon, Karl Thiersch showed that cancers metastasize through the spread of malignant cells and not through some unidentified fluid.

1.2.1.5   Trauma Theory

Trauma was regarded as a factor causing cancer from late the 1800s until the 1920s, in spite of advances in the understanding of cancer. This belief was maintained despite the failure of injury/trauma to cause cancer in experimental animals.

1.2.1.6   Parasite Theory

Cancer was regarded as contagious in the 17th and 18th centuries. In 1779, the first cancer hospital in France was forced to move from the city as people feared that it would spread cancer throughout the city. Current scientific knowledge classifies cancer as noncontagious; however, certain viruses and bacteria increase the risk of developing cancer.

Discovery of the exact chemical structure of DNA by James Watson and Francis Crick can be regarded as a turning point in modern research. DNA was established to carry some sort of genetic code responsible for various activities performed at the molecular level and which often manifests itself in the form of some phenotypic character like the determination of eye color. Scientists were able to translate the information encoded in DNA, which has enabled understanding of biological function or dysfunction in the case of disease as a function of normal or mutant DNA, respectively. Mutation often leads to change in the genetic code, resulting in mistranslation of the coded information and consequent genesis of an abnormal phenotype. Cancer is a genetic disease in which mutations caused by chemicals and radiation are believed to be the causative factor. In the past couple of decades, technology has undergone phenomenal improvement, owing to which now it is possible to pinpoint the exact site of damage of the gene.

Mutations caused by various factors result in defective genes, and these genes are sometimes inherited, making progeny susceptible toward development of cancer. Cancer is believed to be initiated by extrinsic factors like carcinogens such as radiation. These extrinsic factors bring about initial genetic insult, resulting in the development of mutant cells. Under normal conditions, such mutations are detected by cellular regulatory checkpoints and the extent of damage is controlled by forcing such mutant cells to undergo programmed cell death (apoptosis). The subpopulation of mutant cells remains unaffected by the internal control system, and in a course of time, the entire control system is taken hostage and the cellular microenvironment is made conducive for cancerous growth. Such differences in behavior of normal and cancer cells have helped us gain a better understanding of the molecular basis of carcinogenesis.

The genes involved in the development of cancer can be grouped in two categories: (1) oncogenes and (2) tumor suppressor genes (TSGs). Oncogenes are responsible for transforming normal cells into their cancerous counterparts. Oncogenes are a mutated form of otherwise normal genes known as proto-oncogenes. These proto-oncogenes carry out critical functions like cell cycle regulation and differentiation. Mutations in proto-oncogenes compromise these critical functions and thereby lead to uninterrupted growth-promoting activities necessary for tumor formation. Conversely, tumor suppressor genes are responsible for maintaining the normal cellular homeostasis by regulating cell division, repair errors in DNA, and inducing apoptosis in case of accumulation of abnormal signals. Cancer is developed by promoting oncogenes and making tumor suppressor genes dysfunctional by various methods like mutation, epigenetic silencing, etc.

1.2.2 CLASSIFICATION

Cancer can be classified on the basis of the type of cell where tumor induction begins. Following are the various classes.

1.2.2.1 Lymphoma and Leukemia

Leukemia is cancer that affects blood cell production by targeting bone marrow and blood cells. Normal blood cells have a limited period of life, and they need to be constantly replenished by fresh, young cells to support their activity. A stem cell is a kind of cell within the bone marrow that has the capability to mature into various types of blood cells, through a well-regulated process. In leukemia, this controlled process is compromised, which affects development of normal blood cells and consequently results in accumulation of dysfunctional or partially developed cells.

Leukemia can be classified on the basis of speed of development and the persistence of the disorder. Acute leukemia and chronic leukemia are the two types. Acute leukemia is characterized by the presence of immature blood cells, which are not capable of carrying out normal functions. In acute leukemia, the number of blasts of immature cells increases very rapidly, and the condition of the patient deteriorates very fast. Chronic leukemia is characterized by the presence of relatively fewer blasts of immature cells, and these cells are relatively more mature (than their counterparts in acute leukemia), and have the ability to carry out some of their normal functions. In chronic leukemia, the number of blasts increases less rapidly, and therefore leukemia-related abnormalities develop gradually. Leukemia can be further subclassified based on the type of blood cell affected: lymphocytic and myeloid. When leukemia affects lymphoid cells, it is known as lymphocytic, and when it affects myeloid cells, it is known as myeloid leukemia (Figure 1.4). Some of the most common types of leukemia are:

1. **Acute lymphocytic leukemia (ALL):** ALL is an aggressive form of leukemia that has the tendency to metastasize into other organs. It is known to affect both adults and children. It is one of the most common types of cancer in children.
2. **Acute myeloid leukemia (AML):** AML affects people regardless of age and sex. It is known to regress quickly when treated appropriately, and if left untreated, it can be fatal. Its treatment requires regular follow-ups, as it is known to reactivate and redevelop over the time.
3. **Chronic lymphocytic leukemia (CLL):** CLL is the second most common type of leukemia affecting adults.
4. **Chronic myeloid leukemia (CML):** The incidence rate of acute leukemia is approximately 10% more than that of chronic leukemia. The risk of incidence of acute leukemia is comparatively more in older people, with the incidence rate of 60% in people 60 years of age and older. The recovery from leukemia critically depends on early and correct diagnosis and treatment in early stages. There are 245,225 people living with, or in remission from, leukemia in the United States. Thirty-one percent more males are living with leukemia than females. Leukemia causes about one-third of all cancer deaths in children younger than 15 years.

Lymphoma is a cancer of the lymphocytes, which is a constituting unit of the lymphatic system. The lymphatic system consists of a network of lymphatic vessels carrying colorless fluid known as *lymph* (which consists of lymphocytes), with the intervening nodules known as lymph nodes. Clusters of lymph nodes are found in the underarms, groin, neck, chest, and abdomen. Other parts of the lymphatic system are the spleen, thymus, tonsils, and bone marrow. Lymphatic tissue is also found in other parts of the body, including the stomach, intestines, and skin. The lymphatic system is an integral part of the body's immune system. It was first described by Thomas Hodgkin in 1832. The most common form of lymphoma is therefore known as Hodgkin's lymphoma. There are several other forms of lymphomas, together known as non-Hodgkin's lymphomas. The normal appearance of the lymph node in more

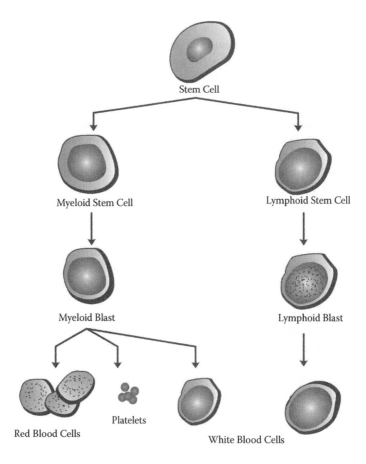

FIGURE 1.4 Division of stem cell to myeloid stem cell and lymphoid stem cell at blast stage.

aggressive lymphomas is replaced by diffused tumors of various sizes. Hodgkin's lymphoma is identified typically by the presence of the Reed-Sternberg cell in the tissue sample from the patient.

1.2.2.2 Carcinoma

Carcinoma is cancer originating from epithelial cells, which lines various organs, like breast, colon, liver, prostate, and stomach. Carcinomas have the tendency to invade neighboring organs and metastasize to remote sites like lymph nodes. Its premalignant form is known as carcinoma in situ (CIS), which has some cytological features pointing toward the presence of malignancy. However, it lacks any histological evidence of invasion through the basement membrane of the epithelium (Figure 1.5). Quite often, CIS will transform into an invasive counterpart, unless it is managed through surgical resection or cryotherapy. It is also observed that CIS can regress or disappear on the withdrawal of carcinogenic stimulus (Ishizumi et al., 2010). It is important to remember that carcinoma in situ is a preinvasive cancer, and

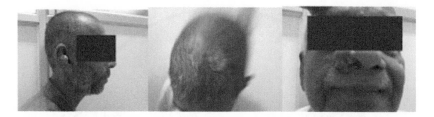

FIGURE 1.5 Squamous cell carcinoma of ear, scalp, and nose. (Courtesy of Dr. Gaurav Agrawal, Cancer Hospital, Birla Institute of Medical Research, Gwalior.)

not a premalignant entity (Banerjee, 2009). Carcinoma is classified on the basis of histopathological appearance (Figure 1.8). The most common carcinomas are adenocarcinoma or squamous cell carcinoma, which have glandular or squamous cell appearances, respectively. Metastatic carcinoma can be diagnosed through biopsy, including fine-needle aspiration, core biopsy, or subtotal removal of a single node.

1.2.2.3 Sarcoma

Cancer with a nonhematopoietic mesenchymal origin is known as sarcoma. Mesenchymal cells are made up of connective tissue, which have the capacity to migrate easily. Mesenchymal cells normally mature into skeletal muscle, smooth muscle, fat, fibrous tissue, bone, and cartilage. Sarcomas derive their name from the type of tissue from which they arise (Table 1.1). For instance, *chondrosarcoma* arises from cartilage, *osteosarcoma* arises from bone, and *leiomyosarcoma* arises from smooth muscle. For the better planning of the therapeutic strategy sarcomas are often assigned a particular grade (low/intermediate/high) on the basis of the presence and severity of certain cellular and molecular characteristics associated with malignancy. Surgical treatment is generally used for low-grade sarcomas, which can be complemented by radiation or chemotherapy. High- and intermediate-grade sarcomas require combinations of various treatment modalities like surgery, chemotherapy, and radiation therapy (Beucker, 2005).

TABLE 1.1

Various Types of Soft Tissue Sarcomas in Adults along with Their Tissue Origin and Body Location

Types of Soft Sarcomas in Adults		
Type of Cancer	**Tissue of Origin**	**Body Location**
Liposarcoma	Fibrous tissue	Arms, legs, and trunk
Fibrosarcoma	Fibrous tissue	Arms, legs, and trunk
Hemangiosarcoma	Blood vessels	Arms, legs, and trunk
Synovial sarcoma	Synovial tissue	Legs
Extraskeletal chondrosarcoma	Cartilage and bone-forming tissue	Legs and trunk

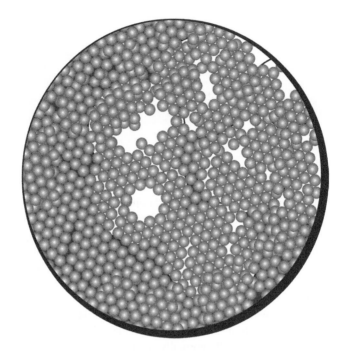

FIGURE 1.6   Rounded cell sarcoma infiltrating adipose tissue.

Sarcomas arising from soft tissue consist of the majority of sarcomas, e.g., *leiomyosarcoma* (smooth muscle), *synovial sarcoma* (synovial tissue), *rhabdomyosarcoma* (striated muscle), *osteosarcoma* (bone), *fibrosarcoma* (fibrous tissue), and *liposarcoma* (adipose tissue) (Figure 1.6). Sarcomas are rare when compared to other cancers, most probably because of the absence of inherent cell growth among constituting cells. In 2006, about 9,500 new cases were diagnosed in the United States (Ries et al., 2006). Soft tissue sarcomas are more commonly found in older patients (>50 years old), although in children and adolescents under age 20, certain histologies are common (rhabdomyosarcoma, synovial sarcoma).

Rhabdomyosarcoma is a relatively rare form of cancer. It affects children aged between 1 and 5 years, and it rarely affects teens and adults. Osteosarcoma is an aggressive cancerous neoplasm that is the most common histological form of primary bone cancer (Ottaviani and Jaffe, 2010). Osteosarcomas may exhibit multinucleated osteoclast-like giant cells (Papalas et al., 2009). The role of cancer stem cells in causing tumors is an active area of research studies (Osuna and de Alava, 2009). The success rate of treating osteosarcoma critically depends upon the correct diagnosis, which should not be confused with other similar bone tumors, like multilobular tumor of bone, or with a range of other lesions (Figure 1.7), such as *osteomyelitis* (Psychas et al., 2009). Leiomyosarcoma is a malignant neoplasm of smooth muscle. Involuntary muscles are made up of smooth muscle cells. Involuntary muscle is found in various organs like the uterus, stomach, intestines, skin, and walls of blood vessels. Leiomyosarcoma can therefore appear in these organs with involuntary

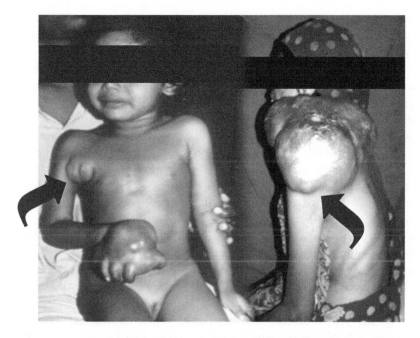

FIGURE 1.7 Osteosarcoma multilobular tumor of bone or with a range of other lesions. (Courtesy of Dr. B.R. Shrivastav, Cancer Hospital and Research Institute, Gwalior.)

muscles; they are most commonly found in the uterus (Arnold et al., 2010), stomach, small intestine, and retroperitoneum (Piovanello et al., 2007).

1.2.2.4 Blastoma

Blastoma is yet another form of the cancer that arises from precursor cells and is usually seen in children. *Blastoma* is added as a suffix to tumors with primitive, incompletely differentiated (or precursor) cells; e.g., *chondroblastoma* consists of cells resembling precursor of chondrocytes. The symptoms of blastomas vary

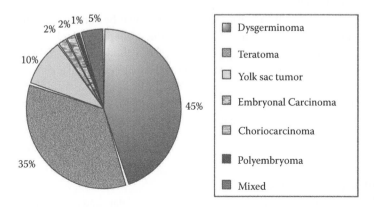

FIGURE 1.8 Various carcinomas and their percentage impact.

greatly and are determined by the affected body parts, like kidneys (*nephroblastoma*), liver (*hepatoblastoma*), nervous system (*neuroblastoma*), retina (*retinoblastoma*), etc.

A mutation in a gene coding for p53 has been linked to *pleuropulmonary blastomas*. Such mutations can vary in individuals, and therefore determine the prognosis. A personalized treatment strategy through genetic profiling of patients should be practiced in the case of dealing with rarer blastomas with the genetic basis (Debnath et al., 2010; Bisen et al., 2012a). Hepatoblastoma originates from liver precursor cells and affects the right lobe of the liver more often than the left lobe. Hepatoblastoma can metastasize to neighboring organs. Hepatoblastomas can also develop as a secondary complication in patients with familial adenomatous polyposis (FAP), a syndrome of early-onset colonic polyps and *adenocarcinoma* (Sanders and Furman, 2006). *Pancreatoblastoma* is a rare tumor of the pancreas, which occurs mainly in childhood (Saif, 2007). Common symptoms include abdominal pain, emesis, and jaundice.

Retinoblastoma (Rb) is another treatable cancer. Retinoblastoma has two forms: (1) a genetic or heritable form and (2) a nongenetic or nonheritable form. Approximately 55% of children with Rb have the nonheritable form. Retinoblastoma is one of the most common types of cancer found in children, with an incident rate of 0.00667%. The disease is labeled as sporadic if there is no trace found within the family history, but this does not imply that it is a nongenetic form. Retinoblastoma affects only one eye (*unilateral retinoblastoma*) in about two-thirds of cases (MacCarthy et al., 2009), and in the other one-third it affects both eyes (*bilateral retinoblastoma*). Tumors in each eye also differ with respect to the size and number. The pineal gland is also affected in certain rare cases of retinoblastoma (*trilateral retinoblastoma*). The factors, which are taken into consideration while treating retinoblastoma, are position, size, and number of tumors. Common symptoms of retinoblastoma are redness, pain, and irritation in the eyes, inflamed tissue around the eye, enlarged pupils, and possibly different-colored eyes.

1.2.2.5 Germ Cell Tumor

Germ cell tumor arises from germ cells, and these can be cancerous or noncancerous. It usually affects ovaries and testes, often due to improper embryonic development. It can affect people in all stages of life, from childhood to adulthood. Germ cell tumors are rare; only 2 to 4 children out of 1 million develop a germ cell tumor each year. Germ cell tumor makes up to 4% of all cancers in children and teens. Germ cell tumors most commonly appear in the gonads (sex organs). However, these tumors can arise in several different places within the body:

- Testes (boys)
- Ovaries (girls)
- Abdomen and pelvis
- Mediastinum (part of the chest between the breastplate and the spinal column)
- Brain

Germ cells are formed during the early developmental period of life. In the growing fetus, the germ cells migrate from point of origin to the gonadal area at about 4 weeks of the embryonic developmental history; however, these germ cells can also reach nonintended destinations. Germ cell tumors can be malignant or benign. The malignant germ cell tumors are life threatening, which includes different types of cancers, such as yolk sac tumor, immature teratoma, and choriocarcinoma. The malignant tumors can destroy gonads (testes or ovaries) and can metastasize to other regions of the body. The benign tumors are nonfatal counterparts, with relatively limited complications (some complications can happen due to size of the tumor). The factors and course of events involved in the development of germ cell tumors are still not fully understood. Some studies have implicated the presence of extra genetic material as the causative factor for germ cell tumors.

Germ cell tumors are broadly divided in two classes:

1. The *germinomatous or seminomatous germ cell tumors* (GGCTs, SGCTs) include only germinoma and its synonyms dysgerminoma and seminoma.
2. The *nongerminomatous or nonseminomatous germ cell tumors* (NGGCTs, NSGCTs) include all other germ cell tumors, pure and mixed.

Testicular cancers, including their metastatic form, show favorable outcomes when treated with chemotherapy. In the United States, the primary testicular tumors are most commonly solid malignant tumors among the age group of 20 to 35 years. The incidence of primary testicular tumors has increased at the rate of 1.2% per year during the past decade; however, the absolute mortality rate has remained more or less constant. According to an estimate by the American Cancer Society, approximately 8,400 new cases of testicular cancer were diagnosed during 2009 in the United States (American Cancer Society, 2009). The exact cause of testicular cancer is still not known. However, some studies have pointed toward the association of the isochromosome of the short arm of chromosome 12 with sporadic cancers, which indicates the importance of this region in the development of germ cell tumors. The genetic basis of testicular cancer is supported by the observation that the risk for developing testicular cancer is higher in first-degree relatives of cancer patients than in the general population. As in the case of other cancers with a genetic link, development of testicular cancer has been linked with multiple genes with varying degrees of association.

1.3 SELECTED TYPES OF CANCERS WITH THEIR SYMPTOMS

1.3.1 BREAST CANCER

Breast cancer has several subtypes, with diverse implications. Complexity among breast cancers can be as simple as ductal carcinoma in situ (DCIS), which has manageable localized growth, or as complex as those with an invasive nature with the ability to metastasize. Somewhere in between this spectrum lies breast cancer types like colloid carcinomas and papillary carcinomas, which have a much more favorable prognosis than invasive breast cancers.

Like other cancers, breast cancer originates in the cells, the basic building block of tissues. Cell growth and death is the continuous and regulated process in normal developing cells. The cellular machinery wears off after some time; therefore, the system needs to replenish exhausted cells with new cells through the regulated process of cell death and cell division. Cancer is invariably characterized by the presence of tumors, which is made possible by altering the normal course of cell regulation through cell death and cell division. Cancer cells promote a factor that allows them to attain tumorous growth, by evading cell death and promoting factors for cell growth. Migrating to other organs is yet another survival strategy of cancer cells that allows them to proliferate (Levy-Lahad and Plon, 2003).

An estimated 192,370 new cases of invasive breast cancer were expected to occur among women in the United States during 2009; about 1,910 new cases were expected in men. An estimated 40,610 breast cancer deaths (40,170 women, 440 men) were expected to occur in 2009. Breast cancer ranks second as a cause of cancer death in women (after lung cancer). Death rates for breast cancer have steadily decreased in women since 1990, with larger decreases in women younger than 50 (a decrease of 3.2% per year) than in those 50 and older (2.0% per year). The decrease in breast cancer death rates can be attributed to progress in both earlier detection and improved treatment (American Cancer Society, 2009).

1.3.1.1   Signs and Symptoms

The early stages of breast cancer are not accompanied by any overt symptoms; however, as the cancer progresses to more advanced stages it manifests itself through various physical symptoms:

- A lump or thickening in or near the breast or in the underarm area.
- A change in the size or shape of the breast.
- Dimpling or puckering in the skin of the breast.
- A nipple turned inward into the breast.
- Discharge (fluid) from the nipple, especially if it's bloody.
- Scaly, red, or swollen skin on the breast, nipple, or *areola* (the dark area of skin at the center of the breast). The skin may have ridges or pitting so that it looks like the skin of an orange (Figure 1.9).

1.3.1.2   Risk Factors

Gender is the strongest risk factor, followed by age, for breast cancer. Females are at greater risk of developing breast cancer than males. Other related risk factors are being overweight or obese after menopause, use of menopasal hormone therapy (MHT) (especially combined estrogen and progestin therapy), an abnormal menstrual cycle, use of oral contraceptives, pregnancy in advanced age, exposure to radiation during a medical procedure, drinking, smoking, and physical inactivity. Some of the pathological and medical parameters useful for profiling patients with

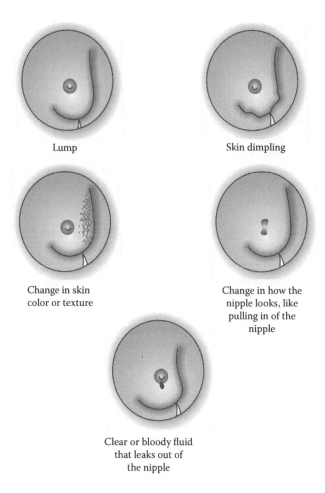

Lump

Skin dimpling

Change in skin
color or texture

Change in how the
nipple looks, like
pulling in of the
nipple

Clear or bloody fluid
that leaks out of
the nipple

FIGURE 1.9 Presentation of various signs and symptoms of breast cancer.

a higher risk of developing breast cancer are (1) high breast density (a mammo-
graphic measure of the amount of glandular tissue relative to fatty tissue in the
breast); (2) biopsy-confirmed hyperplasia, especially atypical hyperplasia; and (3)
high bone mineral density (routinely measured to identify women at increased
risk for osteoporosis). Some genotypic features (like mutations in BRACA1 and
BRACA2) are also known to be associated with increased risk of developing breast
cancer; however, these mutations are very rare and account for only 5 to 10% of
all breast cancers. Therefore, large-scale genetic profiling of patients is still not
practiced in routine clinical settings.

1.3.1.3 Treatment

The aim of treatment for breast cancer and the types of treatment recommended
depend on a number of things:

- The type of breast cancer
- The stage of breast cancer

The various types of breast cancer are:

Ductal carcinoma in situ (DCIS) is localized to ducts of the breast and is a noninvasive breast cancer.

Lobular carcinoma in situ (LCIS) is localized to lobules of the breast and is a noninvasive breast cancer.

Early breast cancer is an invasive breast cancer that is contained in the breast and may or may not have spread to lymph nodes in the breast or armpit. Some cancer cells may have spread outside the breast and armpit area, but cannot be detected.

Paget's disease of the nipple is a rare form of invasive breast cancer that affects the nipple and the area around the nipple (the areola).

Inflammatory breast cancer is a rare form of invasive breast cancer that affects the blood vessels in the skin of the breast, causing the breast to become red and inflamed.

Locally advanced breast cancer is an invasive breast cancer that has spread to areas near the breast, such as the chest wall.

Secondary breast cancer (also called metastatic or advanced breast cancer) is an invasive breast cancer that has spread from the breast to other parts of the body.

Owing to breakthroughs in scientific research, we are now better equipped with the detection and treatment of breast cancer. The molecular understanding of breast cancer has given us promising therapeutic and diagnostic biomarkers, which are in various stages of clinical development. The efficacy of hormone therapy in early-stage breast cancer (estrogen receptor positive) has shown a beneficial effect, and its efficacy is further enhanced when combined with aromatase inhibitors. Targeted therapies like trastuzumab (Herceptin) and lapatinib (Tykerb) are available for HER2/neu positive breast cancer patients.

Early detection and precautionary measures play a critical role in the cure of breast cancer. Analyses of axillary (underarm) lymph nodes help to determine the extent of spread of cancer, and its removal is often recommended as a treatment strategy. Treatment strategy involves a combination of various modalities, like radiation therapy, chemotherapy, hormonal therapy, and targeted drugs. Timely treatment of patients with ductal carcinoma in situ (DCIS) is recommended to avoid its transition into invasive cancer. DCIS can be treated by lumpectomy with radiation therapy or mastectomy; either of these options may be followed by treatment with tamoxifen.

1.3.2 Skin Cancer

Like other cancers, skin cancer originates in cells, the basic building block of tissues. Normal skin cells undergo the regulated and continuous process of cell growth and death. The cellular machinery wears off after some time; therefore, the system needs to replenish exhausted cells with new cells through the regulated process of cell death and cell division. Cancer is invariably characterized by the presence of tumors, which is made possible by altering the normal course of cell regulation through cell death and cell division. Cancer cells promote a factor that allow them to attain tumorous growth, by evading cell death and promoting factors for cell growth. Skin cancer usually occurs at organs exposed to the sun, like the head, face, neck, hands, and arms. Histologically malignant forms of skin cancer can be classified as follows.

1.3.2.1 Basal Cell Skin Cancer

Basal cell skin cancer grows slowly. It usually occurs on areas of the skin that have been exposed to the sun. It is most common on the face. Basal cell cancer rarely spreads to other parts of the body. It is characterized by translucent formation with tiny blood vessels (Figure 1.10).

1.3.2.2 Squamous Cell Skin Cancer

Squamous cell skin cancer also occurs on parts of the skin that may or may not have been exposed to the sun. It can spread to lymph nodes and organs inside the body. It is characterized by a red, crusted, or scaly patch or bump (Figure 1.11).

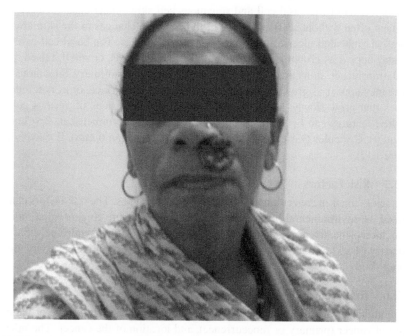

FIGURE 1.10 Patient with basal cell carcinoma near nose/around upper lip region. (Courtesy of Dr. Gaurav Agrawal, Cancer Hospital, Birla Institute of Medical Research, Gwalior.)

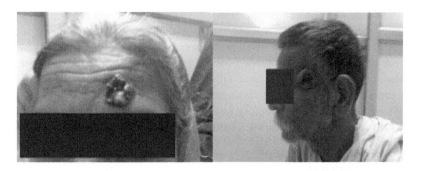

FIGURE 1.11 Basal cell carcinoma on forehead and side head. (Courtesy of Dr. Gaurav Agrawal, Cancer Hospital, Birla Institute of Medical Research, Gwalior.)

1.3.2.3 Melanoma

Melanoma is more rare than basal cell and squamous cell skin cancers. They frequently metastasize and are often fatal in later stages when they have already infiltrated into other body parts. It is characterized by asymmetry with irregular coloration.

Basal cell skin cancer and squamous cell skin cancer are induced by UV-B radiation via direct DNA damage, whereas melanoma is induced by UV-A radiation via free radical-mediated indirect DNA damage.

1.3.2.4 Signs and Symptoms

Early detection of the basal cell and squamous cell skin cancer ensures successful treatment often without scarring. Some of the common changes in the skin are discoloration, unhealed ulcers, and changes in existing moles. The basal cell carcinomas may appear as growths that are flat and firm, pale areas, or small, raised, pink or red, translucent, shiny areas that may bleed following minor injury. Squamous cell cancer may appear as growing lumps, often with a rough surface, or as flat, reddish patches that grow slowly. Melanomas are asymmetrical lesions of various colors, and their features can be summed up as ABCDE (A = asymmetrical, B = border (irregular), C = color (variegated), D = diameter (larger than 6 mm), E = evolving) (Figure 1.12).

1.3.2.5 Risk Factors

Risk factors for skin cancer vary with different cancer types. The risk factors can be grouped as environmental/occupational (prolonged exposure to sun/radiation/coal tar/pitch/creosote/arsenic), genetic (congenital melanocytic nevi syndrome, etc.), and secondary (weak immune system induced by drug or disease, skin sensitivity, light-colored skin).

1.3.2.6 Treatment

The treatment strategy depends on various factors like cancer type, age of patient, nature of cancer (primary or reoccurrence), and location of the cancer. The microscopic examination with the removal of any suspicious lesions in the skin is the first step toward managing skin cancer. Some of the most common methods used for

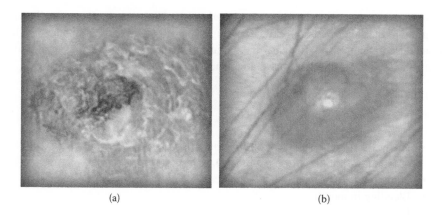

(a)   (b)

FIGURE 1.12   (a) Black rashes on skin presenting skin cancer. (b) Red lumps on skin indicating squamous skin cancer. (Courtesy of Dr. Gaurav Agrawal, Cancer Hospital, Birla Institute of Medical Research, Gwalior.)

the removal of lesions associated with basal cell and squamous cell skin cancers are surgical excision, curettage (tissue destruction by electric current and removal by scraping with a curette), electrodessication, or cryosurgery (tissue destruction by freezing). Other methods involve the use of radiation and application of topical medication. Melanomas are comparatively difficult to treat because of their propensity to metastasize; their treatment usually involves surgical removal of lymph nodes, along with removal of the primary growth and its surrounding region. For more complex melanomas, other treatment modalities, like immunotherapy, chemotherapy, and radiation therapy, are practiced, along with surgery.

1.3.3   ORAL AND PHARYNX CANCER

Cancer of the mouth cavity and pharynx is known as oral cancer. Roughly one-third of oral cancers are found in the pharynx, and the remaining oral cancers are found across other parts of the oral cavity. According to an estimate, around 35,000 Americans are diagnosed with oral cancer every year, and approximately 7,500 die due to oral cancer every year. The incidence of oral cancer is associated with factors like age and gender; it occurs in people over 40 years of age and affects more than twice as many men as women. However, this gap in incidence rate between genders is gradually narrowing, thanks to change in lifestyle, such as use of alcoholic drinks and tobacco products by females. The majority of oral cancers start in the flat cells (*squamous cells*) that cover the surface of the mouth, tongue, and lips. These cancers are known as *oral squamous cell carcinomas* (OSCCs). An estimated 7,600 deaths from the oral cavity and pharynx cancer were expected in 2009. Death rates from oral cancer have decreased by more than 2% per year since 1980 in men, and since 1990 in women (American Cancer Society, 2009).

Higher-grade oral cancers are known to metastasize to other organs, mostly via the lymphatic system. The lymphatic system consists of a complex web of lymphatic vessels carrying *lymph* (a clear, watery fluid). Cancer cells are carried along with lymph and appear initially in nearby *lymph nodes* in the neck. Cancer cells can migrate to the other parts of body, like lungs; in such cases the malignancy is still treated as oral cancer (malignant), and not as lung cancer. Such a medical condition is also known as "distant" or metastatic cancer.

1.3.3.1   Signs and Symptoms

Common symptoms of oral cancer include:

- Patches inside the mouth or on lips that are white (*leukoplakia*), a mixture of red and white (*erythroleukoplakia*), or red (*erythroplakia*)
- Sore on the lip or mouth that won't heal
- Bleeding in mouth
- Loose teeth
- Difficulty or pain when swallowing
- Difficulty wearing dentures
- Lump in neck
- Earache

1.3.3.2   Risk Factors

Well-established risk factors of oral cancer are tobacco products in any form and consumption of alcohol, which are also reported to act synergistically to aggravate oral carcinogenesis. Risk of developing oral cancer is increased by over 30-fold in individuals who are heavy smokers and drinkers. Human papillomavirus (HPV) infection is also associated with increasing risk in a certain kind of oropharyngeal cancer.

1.3.3.3   Treatment

The treatment strategy of oral cancer involves the combination of various methods, such as surgery, chemotherapy, radiation therapy, targeted therapy, and hormonal therapy. Targeted therapy with cetuximab (Erbitux) may be combined with radiation in the initial treatment or used alone to treat recurrent cancer.

1.3.4   Lung and Bronchus Cancer

The lungs are a pair of sponge-like organs that are an indispensable part of the respiratory system. Histologically, lung cancers can be divided into two types: (1) *non-small cell lung cancer* and (2) *small cell lung cancer*. These cancer subtypes have different development patterns, and therefore are treated differently. Lung cancer is one of the major killers across the globe, with a poor survival rate. It is usually detected in the advanced stages, and by that time has spread to distant organs, making its management very complex. Higher-grade lung cancers are known to metastasize to other organs, mostly via the lymphatic system. The lymphatic system consists of a complex web of lymphatic vessels carrying lymph (a clear, watery fluid). Cancer cells are carried along with lymph and appear initially in nearby lymph nodes in the lungs and in the center of the chest. The lung cancer often spreads toward the center of the chest because the natural flow of lymph out of the lungs is toward the center of the chest. Metastatic lung cancer has very poor prognosis.

Lung cancer accounts for the most cancer-related deaths. An estimated 219,440 new cases of lung cancer were expected in 2009, accounting for about 15% of cancer diagnoses. The incidence rate is declining significantly in men, from a high of 102.1 cases per 100,000 in 1984 to 73.2 in 2005. In women, the rate is approaching a plateau after a long period of increase. These trends in lung cancer mortality reflect historical differences in cigarette smoking between men and women and the decrease in smoking rates over the past 40 years. Lung cancer is classified clinically as small cell (14% of all lung cancers) or non-small cell (85% of all lung cancers) for the purpose of treatment.

1.3.4.1 Signs and Symptoms

Persistent coughing is one of the most noticeable symptoms of lung cancer, which indicates malfunctioning of the lungs due to cancer formation. Some other symptoms involve dysfunction of metabolic or other organ systems, which are not directly attributable to lung function. Some of the symptoms that can be directly attributed to lung function are:

- Coughing, the most common symptom, experienced by 74% of patients
- Bloody sputum (phlegm; 57%)
- Shortness of breath (37%)
- Chest pain (25%)
- Hoarseness (18%)
- Paralysis of the diaphragm, either symptomless or perceived as shortness of breath
- Wheezing or vibrating breathing noises
- Recurrent pneumonia or bronchitis
- Difficulty swallowing (dysphagia)

The formation of a tumor puts pressure on other neighboring organs, which is responsible for other nonrespiratory symptoms associated with lung cancer (Figure 1.13). Metastasis is also responsible for nonrespiratory symptoms. Some of the nonrespiratory symptoms are swelling of the face, arms, and neck, visible veins in the chest (superior vena cava syndrome (SVCS)), and pancoast syndrome. Pancoast syndrome is caused by the pressure created by a tumor on a nerve in the superior sulcus, a groove in the upper lung, and its sac, through which runs a major artery. Pancoast syndrome is associated with symptoms like weak or drooping eyelid, lessened or no perspiration on one side of the face, smaller pupil in one eye, pain in the shoulder, weakening of hand muscles, destruction of bone, which might be perceived as bone pain, headache, weakness, numbness, or paralysis, dizziness, partial loss of vision, bone or joint pain, abdominal pain upon probing, unexplained weight loss, loss of appetite, unexplained fever, yellowing of the skin (jaundice), accumulation of fluid in the chest or abdomen (effusion, ascites), and cardiac symptoms (irregular pulse and difficulty breathing).

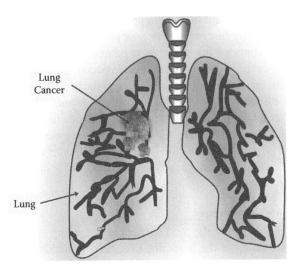

FIGURE 1.13   Illustrating cancerous cell growth on lungs.

1.3.4.2   Risk Factors

Smoking is the strongest risk factor among all that causes lung cancer. Carcinogens and other ingredients present in tobacco smoke cause direct (physical damage to lung tissue) and indirect (DNA damage, etc.) harm leading to initiation of lung cancer. The risk of developing lung cancer increases with the frequency of smoking, and the damage done by smoking is often irreversible (i.e., the person remains at risk even after quitting smoking, albeit with lesser risk). Exposure to smoke from someone else's cigarettes (passive smoking) also puts the person at risk of developing lung cancer. On average, smokers have a 15-fold higher risk of dying from lung cancer than nonsmokers. Frequency and duration of smoking determine the risk level of lung cancer in smokers. Compared with nonsmokers, those who smoke between 1 and 14 cigarettes a day have 8 times the risk of dying from lung cancer, and those who smoke 25 or more cigarettes a day have 25 times the risk. However, risk is more dependent on duration of smoking than consumption: smoking one pack of cigarettes a day for 40 years is more hazardous than smoking two packs a day for 20 years (Lubin et al., 2007). The risk of developing lung cancer also increases upon accidental/occupational exposure to other substances, like asbestos (asbestos is hair-like crystals found in many types of rock and is often used as fireproof insulation in buildings). The combination of smoking and asbestos exposure has been identified in several studies as a risk factor for lung cancer. Workers in asbestos-related industries (shipbuilding, asbestos mining, insulation, or automotive brake repair) are advised to use protective breathing equipment to reduce the risk of developing lung cancer. Radon is another element that increases the risk of some types of cancer, including lung cancer. Radon is an invisible, odorless gas naturally released by some soils and rocks. Radon levels can be detected with kits available in hardware stores, and proper ventilation of basements can reduce chances of radon exposure. Some other risk factors for lung cancer are occupational or environmental exposure to certain

metals (chromium, cadmium, arsenic), some organic chemicals, radiation, air pollu-
tion, and a history of tuberculosis. Genetic susceptibility puts the person of a younger
age at higher risk of lung cancer.

1.3.4.3   Treatment

The treatment plan of lung cancer traditionally involves the stages listed below. The
stage of both small cell and non-small cell lung cancer is described by a number,
zero (0) through four (Roman numerals I through IV). Cancer is associated with a
particular stage on the basis of factors like effort required in removal of the tumor,
the size of the tumor, the presence or absence of metastasis, etc.

Stage 0: This is called in situ disease, meaning the cancer is "in place" and has
not invaded nearby tissues and spread outside the lung.

Stage I: A stage I lung cancer is a small tumor that has not spread to any
lymph nodes, making it possible for a surgeon to remove it completely.
Stage I is divided into two substages, stage IA and stage IB, based on the
size of the tumor. Smaller tumors are stage IA, and slightly larger ones are
stage IB.

Stage II: Stage II lung cancer is divided into two substages: stage IIA and
IIB. A stage IIA cancer describes a slightly larger tumor that has not spread
to the nearby lymph nodes or a small tumor that has spread to the nearby
lymph nodes. Stage IIB lung cancer describes a slightly larger tumor that
has spread to the lymph nodes or a larger tumor that may or may not have
invaded nearby structures in the lung, but has not spread to the lymph
nodes. Sometimes, stage II tumors can be removed with surgery, and at
times, other treatments are needed.

Stage III: Stage III lung cancers are classified as stages IIIA and IIIB. For
many stage IIIA cancers and nearly all stage IIIB cancers, the tumor is
difficult, and sometimes impossible, to remove. For example, lung cancer
may spread to the lymph nodes located in the center of the chest, which is
outside the lung, or the tumor may have invaded nearby structures in the
lung. In either situation, it is less likely that the surgeon can completely
remove the cancer because removal of cancer has to be performed in steps.

Stage IV: Stage IV means the lung cancer has spread to multiple sites, in addition to the lung, the fluid surrounding the lung or the heart, or distant sites
within the body by way of the bloodstream. Once released in the blood,
cancer can spread anywhere in the body, but has a tendency to spread to the
brain, bones, liver, and adrenal glands. In general, surgery is not successful
for any stage III or IV lung cancer. Other situations that make lung cancer
impossible to remove are if it has spread to the lymph nodes above the col-
larbone, or if the cancer grows into vital structures within the chest, such as
the heart, large blood vessels, or main breathing tubes leading to the lungs.

The treatment strategy depends on the histological type of cancer (small cell or
non-small cell) and stage of cancer. Lung cancer treatment includes surgery, radiation therapy, chemotherapy, and targeted biological therapies, such as erlotinib

(Tarceva) and bevacizumab (Avastin). Surgery is a method of choice to treat localized cancers. Surgery is usually followed by a chemotherapeutic regime, to avoid any recurrence. In case of advanced stage cancers, surgery alone is usually not sufficient to treat cancer; therefore, other methods, like chemotherapy, radiation therapy, and targeted therapy, are practiced in combination with the surgery.

1.3.5 COLON AND RECTUM CANCER

Colorectal cancer is initiated in the colon or rectum. The colon and rectum make the terminating portion of the alimentary canal, which is involved in the absorption of fluid from solid waste (fecal matter or stool), which is extracted from the body. Cancer of the colon and rectum has many features in common, and is therefore often referred to as colorectal cancer. Colorectal cancer develops gradually over a period of several years. Development of a cancerous tumor is preceded by a growth of a noncancerous *polyp* on the inner lining of the colon or rectum. Some polyps transform into cancerous tumors.

The chance of changing into cancer depends upon the kind of polyps:

- **Adenomatous polyps (adenomas):** These are polyps that have the potential to become cancerous. Because of this, adenomas are called a precancerous condition.
- **Hyperplastic polyps and inflammatory polyps:** In general, these are not precancerous; however, few medical professionals believe that some hyperplastic polyps can become precancerous or might be a sign of having a greater risk of developing adenomas and cancer, particularly when these polyps grow in the ascending colon. *Dysplasia* is another kind of precancerous condition, which refers to the presence of abnormal cells (different from true cancer cells) in the lining of the colon or rectum. These histologically abnormal cells can transform into cancer cells over time. Dysplasia is usually seen as a secondary complication in people suffering from ulcerative colitis or Crohn's disease for many years. Chronic inflammation in the colon is caused by ulcerative colitis and Crohn's disease. A cancerous polyp eventually invaginates into the wall of the colon or rectum. Once cancer cells get inside the wall of the colon, they are able to access the blood or lymph vessels. Cancer cells spread throughout the body and reach distant organs through the circulatory or lymphatic system. Such cells can be detected in nearby lymph nodes, or in distant organs such as the liver. Colorectal cancer can be of the following types:
- **Adenocarcinomas:** The majority of colorectal cancers are adenocarcinomas. These cancers originate from glandular cells, which make mucus to lubricate the inside of the colon and rectum.
- **Carcinoid tumors:** These tumors originate from specialized hormone-producing cells of the intestine.
- **Gastrointestinal stromal tumors (GISTs):** These tumors start from specialized cells in the wall of the colon called the *interstitial cells of Cajal*. They can be benign or malignant. These tumors can be found anywhere in the digestive tract, but they are usually found in the colon.

- **Lymphomas:** These are cancers of the immune system cells that typically start in the lymph nodes, but they may also start in the colon, rectum, or other organs.

An estimated 49,920 deaths from colorectal cancer were expected to occur in 2009, accounting for almost 9% of all cancer deaths. Mortality rates for colorectal cancer have declined in both men and women over the past two decades, with a steeper decline since 2002 (4.3% per year from 2002 to 2005 in both men and women, compared to 2.0% per year from 1990 to 2002 in men and 1.8% per year from 1984 to 2002 in women). Improvements in detection techniques and treatment are the main reason behind the decrease of mortality and incident rates of colorectal cancer (American Cancer Society, 2009).

1.3.5.1 Signs and Symptoms

Screening tests are usually recommended to detect colorectal cancer in its early stage, as early-stage cancers are devoid of any overt symptoms. Some of the symptoms in the advanced stages of colorectal cancer are rectal bleeding, blood in the stool, a change in bowel habits, and cramping pain in the lower abdomen. Blood loss due to colorectal cancer leads to anemia, causing symptoms like weakness and excessive fatigue. The nature of symptoms of colorectal cancer depends on the location of the tumor and extent of spread of the tumor (metastasis) (Figure 1.14).

Colorectal cancer is usually associated with nonspecific symptoms, like change in bowel habit and a feeling of incomplete defecation (*rectal tenesmus*) and reduction in the diameter of the stool; tenesmus and change in stool shape are both characteristic of rectal cancer. Lower gastrointestinal bleeding, including the passage of bright red blood in the stool, and the increased presence of mucus may indicate colorectal cancer. Melena, black stool with a tarry appearance, normally occurs in upper gastrointestinal bleeding (such as from a duodenal ulcer), but is sometimes encountered in colorectal cancer when the disease is located in the beginning of the large bowel. Bowel obstruction is caused by a large-sized tumor, which causes constipation, abdominal pain, abdominal distension, and vomiting. This may also lead to the obstructed and distended bowel. Hydronephrosis may be caused through

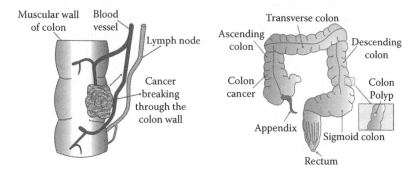

FIGURE 1.14 Cancerous cell breaking out through the colon wall with brief disruption of colon parts.

compression of the left ureter by a large left colonic tumor. Certain local effects of colorectal cancer occur when the disease has become more advanced. A large tumor is more likely to be noticed on feeling the abdomen, and it may be noticed by a doctor on physical examination. The disease may invade other organs and may cause blood or air in the urine (invasion of the bladder) or vaginal discharge (invasion of the female reproductive tract).

1.3.5.2   Risk Factors

Colorectal cancer can be regarded as an age-related malignancy because almost 91% of people diagnosed with colorectal cancer are aged over 50 years. Colorectal cancer is associated with several modifiable factors, like obesity, physical inactivity, alcohol consumption, smoking, diet high in red or processed meat, and inadequate intake of fruits and vegetables. Colorectal cancer risk is also increased by certain inherited genetic mutations (familial adenomatous polyposis (FAP) and hereditary nonpolyposis colorectal cancer (HNPCC), also known as Lynch syndrome), a personal or family history of colorectal cancer or polyps, or a personal history of chronic inflammatory bowel disease. Diabetes is also known to increase the risk of colorectal cancer.

Consumption of calcium and milk appears to decrease the risk of developing colorectal cancer. The intake of vitamin B6 has been reported to reduce the risk of colorectal cancer (Larsson et al., 2010). Some studies suggest that prophylactic use of nonsteroid anti-inflammatory drugs (NSAIDs) such as aspirin may reduce the risk of colorectal cancer.

1.3.5.3   Treatment

Colorectal cancer is usually treated with surgery, which can be combined with chemotherapy or radiation therapy, depending on factors like cancer stage, metastasis, age of patient, etc. Fitness level should be considered for the elderly patient before administering an aggressive chemotherapeutic regime for treating colorectal cancer (Ades, 2009).

The treatment of small cancerous polyps is usually done locally through a tube inserted through the rectum; however, removal of part of the intestine becomes mandatory while dealing with large tumors. The healthy parts of the intestine are sewn together to reconstitute the intestine after surgical excision of the tumor-infested region. In the case of complications in reconstituting the intestine, a temporary colostomy can be performed, in which the colon would be pulled out through the opening made in the abdomen, which would ensure passing of waste directly from the colon through the abdominal opening. A colostomy is a temporary workaround to ensure removal of waste and is rarely required as a permanent solution in colorectal cancer. The metastatic colorectal cancer is treated with surgery in combination with chemotherapy or radiation therapy, administered before and after surgery. A person with metastatic colorectal cancer is given oxaliplatin along with 5-fluorouracil (5-FU), followed by leucovorin (LV). The targeted therapies approved by the Food and Drug Administration (FDA) to treat metastatic colorectal cancer are:

Bevacizumab (Avastin): It blocks the growth of blood vessels to the tumor.
Cetuximab (Erbitux): It's an EGFR inhibitor, which inhibits growth of the tumor.

Pantitumumab (Vectibix): It's an EGFR inhibitor, which inhibits progression of the cancer.

1.3.6 OVARIAN CANCER

The pair of ovaries make up gonads in the female; they are located in the pelvic region (lower abdomen), on either side of the uterus (womb). The ovaries release ova, which reach the uterus through the fallopian tube. Once they reach the uterus, they may or may not get fertilized by the male sperm. The ovaries are also involved in production of critical hormones like estrogen and progesterone. Tumorous formation in one or both ovaries is known as ovarian cancer. Similar to other cancers, in ovarian cancer too, some normal cells undergo transformation into a cancerous cell; the process is induced by various extrinsic and intrinsic factors. A cancerous cell has the tendency to overpower normal cellular machinery and make conditions conducive for tumorous growth. In order to attain tumorous growth the cancer cell adapts various strategies to survive and thrive at the expense of the normal cellular machinery. Cell division and cell death are normal physiological phenomena by which the system replaces new/fresh cells with old cells; in cancer cells this phenomenon of cell death is compromised by activation of survival pathways, which ensures that cells keep on growing without being replaced by new cells (evading *apoptosis* or programmed cell death). Nutrition is vital for any cell, including cancer cells, which make up tumors, and when a tumor grows in size, it needs nutrition for the cancer cells residing in its core. A tumor induces formation of new vessels to ensure a nutrient supply to cancer cells (angiogenesis). Cancer cells increase their chance of survival by spreading to other organ systems, such as the liver or lungs (metastasis).

Ovarian cancer can be classified based on the type of cell where the cancer originated:

Epithelial tumors: This class of tumor arises from epithelial cells, which make the outer layer of the ovary. Around 70 to 80% of all ovarian cancers belong to this class. It occurs predominantly in women who have been through menopause (aged 45 to 70 years).

Stromal tumors: This class of tumor develops from stroma or connective tissue cells that form the structure of the ovary and are responsible for hormonal production. Usually a tumor develops in only one ovary. Around 5 to 10% of all ovarian cancers belong to this class. It occurs in women aged 40 to 60 years. Surgical removal of the tumor is sufficient as a treatment method, unless the tumor has spread; in such cases chemotherapy becomes necessary along with surgery.

Germ cell tumors: This class of tumor develops for germ cells, which are responsible for the production of ova. Around 15% of all ovarian cancers belong to this class. It develops most often in young women/girls. The success rate of treatment for this class of tumor is around 90%; however, many patients become infertile after treatment because of removal of germ cells that make eggs or ova.

Early detection of ovarian cancer is the key to attain successful treatment; however, symptoms from ovarian cancer start appearing when it has already spread to the abdomen, which is the main cause of its poor survival rate. Ovarian cancer is the biggest killer among cancers of the female reproductive system. The death rate for ovarian cancer has been stable since 1998 (American Cancer Society, 2009).

1.3.6.1   Signs and Symptoms

Ovarian cancer is associated with nonspecific symptoms such as abdominal pain or discomfort, back pain, urinary urgency, abdominal mass, bloating, constipation, tiredness, and a few more specific symptoms, such as abnormal vaginal bleeding, involuntary weight loss, or pelvic pain (Ryerson et al., 2007). Ovarian cancer can lead to accumulation of fluid (ascites) in the abdominal cavity, which causes enlargement of the abdomen (Figure 1.15).

1.3.6.2   Risk Factors

Advanced age is associated with the elevated risk of developing ovarian cancer. The long-term use of oral contraceptives and pregnancy are well-established factors that reduce the risk of developing ovarian cancer. Hysterectomy (removal of the uterus) and tubal ligation are believed to decrease the risk of ovarian cancer. Use of estrogen as a postmenopausal hormone therapy is associated with the increased risk of developing ovarian cancer. Obesity is also associated with an increased risk of ovarian cancer. Women with breast cancer or with a family history of breast or ovarian cancer, or with another genetic predisposition for cancer (BRCA1 and BRCA2 mutations), are at greater risk of developing ovarian cancer. A familial nonpolyposis colon cancer has also been associated with ovarian cancer.

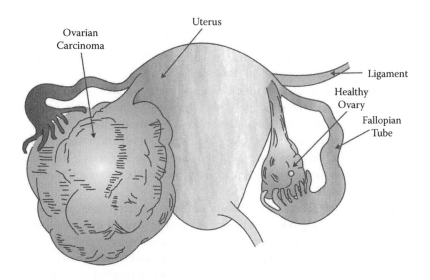

FIGURE 1.15   Illustrates developed ovarian carcinoma.

1.3.6.3 Treatment

The surgical removal of the affected region is usually the first treatment option for ovarian cancer. The surgery is carried out in close association with a pathologist, who confirms whether various regions show signs of cancer and should be removed accordingly. The treatment strategy depends on factors like staging, and possibility retaining fertility of the patient.

In case of stage I ovarian cancer, only the involved ovary and fallopian tube may be removed in women who want to conceive in the future. For those women who don't want to conceive in the future, both ovaries along with the fallopian tubes (*bilateral salpingo-oophorectomy*) and uterus (*hysterectomy*) are removed, and as a further precautionary measure, lymph nodes present in neighboring organs and the omentum are also removed. Chemotherapy can also be given after surgery.

In case of stage II ovarian cancer, both ovaries, along with the fallopian tubes and uterus, are removed. Tumor present in other body parts (usually in the pelvic region) is resected. Chemotherapy is highly recommended; it usually involves administration of carboplatin and paclitaxel in six cycles of 3 weeks each.

The treatment of stage III ovarian cancer is similar to that of stage II, with the only difference in aggressiveness of chemotherapeutic regime for the more advanced stage. Additional experimental treatments like peritoneal therapy may also be required for some women.

In case of stage IV ovarian cancer, extensive debulking and multiagent chemotherapy, and possibly secondary surgery, are practiced.

1.3.7 PANCREATIC CANCER

Cancer, which originates from the pancreas, is called pancreatic cancer (Figure 1.16). The pancreas is an organ located behind the stomach, and it has a shape similar to a fish (wide head with a tapering body and a pointing tail). It extends horizontally across the abdomen, with its head on the right side of the abdomen, body behind the stomach, and tail on the left side of the abdomen next to the spleen. The pancreas is a mixed gland; i.e., it consists of both endocrine and exocrine glands. Approximately 95% of the pancreas is made up of cells dedicated to carrying on exocrine function. The exocrine glands are involved in production of "pancreatic juice," which consists of a cocktail of enzymes involved in the digestion of fats, proteins, and carbohydrates present in food. These enzymes are collected by tiny tubes called ducts, which merge to form a larger duct that empties into the pancreatic duct. The pancreatic duct merges with the bile duct from the liver and pours pancreatic juice into the duodenum (the first part of the small intestine).

Small fractions of cells of the pancreas are dedicated to carrying on endocrine function, which involves production of two critical hormones: (1) insulin and (2) glucagon. Endocrine cells of the pancreas are arranged in small groups/clusters known as islets (or *islets of Langerhans*). Insulin reduces the amount of sugar in the blood, whereas glucagon is involved in elevation of the glucose level in blood. A metabolic disease like diabetes develops when insulin production is affected. Similar to other

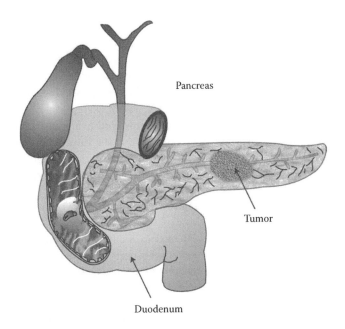

Pancreas

Tumor

Duodenum

FIGURE 1.16   Pancreatic cancer.

cancers, pancreatic cancer can also spread to the other organs (metastatic pancreatic cancer) in more advanced stages, making its management more complex with poor chances of survival.

1.3.7.1   Types of Pancreatic Tumors

The nature of a tumor formed by endocrine cells and exocrine cells is completely different.

Exocrine tumors: They are the most common type of pancreatic cancer. They can be benign or malignant, but mostly they are malignant in nature. Histologically pancreatic cancers are of different types, like adenocarcinoma, adenosquamous carcinoma, squamous cell carcinoma, and giant cell carcinoma. Adenocarcinomas form the majority of exocrine pancreatic cancer; they originate from glandular cells present in the pancreatic duct and sometimes from enzyme-forming cells of the pancreas.

Ampullary cancer is a special type of cancer related to the pancreas. It is formed in the area known as *amulla of Vater* (which is located in the place where the bile duct and pancreatic duct come together and open into the duodenum). Even in the initial stages ampullary cancer blocks the bile duct, leading to accumulation of bile in the body. This buildup of bile in the body leads to well distinct external features like yellowing of the skin and eye (jaundice). The presence of such overt symptoms makes its detection easy, and therefore is often associated with successful treatment.

Endocrine tumors: These tumors are not as common as exocrine pancreatic cancer. The tumors belonging to this group are known as pancreatic neuroendocrine tumors (NETs), which are sometimes also known as islet cell tumors. There are several subtypes of islet cell tumors. Each is named according to the type of hormone-making cell it starts in:

- *Insulinomas* come from cells that make insulin.
- *Glucagonomas* come from cells that make glucagon.
- *Gastrinomas* come from cells that make gastrin.
- *Somatostatinomas* come from cells that make somatostatin.
- *VIPomas* come from cells that make vasoactive intestinal peptide (VIP).
- *Pomas* come from cells that make pancreatic polypeptide.

NETs can be further classified as functioning or nonfunctioning, for their ability or inability to produce hormones, respectively. They can be either benign or malignant in nature. The treatment strategy and prognosis depend on the specific tumor type and stage of tumor; however, its prognosis is better than that of pancreatic exocrine cancers. Gastrinomas and insulinomas are the most common type of endocrine tumors.

Accurate identification of the type of pancreatic cancer is critically linked with the success rate of treatment, as exocrine and endocrine cancers have very distinct risk factors, etiology, symptoms, and prognosis.

1.3.7.2 Signs and Symptoms

- **Jaundice:** It is usually characterized by the yellowing of the skin and eye. Jaundice occurs in around 50% of patients with pancreatic cancer and 100% of those with ampullary cancer. It is caused by accumulation of bilirubin in the body. Bilirubin is produced in the liver and is excreted as bile. It is passed on by the liver into the intestine through the common bile duct. Buildup of bilirubin in the body occurs when the passage of the common bile duct is obstructed/blocked and bilirubin is not able to reach the intestine. Cancers located in the head of the pancreas have physical proximity to the common bile duct, and thus have a strong tendency to block the common bile duct, even when a tumor is small in its initial stages. This blockage of the common bile duct leads to buildup of the bilirubin, and consequently leads to the development of jaundice. Such overt symptoms act as a blessing in disguise, as they help in the detection of cancer, in its nascent stage. Other pancreatic cancers, which develop in the body or tail portion of the pancreas, are located away from the common bile duct, and hence do not compress the bile duct in its initial or localized stages.
- **Abdomen pain:** Abdominal or back pain is the common symptom of advanced pancreatic cancer. The abdominal pain is caused by compression from large cancerous development in the body or tail of the pancreas. Back pain is also caused by the spread of cancer to the nerves surrounding the pancreas.

- **Weight loss and poor appetite:** Abnormal weight loss is one of the very common symptoms in pancreatic cancer patients. It is also accompanied by related symptoms, like tiredness and loss of appetite.
- **Digestive problems:** Pancreatic cancer can cause blockage of the release of pancreatic juice into the intestine. Since pancreatic juice contains enzymes involved in the digestion of fatty food, its blockage leads to the inability of the system to digest fatty food. The undigested fat may cause stools to be unusually pale, bulky, greasy, and to float in the toilet. When cancer grows larger, it can wrap around the far end of the stomach and obstruct it, causing symptoms like nausea, vomiting, and pain, which aggravate.
- **Gallbladder enlargement:** The blockage of bile duct can lead to buildup of bile in the gallbladder. This buildup of bile inside the gallbladder leads to enlargement of the gallbladder, which can be felt by a physician during a physical checkup. The enlargement of the gallbladder can also be detected by imaging studies.
- **Blood clots or fatty tissue abnormalities:** Some other nonspecific symptoms of pancreatic cancer are the presence of blood clots and the abnormal texture of fatty tissue. Blood clots can occur in veins, more often in large veins in specific regions/organs of the body, like leg (deep venous thrombosis (DVT)) or lungs (pulmonary embolism (PE)). Sometimes the presence of pancreatic cancer can be suspected due to development of an uneven texture of the fatty tissue underneath the skin.
- **Diabetes:** In some rare cases pancreatic cancer can also promote the development of diabetes, especially when cancer destroys insulin-making cells of the pancreas.

1.3.7.3 Risk Factors

Tobacco smoking is known to increase the risk of developing pancreatic cancer by more than twofold when compared with nonsmokers. Other risk factors for pancreatic cancer are obesity, gender (males are more prone to pancreatic cancer), family history, cirrhosis, diabetes, chronic pancreatitis, and use of smokeless tobacco. Consumption of red meat is also believed to increase the risk of developing pancreatic cancer, whereas physical activity may decrease the risk.

1.3.7.4 Treatment

It's often difficult to treat pancreatic cancer, because of its agility to spread to neighboring organs. Several treatment options, like surgery, radiotherapy, and chemotherapy, are suggested to improve survival and relieve patients from symptoms produced by pancreatic cancer. Targeted therapies like erlotinib (Tarceva) have been approved by the FDA for treatment of advanced pancreatic cancer.

The surgical procedures undertaken in the case of pancreatic cancer can be best described as palliative with an objective of relieving patients from pain or symptoms caused by the cancer, rather than with the intention to cure pancreatic cancer. Several surgical strategies are adopted to circumvent symptoms caused by blockage of the bile duct. A surgical procedure of rerouting the common bile duct directly into the intestine by bypassing the pancreas is carried out.

1.3.8 PROSTATE CANCER

The Greek meaning of *prostate* literally means "to stand in front of," which is a name suggestive of its location as a small organ placed in front of the bladder. It produces slightly acidic and whitish fluid, which form mixes with constituents from the seminal gland (seminal fluid) and testis (spermatozoa), to form semen. The prostate gland contracts with ejaculation to provide enzymes to nourish sperm and increases their survival in the alkaline environment presented through the female reproductive organ system. Prostate cancer is traditionally considered a disease of older men, and its treatment could be best described as palliative rather than curative. Digital rectal examination (DRE), along with the presence of overt symptoms like experiencing pain during urination, probably due to obstruction of the bladder, was a standard for suspecting prostate cancer. The downside of reliance on the presence of these symptoms to infer prostate cancer is that by this time the cancer would have already metastasized to other organs, making it incurable.

With advancements in diagnostic techniques, the prostate-specific antigen (PSA) blood test has become a standard method to ascertain the presence of prostate cancer. Persons greater than 50 years of age having PSA in the range of 4 to 10 ng/ml should be closely monitored for the presence of prostate cancer. PSA has definitely helped in the early identification of cases of prostate cancer, but at times its results are not conclusive enough. Along with prostate cancer development, the level of PSA is known to also be correlated with other conditions, like obesity, urinary infection, irritation, etc.

The stromal component of the prostate consists of smooth muscle cells, fibroblasts, and endothelial cells, and the epithelium consists of the basal epithelium, secretory epithelium, transit amplifying cells, neuroendocrine cells, and stem cells. Dysregulated proliferation of constituting cells, especially those that make the epithelium, leads to development of prostate cancer. The cancerous development is believed to be triggered by cytokines and growth factors secreted by stromal and epithelial cell types (Chung et al., 2005). However, triggers and the sequence of development of prostate cancer are still not well understood. For the past several years efforts have been on screening targets (therapeutic and biomarkers) for prostate cancer by utilizing high-throughput *omics* technologies. These techniques have suggested several proteins and genes with differential behaviors in benign, malignant, and normal prostate tissue. Engrailed-2 (EN-2) is a one of the recent prostate-specific biomarkers detectable in the urine of a person with prostate cancer, which has the potential to be used as an independent diagnostic biomarker for prostate cancer (Morgan et al., 2011).

Prostate cancer is the second leading cause of cancer deaths in men. The prostate cancer-related mortality rate has declined more rapidly among African Americans than among white men; since the early 1990s, rates in African

Americans remain more than twice as high as those in whites (American Cancer Society, 2009).

The types of prostate cancer are clinically categorized as

- **Adenocarcinomas:** Over 95% of prostate cancer falls under this category. It starts in glandular tissue. The types of prostate cancer include *leiomyosarcoma* and *rhabdomyosarcoma*.
- **Prostatic intraepithelial neoplasia (PIN):** It's premalignant formation, which eventually becomes malignant.
- **Prostate carcinoma:** It usually develops near the surface of the gland.
- **Benign prostatic hyperplasia (BPH):** The abnormal growth of benign prostate cells is called BPH. In BPH, the prostate enlarges and blocks the normal flow of urine.

An advanced stage of prostate cancer is involved in the metastasis and invasion. It can invade to fat, tissue, seminal vesicles, and the neck of the bladder. It can spread across lymph nodes situated in the pelvis, bones of the spine, hip, and chest (Figure 1.17). Lungs, liver, and adrenal glands are other organs that are invaded by prostate cancer.

1.3.8.1 Signs and Symptoms

Prostate cancer does not have any overt symptom in its early stages. These cancers are detected by the higher level of prostate-specific antigen (PSA) or as a hard nodule (lump) in the prostate gland that can be detected during routine digital rectal examination (DRE). The prostate gland is located immediately in front of the rectum.

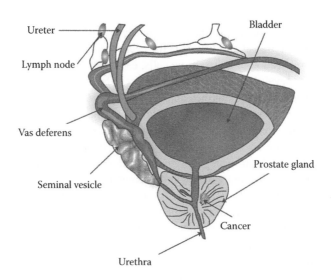

FIGURE 1.17 Stage I of prostate cancer.

In rare cases of advanced prostate cancer, the cancer could enlarge and press the urethra, which can result in diminished flow of urine and difficulty in urination. Consequently, patients also complain of a burning sensation or blood in the urine. With the progression of cancer, the urethra can be completely blocked, leading to an obstructed and enlarged urinary bladder. These symptoms could also be due to noncancerous enlargement of the prostate; therefore, their presence does not necessarily confirm the presence of the prostate cancer, but does warrant a complete diagnosis followed by timely and appropriate treatment.

The later stage of prostate cancer can invade to neighboring tissue/organs/lymph nodes (mostly in the pelvic region). Metastases to the liver can cause abdominal pain and jaundice. Metastases to the lungs can cause chest pain and coughing.

1.3.8.2 Risk Factors

Well-established risk factors of prostate cancer are age, ethnicity, and family history of the disease. It occurs predominantly in men aged 65 and older (more than 63%). Jamaican men of African descent and African American men have the highest incident rate of prostate cancer in the world. Dietary factors like high intake of animal fat may also predispose men toward developing prostate cancer. A diet rich in vegetables and fruit is known to reduce the risk of developing many cancers, including prostate cancer.

1.3.8.3 Treatment

The treatment strategy for prostate cancer depends on various factors like the patient's age and the stage and grade of the cancer. The Gleason score is used to formally grade tumors, which typically indicates the aggressiveness of the cancer (it ranges from 2 (nonaggressive) to 10 (very aggressive)). Early-stage cancer is often treated with surgery, hormonal therapy, radiation, or radiation seed implants (*brachytherapy*). The patient is put on careful observation rather than on an aggressive treatment regime in the case of old-age-related complications. More advanced therapy requires multiple treatment options, like hormonal therapy, chemotherapy, and radiation. Finasteride and dutasteride are used as a chemopreventive agent for prostate cancer, because of their ability to reduce the amount of male sex hormone (testosterone).

1.4   MAJOR CAUSES BEHIND CANCER

Cancers are multifactorial in nature and caused by extrinsic factors like the environment, lifestyle, etc. (constitutes around 90 to 95% of all casual factors), and intrinsic factors, which constitutes 5 to 10% of all causes (Table 1.2). The most common extrinsic factors include tobacco (25 to 20%), diet and obesity (30 to 35%), infections (15 to 20%), radiation, radon exposure, stress, lack of physical activity, and environmental pollutants (Figure 1.18).

1.4.1   GENETICAL CAUSES

Cancer is characterized by abnormal cell growth. Cancer cells with an abnormal number of chromosomes were observed as early as 1914. Discovery of oncogenes

TABLE 1.2

Ten Leading Cancer Types for the Estimated New Cancer Cases and Deaths by Sex, 2010

Males		Females	
Cancer	**Deaths**	**Cancer**	**Deaths**
Lung and bronchus	29%	Lung and bronchus	26%
Prostate	11%	Breast	15%
Colon and rectum	9%	Colon and rectum	9%
Pancreas	6%	Pancreas	7%
Liver and intrahepatic bile duct	4%	Ovary	5%
Leukemia	4%	Non-Hodgkin's lymphoma	4%
Esophagus	4%	Leukemia	3%
Non-Hodgkin's lymphoma	4%	Uterine corpus	3%
Urinary bladder	3%	Liver and intrahepatic bile duct	2%
Kidney and renal pelvis	3%	Brain and other nervous system	2%

Source: Jemal, A., et al., *CA Cancer J Clin* 60: 277–300, 2010. With permission.

and tumor suppressor genes, in the context of development of cancer, paved the way for understanding cancer at the molecular level.

Cells in human beings and other animals can be broadly grouped as (1) germ cells or (2) somatic cells. A germ cell (ova in female and sperm in male) contains only one set of each chromosome. The germ cell plays a critical role in sexual reproduction and acts as a carrier of genetic material contributed by parents to their offspring. Fertilization is a process in which sperm and ova fuse/combine to form a fertilized egg, which undergoes a series of divisions to become an embryo, and later undergoes differentiation and cell multiplication to become a full-grown adult. Somatic

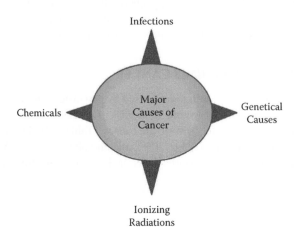

FIGURE 1.18 Major causes of cancer.

cells form the body of an organism and have the same genes, which were passed by parents, via fertilized egg. Any defective gene is thus passed on to offspring, which predisposes them to cancers that may appear in childhood or in the later part of the life. Few genes put the person at a greater risk of developing cancer, and the development of cancer depends critically on a few additional triggers (like smoking, alcohol consumption, infection, carcinogens, etc.). The BRCA1 and BRCA2 mutations are known to predispose women to develop breast cancer.

The BRCA1 and BRCA2 mutations increase the risk of development of breast cancer in women by eightfold compared to the normal population, and women with such mutations have an 80% lifetime risk of developing breast cancer. The level of predisposition varies from one gene to another, and most of the mutations are weakly associated with cancer, but when considered together, the presence of multiple weak predispositions puts the person at greater risk of developing cancer. Around 5 to 10% of breast cancer cases can be attributed to strong and weak mutations inherited from parents (Fearon et al., 2003). These mutations can be confined to certain families or ethnic groups, which explains why they are more susceptible in developing certain cancers.

Cancer is caused by the interaction of multiple genes and factors. Cellular activity is tightly a regulated phenomenon in the normal cells. In cells with mutated genes, normal cellular regulation is compromised, which sometimes leads to the formation of carcinogens due to incomplete metabolism of chemicals. These carcinogens cause DNA damage, which under normal conditions is taken care of by DNA repair enzymes, or in some cases, the cell is pushed to controlled cell death (apoptosis). In the case of cancer, the repair mechanism and apoptotic mechanism become dysfunctional to promote abnormal cell growth. A complex interaction among the presence of a gene mutation and environmental toxicants exists that predisposes some persons to the risk of developing cancer over time. The study that aims to understand this interaction among genetic risk factors and environmental factors is known as molecular epidemiology.

Cancers are called sporadic when they are not inherited. They occur in somatic cells, which make the bulk of the body (Figure 1.19). Somatic cells are in the process of continuous replenishment through the process of cell division and cell death. A few somatic cells, like muscle cells or neurons in the brain, stop dividing after reaching maturity; cancer rarely occurs in such nondividing cells. Cancer occurs in somatic cells that are in the cyclic process of cell division and cell death.

Cell division is a well-controlled process in which the cell passes its genes to two daughter cells, which receives the same copy of genes as were present in the parent cell. However, there is a rare possibility of an error in this process, which could result in the daughter cell(s) receiving a wrong (mutated) copy of the gene(s). Our cellular system is well equipped to detect such mutations, and mark such cells with an abnormal genetic material for controlled death (apoptosis or programmed cell death). However, in rare instances a few genetic mutations activate the survival mechanism in somatic cells, and thus apoptosis is skillfully evaded in such instances. Cells with mutated genes proliferate and create an environment conducive for the second set of mutations, which would give them potential to replicate and grow without any restrictions. In order to achieve malignancy cancers, mutated series of genes, these mutations activate a group of genes (oncogenes) that would help them to survive and

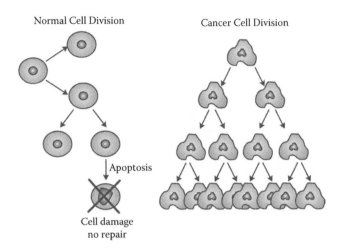

FIGURE 1.19   Illustrates uncontrolled cell division.

thrive and deactivate or silence another group of genes (tumor suppressor genes) that have the potential to check cancer growth. Our body is equipped with countermechanisms to suppress the development of cancer. In fact, most of the tumor suppressor genes are inherited dominantly, and there are only three oncogenes (RET, MET, and CDK4) that are known to be inherited in different types of cancer. The development of cancer is a rare event, which requires the concomitant presence of environmental, genetic, and mutational factors.

1.4.2   INFECTION

Cancer can also be caused by infection. Viral infection accounts for up to 20% of the human cancer incidences worldwide. Some of the cancer caused by viruses in humans are cervical carcinoma (human papillomavirus), mesothelioma (human polyomavirus), B-cell lymphoproliferative disease and nasopharyngeal carcinoma (Epstein-Barr virus), Kaposi's sarcoma and primary effusion lymphomas (Kaposi's sarcoma herpesvirus), hepatocellular carcinoma (hepatitis B and hepatitis C viruses), T-cell leukemia (human T-cell leukemia virus-1), and gastric carcinoma (*Helicobacter pylori*). We can expect more viruses to be added in this group to cause cancer (Pagano et al., 2004). Viruses are rarely causally associated with cancer; more often they help in tumor progression.

Hepatitis viruses (hepatitis B and hepatitis C) are known to induce chronic viral infection, which can lead to development of liver cancer. The risk of development of liver cancer is increased by the concomitant presence of viral hepatitis and cirrhosis.

1.4.3   IONIZING RADIATIONS

Prolonged exposure to UV radiation from the sun can lead to development of skin cancer. Ionizing radiation is a well-established carcinogen that is responsible for causing over 3% of all cancers. At a molecular level, ionizing radiation causes abnormalities by damaging nuclear DNA. Radiation from natural sources like cosmic rays and radioactive substances in the earth's crust can also cause cancer. According to an estimate, we are exposed to this "background" radiation at a rate of about 1 to 2 mGy per year. The gray (Gy) is a unit of measurement for the amount of radiation energy absorbed by body tissues. It is equal to 100 rad and is now the unit of dose. Radiation of a very high dose is often fatal, especially if received all at once, but is tolerated when received in batches over a period of time. For instance, whole-body irradiation of a 5 Gy dose would kill about 50% of people exposed within 30 days, whereas patients on radiation therapy receive around 2 Gy per session and are exposed to tens of gray over a period of weeks.

The role of radiation in causing cancer came from observations of patients enrolled for radiation therapy for other ailments, like ankylosing spondylitis (a spinal disorder). Cases of leukemia and cancers of the lung, esophagus, bone, and other organs exposed to radiation were reported relatively more from these patients. Since the dangers of radiation exposure were not known before the 1920s, many radiologists who were extensively exposed to x-rays without shielding developed leukemia. There is also some evidence that suggests radiation therapy given for cancer treatment can also be deleterious to health. Some organs, like relatively the breast, thyroid, and bone marrow, are highly sensitive to the effects of ionizing radiation. Radiation-induced cancer develops over a period of time; it takes around 2 years for leukemia to develop and around 10 to 15 years for other cancers to develop. It is best to avoid unnecessary medical x-rays; however, sometimes the benefits outweigh the risks, and early detection by the x-ray technique plays a critical role in the treatment of cancers.

1.4.4 CHEMICALS

Synthetic products like plastics, insulators, detergents, refrigerants, paints, deodorants, dyes, and cosmetics are just a few names that have become an integral part of our house and life. We have not only embraced them as a part of necessity or luxury, but also gone a step beyond by making our body a junkyard of various chemicals. Dr. Theo Colborn, of World Wildlife Fund and coauthor of the 1996 bestseller *Our Stolen Future*, has aptly described the current scenario: "Every one of you sitting here today is carrying at least 500 measurable chemicals in your body that were never in anyone's body before the 1920s." One study conducted in the United States regarding the "body burden" shows that a typical nonsmoking, middle-aged person harbors around 53 known carcinogens in his or her body fluids (urine and blood) (Colborn et al., 2006).

The causal link between environmental exposures and childhood cancer is well documented. There is growing evidence that confirms the role of parental or childhood exposures to certain toxic chemicals, including solvents, pesticides, petrochemicals, and certain industrial by-products (dioxins and polycyclic aromatic hydrocarbons) in causing childhood cancer. The risk of the development of breast cancer in women increases by twofold when they are occupationally exposed to deleterious chemicals while working in the wood, furniture, printing, or chemical industry.

The absence of natural pathways to break down synthetic chemicals is the root cause of the problem. The toxic buildup of these synthetic chemicals happens because of the absence of natural ways to eliminate or recycle these chemicals. We have not only surrounded ourselves with these chemicals, but have also kept them inside our body, usually tucked inside/between fatty tissues.

1.5   EPIDEMIOLOGY

Epidemiology is a study to identify causal factors of disease through large-scale studies conducted at the population level. It helps in drafting policies for disease prevention, dedicating effort toward susceptible subgroup(s), identifying risk factors, identifying emerging diseases, and personalizing treatment options. Current knowledge tells us that cancer is a multifactorial disease with some common factors shared by all cancers, and few factors specific to a particular cancer. Association of causal factors with cancer varies in degree of strength. For any cancer there are strong factors—smoking, tobacco etc.—along with weak factors. The coexistence of various factors puts persons or a population at greater risk of developing cancer. Establishing causality of some slow-acting factors with cancer requires designing long-term epidemiological studies. For instance, it took nearly 10 years for Hiroshima survivors to develop cancer because of brief exposure to intense radiation caused by the atomic bomb. Precise identification of causal factors in long-term studies becomes challenging due to the emergence of additional parameters, like age, personal health, migration, etc. To disambiguate causality in epidemiological studies, it is a common practice to form homogenous groups and compare results across such groups.

Epidemiological studies provide a bird's-eye view of causal factors, but lack the underpinning mechanism by which these risk factors cause/aid phenotypic behavior like tumor growth. Epidemiological studies form the basis for bench studies in which mechanistic details of these factors are worked out at cellular and molecular levels.

Causal factors can be broadly grouped as genetic or environmental. A study of immigrants helps in identifying whether the factor is genetic or environmental. For instance, Japan has a high incidence rate of stomach cancer when compared to colon cancer, whereas Hawaii has a higher incidence rate of colon cancer when compared to stomach cancer. The cancer pattern of second-generation Japanese immigrants in Hawaii is similar to that of natives of Hawaii, which suggests the nongenetic basis for most cancers.

Cancer places patients in a vulnerable position even after their primary treatment. Cancer survivors have a 20% higher risk of developing another primary cancer than the baseline normal population. This increased risk among survivors can be attributed to the same risk factors (like genetics, tobacco and alcohol consumption, obesity, exposure to radiation, etc.) that first initiated cancer. Sometimes even treatment options (like mutagenic chemotherapeutic drugs or radiation) also cause development of a new primary cancer. It can also be attributed to the higher detection rate, because of patients' compliance to follow-up visits for regular screening.

Molecular epidemiology is the current area of research in which studies are conducted to establish connections between genetic and environmental factors. In various observational studies we have seen that not all persons exposed to similar environmental conditions develop cancer, and it has also been observed that there is diversity in cancer type/subtype among patients exposed to similar conditions. Such findings suggest an interplay between genes and environmental factors (possible carcinogens). Chemicals present in the environment are metabolized into innocuous substances, and by certain genes, if this activity is compromised due to certain genetic reasons, it leads to buildup of toxicants within the body and consequently development of cancer. Molecular epidemiological studies are conducted with the objective to pinpoint such a risk genotype that makes the person more susceptible toward developing cancer when exposed to a particular environmental factor.

1.6   DIAGNOSIS AND PATHOLOGY

Pathology is the study of tissues and cells using the microscope, in order to ascertain the presence or absence of the diseased condition. Pathology is regarded as a gold standard for cancer diagnosis and classification. Downstream treatment options like surgery and chemotherapy depend heavily on the result of pathological examination; therefore, any misdiagnosis can cause great damage to subjects' well-being. The reliability of pathological examination is at stake, at times because of its very nature to be subjective in nature; the same sample can be labeled differently by an independent pathologist, and sometimes the results are totally in contrast.

Following are the tests in practice for cancer diagnosis:

- Physical examination
- Imaging—used to see the tumor or its related damage
  - Computed tomography (CT)
  - Magnetic resonance imaging (MRI)
  - X-rays
  - Ultrasonography
  - Radionuclide scanning
- Biopsy
  - Needle biopsy
  - Surgical biopsy
- Pathology tests—the analysis of biopsy tissue
- Other tests, depending on the specific type of cancer

Some more specific tests are:

Antimalignin antibody screen (AMAS): Capable of detecting cancers much in advance compared with traditional tests. It can detect cancer months before the appearance of signs and symptoms. This test is not useful for such cases of advanced cancer where antimalignin antibody is absent from the system.

Biological terrain assessment (BTA): A computerized device to estimate the health status of the cell. It simultaneously measures body fluids like blood, saliva, and urine for the pH balance, amount of electrons present, and minerals. It shows if the cellular environment is too acidic, if there are too few electrons to combat free radicals or produce energy, or if there are too few minerals to buffer the acids.

Cancer marker tests: When normal cells transform progressively into a cancerous state, there is change in the cellular environment, and this change is also reflected at the level of biomolecules (proteins/genes), which act as cancer markers. Properly designed marker-based tests have the ability to detect cancer in its nascent stages, and thereby increase the chance of recovery by manyfold. Some of the commonly used cancer marker tests are:

- *Alpha fetoprotein* (AFP) is a cancer marker that is elevated in testicular and liver cancers. Its level is also reported to be elevated in cases of pregnancy or some gastrointestinal cancers.
- *Cancer antigen 15.3* (CA 15.3) is known to be elevated in cases of breast cancer. Its level is also reported to be elevated in cases of cirrhosis and benign diseases of ovaries and breast.
- *Cancer antigen 19.9* (CA 19.9) is a cancer marker for gastric/pancreatic or stomach cancer. Its level is best evaluated along with the CEA marker test.
- *Cancer antigen 125* (CA 125) is a widely used tumor marker for cancers of the female reproductive system, including the uterus, fallopian tubes, and ovaries. Its level is also reported to be elevated in cancers of the pancreas, lungs, breast, and colon. It is also elevated in some noncancerous conditions: during menstruation, pregnancy, or in individuals with ovarian cysts, pericarditis, hepatitis, cirrhosis of the liver, or peritonitis, an infection of the lining of the abdomen, and even in 1 to 2% of healthy individuals. It also acts as a prognostic marker, and its reduced level after treatment indicates good prognosis.
- *Carcinoembryonic antigen* (CEA) is commonly used as a cancer marker for colorectal cancer. The medullary thyroid cancer (MTC) can also be detected with this cancer marker.
- *Epstein-Barr virus protein* (EVP) is a cancer marker elevated in nasopharyngeal cancer.
- *DR-70* is a blood test designed to detect the presence of fibrin and fibrinogen degradation products. The level of fibrin and fibrinogen degradation products is elevated in the presence of malignant tumor and increases as malignancy grows. It's a versatile blood test that can be used to screen for 14 different types of cancers, including lung, breast, stomach, liver, colon, rectal, ovarian, esophageal, cervical, trophoblastic, thyroid, malignant lymphoma, brain, and pancreatic cancers.

Pylori probe: Detects the presence of *Helicobacter pylori* in the stomach, which is the causative agent for ulcers and possibly stomach cancer.

Darkfield microscopy: A technique that allows monitoring of blood cells in continuum. It can indicate the presence of cancer in the body. It is based on the general observation that a tumor exists under conditions where the oxygen supply is low. Although this technique is not approved by the FDA, it's been used by a number of physicians practicing alternative medicine or naturopathy.

Electrodermal screening (EDS): An acupuncture-based computerized screening test. It works by taking readings at different acupuncture points. It gives an overview about the health of various organs. It can help in the detection of various diseases and the presence of toxins, allergens, and irritants in the body. It is used by a number of physicians practicing alternative medicine or naturopathy as a method to monitor or detect cancer.

Endoscopic ultrasound: Used to detect tumors associated with GI cancers. It's a noninvasive method that leverages an instrument's ability to reach into the GI tract and see through the tissue into surrounding organs.

Lymphocyte size analysis: Developed by Valentin Govallo, a Russian immunologist, this test is based on size, physical attributes, and count of lymphocytes. It is based on the hypothesis that the number of swollen lymphocytes increases in malignant conditions like cancer.

Maverick monitoring test (MMT): Monitors the level of malondialdehyde (MDA) in the urine or blood. The level of malondialdehyde is indicative of oxidative stress in the body. Malondialdehyde is a reactive oxygen species (ROS), which is one of the factors that causes DNA damage and pushes cells toward either apoptosis (in normal cells) or proliferation (under cancerous conditions). It can be used as a tool to check the effectiveness of natural chemopreventive agents, which contains antioxidants.

Positron emission tomography (PET) scan: It gives a dynamic image of the body's interior. It helps doctors visualize the body's metabolism and goes beyond standard techniques like x-ray (which gives a picture of bones) and MRI (which gives a picture of internal organs and soft tissues). PET enables physicians to monitor the energy consumption level in various organs and body parts. Sugar acts as a source of fuel to cells in our body. The structure of sugar is modified to attach radioactive tags. Cancer cells are characterized by high proliferative growth and require relatively high quantities of glucose as a source of energy to fuel this growth. Radiolabeled sugar allows physicians to detect such *hot spots* that are suspected tumor sites. It comes in handy to detect localized as well as metastatic cancer. It has the ability to overpower CAT in terms of detection power, as it is based on the principle of energy consumption at tumor sites, rather than ambiguous physical attributes like tumor size.

Thermography: Uses thermal imaging to detect new blood vessels and chemical changes associated with a tumor's origin and growth. It uses the radiation of infrared heat from our body and translates this information into anatomical images.

T/Tn antigen test: Based on the presence of T/Tn antigens and was developed by Dr. Georg Springer. The T and Tn antigens are surface proteins present on blood and skin cells and can be identified by the immune system antibodies. The concentration of these antigens varies depending on the cancer type and stage.

Whole-body CT scans: Radiation-based diagnostic method that is used for screening patients for the presence of any tumor in the body. Radiation is already known to induce the development of cancer; therefore, this test should be used only as an alternative to a traditional x-ray-based examination.

1.7   CONCLUSION

Cancer has existed probably since the dawn of civilization; however, it came into prominence during the 19th century. There could be many reasons why cancer remained out of public attention. Age is one of the strongest causal factors in the case of most of the cancers, and symptoms of cancer manifest or become more prominent in old age; therefore, it was regarded as just another old age-related health complexity. Before the 19th century, there were other diseases, like cholera, plague, etc., that took the lives of millions of people around the world, because of the absence of proper medical facilities. The life expectancy in those times was much less than that of today because of the absence of proper medical facilities, and other factors, like war. Simplicity in lifestyle and food habits was also responsible for keeping a check on cancer-related incidences. In the last couple of centuries we have witnessed tremendous progress in medical science and healthcare, which is why the average life expectancy has increased. Owing to increased life expectancy, there are more chances for age-related malignancies like cancer to appear in the lifetime of an individual. The so-called modern life is often accompanied by various vices like processed food, sedentary lifestyle, smoking, dinking, etc., which are known to trigger cancer initiation.

Cancer has become one of the biggest killers of our time, ranking second only to deaths caused by cardiovascular disease. The huge mortality and economic burden associated with cancer has brought it into prominence for policy makers, pharmaceutical companies, and researchers with the common objective to control mortality and improve the survival of affected persons. The economic burden due to cancer incidences is increasing every day; according to an estimate, the amount of money spent on cancer treatment has doubled in the past two decades. The mental agony it puts on cancer patients, and their near and dear ones, cannot be enumerated by any statistics.

The multifactorial role of cancer is established beyond any doubt, which necessitates devising treatment and detection strategies that address these factors. Cancer research is moving toward establishing strength of causality for various factors that contribute in various phases of cancer (initiation, progression, tumorigenesis). Cancer research started with a reductionist approach, wherein scientific experiments were designed to understand the role of individual factors (gene/protein) in causing cancer. The knowledge gained through such a reductionist approach has helped us to get a basic understanding of the biological complexity of cancer. Since the nature of cancer is multifactorial, it is logical to search for a medicinal compound that has the capability of modulating multiple factors at the same time. It also explains why combination therapy works in the case of cancer treatment. Combination therapy is designed to address these multiple factors and bring about a synergistic effect.

Timely detection of cancer in its nascent stages can be regarded as one of the most effective methods to control mortalities due to cancer. The majority of cancers are detected on the basis of appearance of some overt symptoms. The detection of cancer at more advanced stages makes management of cancer more complex, with little hope of recovery with existing treatment options. Incorrect identification of a cancer subtype is also one of the reasons why a cancer patient doesn't respond to drugs designed for a specific class of cancer. With the current advances in technologies, like next-generation sequencing, now it is possible to accurately characterize cancer-related features like stage, subtype, resistance to certain drugs, etc. It is more than obvious that cancer is predominantly a man-made malignancy, which means our concerted and sincere efforts can bear fruit by eradication of causal factors like tobacco, alcohol, harmful chemicals, and other carcinogens. Adapting a healthy lifestyle and diet can also help tremendously in preventing cancer.

We are living in a fast-changing world with path-breaking technological advancements as the norm of the day. With these advancements, it should not be an overoptimistic expectation to believe that cancer would be treated as routine diseases and viruses like flu, infection, etc. We firmly believe that this would be possible by an effective diagnosis and treatment strategy tailored to address phenotypic and genotypic heterogeneity associated with cancer. We envision the following milestones to be achieved by the research community in the near future:

- Development of noninvasive, handy, simple yet efficient diagnostic kit for cancer detection
- Characterization of diagnostic, prognostic efficacy biomarkers
- Unraveling of molecular events associated with disease pathogenesis
- Development of rule-based therapy to deal with heterogeneity in cancer (personalized cancer treatment)
- Clinical use of effective therapies like nanomedicine, oncolytic viruses, immunotherapy, and network medicine

REFERENCES

Ades, S. 2009. Adjuvant chemotherapy for colon cancer in the elderly: Moving from evidence to practice. *Oncology* 23: 1–3.

American Cancer Society. 2009. Cancer facts and figures 2009. No. 500809, pp. 1–68.

Arnold, L.M., 3rd, S.D. Burman, and A.H. O-Yurvati. 2010. Diagnosis and management of primary pulmonary leiomyosarcoma. *J Am Osteopath Assoc* 110: 244–246.

Banerjee, A.K. 2009. Preinvasive lesions of the bronchus. *J Thorac Oncol* 4: 545–551.

Bisen, P.S. 2012a. Editorial on enabling oral cancer translational research in India. *J Cancer Sci Ther* 4: i. doi:10.4172/1948-5956.1000e103.

Bisen, P.S. 2012b. Editorial on marine microbes: Unexplored therapeutic bio-mine. *J Cancer Sci Ther* 4: xviii–xix. doi:10.4172/1948-5956.1000e112.

Bisen, P.S., S. Bundela, and A. Sharma. 2012b. Ellagic acid—Chemoprotective role in oral cancer. *J Cancer Sci Ther* 4: 23–30. doi:10.4172/1948-5956.1000106.

Bisen, P.S., M. Debnath, and G.B.K.S. Prasad. 2012a. Human and microbial world. In *Microbes concepts and applications*, 1–64. Wiley-Blackwell, Hoboken, NJ.

Brayand, F., and B. Moller. 2006. Predicting the future burden of cancer. *Nat Rev Cancer* 6: 63–74.

Buecker, P. 2005. Sarcoma: A diagnosis of patience. ESUN 2(5). http://sarcomahelp.org/learning_center/patience.html.

Chung, L.W., A. Baseman, V. Assikis, et al. 2005. Molecular insights into prostate cancer progression: The missing link of tumor microenvironment. *J Urol* 173: 10–20.

Colborn, T., D. Dumanoski, and Meyers, J.P., *Our stolen future: Are we threatening our fertility, intelligence, and survival?* Dutton, 1996, U.S.

Debnath, M., P.S. Bisen, and G.B.K.S. Prasad. 2010. Personalised medicine. In *Molecular diagnostics: Promises and possibilities*, 393–412. Springer, Dordrecht.

Dikshit, R. Gupta, P.C., Ramasundarahettige, C., Gajalakshmi, V., Aleksandrowicz, L. et al. (2012). Cancer mortality in India: A nationally representative survey. *Lancet*. 379(9828): 1807–16.

Fearon, K.C.H., M.F. von Meyenfeldt, A.G.W. Mosses, et al. 2003. Effect of a protein and energy dense n-3 fatty acid enriched oral supplement on loss of weight and lean tissue in cancer cachexia: A randomised double blind trial. *Gut* 52: 1479–1486.

Ishizumi, T., A. McWilliams, C. MacAulay, et al. 2010. Natural history of bronchial preinvasive lesions. *Cancer Metastasis Rev* 29: 5–14.

Jemal, A., R. Siegel, E. Ward, et al. 2007. Cancer statistics, 2007. *CA Cancer J Clin* 57: 43–66.

Jemal, A., R. Siegel, J. Xu, et al. 2010. Cancer statistics, 2010. *CA Cancer J Clin* 60: 277–300.

Larsson, S., N. Orsini, and A. Wolk. 2010. Vitamin B6 and risk of colorectal cancer: A meta-analysis of prospective studies. *JAMA* 303: 1077–1083.

Levy-Lahad, E., and S.E. Plon. 2003. A risky business—Accessing breast cancer risks. *Science* 302: 574–575.

Lubin, J.H., M.C.R. Alavanja, N. Caporaso, et al. 2007. Cigarette smoking and cancer risk: Modeling total exposure and intensity. *Am J Epidemiol* 166: 479–489.

MacCarthy, A., J.M. Birch, G.J. Draper, et al. 2009. Retinoblastoma in Great Britain 1963–2002. *Br J Ophthalmol* 93: 33–37.

Morgan, R., A. Boxall, A. Bhatt, et al. 2011. Engrailed-2 (EN2): A tumor specific urinary biomarker for the early diagnosis of prostate cancer. *Clin Cancer Res* 17: 1090–1098.

Osuna, D., and E. de Alava. 2009. Molecular pathology of sarcomas. *Rev Recent Clin Trials* 4: 12–26.

Ottaviani, G., and N. Jaffe. 2010. The epidemiology of osteosarcoma. In *Pediatric and adolescent osteosarcoma*, ed. N. Jaffe, O.S. Bruland, and S. Bielack, 3–13. Vol. 152. Springer, New York.

Pagano, J.S., M. Blaser, M.A. Buendia, et al. 2004. Infectious agents and cancer: Criteria for a causal relation. *Semin Cancer Biol* 14: 453–471.

Papalas, J.A., N.N. Balmer, C. Wallace, et al. 2009. Ossifying dermatofibroma with osteoclast-like giant cells: Report of a case and literature review. *Am J Dermatopathol* 31: 379–383.

Piovanello, P., V. Viola, G. Costa, et al. 2007. Locally advanced leiomyosarcoma of the spleen: A case report and review of the literature. *World J Surg Oncol* 5: 135.

Psychas, V., P. Loukopoulos, Z.S. Polizopoulou, et al. 2009. Multilobular tumour of the caudal cranium causing severe cerebral and cerebellar compression in a dog. *J Vet Sci* 10: 81–83.

Ries, L.A.G., D. Harkins, M. Krapcho, et al., eds. 2006. SEER cancer statistics review, 1975–2003. National Cancer Institute, Bethesda, MD. http://seer.cancer.gov/csr/1975_2003/ (based on November 2005 SEER data submission, posted to the SEER website 2006).

Ryerson, A.B., C. Eheman, J. Burton, et al. 2007. Symptoms, diagnoses, and time to key diagnostic procedures among older U.S. women with ovarian cancer. *Obstet Gynecol* 109: 1053–1061.

Saif, M.W. 2007. Pancreatoblastoma. *J Pancreas* 8: 55–63.

Sanders, R.P., and W.L. Furman. 2006. Familial adenomatous polyposis in two brothers with hepatoblastoma: Implications for diagnosis and screening. *Pediatr Blood Cancer* 47: 851–854.

2 Oral Cancer

KEY WORDS

Human papillomavirus
Oral cancer
Premalignant lesions
Squamous cell carcinoma
Tobacco consumption

2.1 INTRODUCTION

The oral cavity is a part of the upper aerodigestive tract that begins at the lips and ends at the anterior surface of the faucial arch. Oral cancer is an umbrella term that consists of cancers that originate in the oral tissues. The primary site of origin of oral cancers is submucous tissues, the epithelium, and minor salivary gland. Some of the other common subsites of oral carcinoma are the alveolus, tongue, buccal mucosa, and gingivobuccal sulcus.

Oropharyngeal cancer is responsible for the deaths of thousands of people every day (Menck et al., 1991). It is very surprising that cancers of the oral cavity remain undetected in its nascent stages, despite easy accessibility of the oral cavity. Approximately 40,000 new cases of oral cancer are reported every year in the United States. Oral cancer occurs predominantly in males with a frequency of incidence over two times that of females (Neville et al., 2002; Swango, 1996; Ries et al., 1991). The incidence rate has declined annually by 1.4% in men and 1.1% in women during the past couple of decades. Age was previously regarded as one of the important causal factors; however, it has been increasingly reported among men younger than 50, which is associated with human papillomavirus (HPV) infection (WHO Fact Sheet, 2011).

2.2 EPIDEMIOLOGY

Oral cancer commonly occurs in middle-aged and older individuals, although a disturbing number of these malignancies have also been documented in younger adults in recent years (Chen et al., 1990; Llewellyn et al., 2001; Schantz and Yu, 2002). From an epidemiological and clinicopathological perspective, oral cancer can be divided into three categories: (1) carcinomas of the oral cavity proper, (2) carcinomas of the lip vermilion, and (3) carcinomas arising in the oropharynx. Males were more susceptible for oral cancer than females; however, now relatively more females are contributing toward the global oral cancer population, which can be attributed to lifestyle changes like exposure to carcinogens such as tobacco and alcohol (Silverman, 1998; Chen et al., 1990; Khan and Bisen, 2013).

The distribution of oral cancer incidence rates varies in different geographical regions. The annual age-adjusted incidence rates per 100,000 in several European countries vary from 2.0 (UK, South Thames region) to 9.4 in France. In the Americas the incidence rates vary from 4.4 (Cali, Colombia) to 13.4 in Canada. In Asia, it ranges from 1.6 (Japan) to 16.0 (India). In Australia and New Zealand, it varies from 2.6 (New Zealand, Maori) to 7.5 in South Australia. In Papua, New Guinea, and in the Lowlands and Highlands the incidences per 100,000 among men were 6.8 and 1.0 and among women 3 and 0.4, respectively. In Iran the incidence was reported to be 1.1 per 100,000 per year (Atkinson et al., 1982; Fahmy et al., 1983). India is one of the countries with a high oral cancer incidence rate (~16 to 18 per 100,000), which accounts for about 200,000 cases of oral cancer reported every year in India (study by Indian Council of Medical Research). Lips and the floor of the mouth are the most common sites of oral cancer in Indians, which together constitute over 30% of reported cases of oral cancer. According to an observational study among 150,000 villagers aged 15 years and above across six districts in India, an oral cancer prevalence rate of 0.1% was reported (Mehta et al., 1972; Mehta, 1982).

In contrast to intraoral and oropharyngeal carcinomas, cancers of the lip vermilion are more akin epidemiologically to squamous cell carcinoma of the skin and occur primarily in white men (Neville et al., 2002). Some of the strong causal factors of lip tumors are chronic sun exposure, cigarettes, or pipestems held habitually near the lip vermilion (Silverman, 1998). Cancers of the lip are more common in men, who are relatively more exposed to sun due to their vocational constraints. However, the incidence rate has declined significantly in the past half century, mostly because of changes in working conditions that seldomly require men to work outdoors (Neville et al., 2002, Silverman, 1998).

The 5-year survival rate for oral cancer has remained constant at about 50 to 55% in the past several decades, in spite of advancements in the field of surgery, radiation, and chemotherapy (Silverman, 2001; Ries et al., 1991). African Americans have a significantly higher mortality rate than whites (4.4 vs. 2.4 per 100,000 populations), most probably because of detection of tumors at an advanced stage (Silverman, 2001; Ries et al., 1991; Goldberg et al., 1994; Caplan and Hertz-Picciotto, 1998). The 5-year survival rate for lip cancer is 95% (Neville et al., 2002; Silverman, 2001). According to mortality statistics and projections for oral cancer, it isn't among the top 5 causes of death due to cancer; however, analysis of global trends projections for 2002 to 2030 makes some grim predictions about oral cancer mortality. Deaths due to oral cancer are expected to grow by almost 60% (Matters and Loncar, 2006).

2.3   SIGNS AND SYMPTOMS

Symptoms of oral cancer include the following:

- A sore, irritation, lump, or thick patch in the mouth, lip, or throat
- A white (leukoplakia) or red (erythroplakia) patch in the mouth
- Difficulty in chewing, swallowing, and moving the jaw or tongue
- Lumps in the neck or cheek
- Numbness or bleeding in the mouth

- Loose teeth
- Pain in one ear without hearing loss

The above-mentioned symptoms could be due to primary or secondary infection in the mouth cavity; however, it is suggested to get screened for oral cancer if these symptoms persist for more than 2 weeks.

2.4 SEX AND AGE DISTRIBUTION

Oral cancer incidences have a well-defined gender pattern; they predominantly affect men (Fleming et al., 1982; Fahmy et al., 1983). However, association strength of gender as an independent causative factor is questionable; for instance, the distribution of oral cancer incidences among males and females in India was found to be proportional to the prevalence of tobacco habits among men and women in the general population (Mehta, 1982). Age is another important factor associated with risk of development of oral cancer. The peak occurrence of oral cancer varies between different geographical locations or population groups; for instance, in Western countries it is around 60 to 70, whereas in Asian countries it is usually around 50 to 60 (Paymaster, 1962; Fahmy et al., 1983; Sanghvi et al., 1986).

2.5 ETIOLOGY

Oral cancer has been associated with various risk factors identified through observational and epidemiological studies. Some of the important risk factors follow.

2.5.1 Tobacco and Alcohol Consumption

Tobacco use is one of the strongest causal factors for cancers of the oral cavity and pharynx. The associative link between oral carcinogenesis and extensive use of tobacco, in the form of either smokeless tobacco or cigarette smoking, has been established through several epidemiological studies (Winn, 1984; Preston-Martin and Correa, 1989; Hoffmann et al., 1991; Stich et al., 1992). The risk of developing oral cancer increases by over 17-fold in smokers compared with nonsmokers, and this risk may increase as much as 17-fold in heavy smokers, who smoke more than 80 cigarettes per day. Smoking is also associated with a poor prognosis among oral cancer patients who continue to smoke even after treatment. Risk of developing secondary cancer (primarily in the upper aerodigestive tract) raises by two- to threefold in treated oral cancer patients who continue to smoke, in comparison to patients who quit smoking after treatment.

DNA damage is known to be induced by tobacco use, more often through direct damage by carcinogens like benzo(a)pyrene (B(a)P) and tobacco-specific N'-nitrosamines (TSNAs) present in tobacco smoke (Preston-Martin and Correa, 1989; De Stefani et al., 1990; Stich et al., 1992; Martin et al., 1996; Baan et al., 1988; Pulera et al., 1997). TSNAs like N-nitrosonornicotine (NNN) and 4-(methylnitrosoamino)-1-(3-pyridyl)-1-butanone (NNK), have exhibited carcinogenicity in animals (Huberman et al., 1976; Hoffmann et al., 1982; Preston-Martin and

Correa, 1989). Some other factors along with tobacco are required to induce oral carcinogenesis, since continued intraoral placement of smokeless tobacco failed to evoke malignant conversion of oral mucosal cells of animals in vivo (Shklar et al., 1985; Park et al., 1986). Therefore, it is logical to believe that other factors, like alcohol consumption, nutritional deficiencies, and human papillomavirus (HPV) (zur Hausen, 1996; Khan and Bisen, 2013), along with smoking, cause oral cancer.

Consumption of tobacco plus alcohol poses a much greater risk than using either substance alone (Baan et al., 2007; Day et al., 1993). Betel quid or paan is consumed primarily in South Asian countries, with or without tobacco. In both forms it has been also associated with an etiology of oral cancer (Merchant et al., 2000). The genetic predisposition of an individual makes him or her more susceptible to oral cancer when exposed to carcinogens present in tobacco or alcohol. Most tobacco carcinogens are metabolized via complex enzymatic mechanisms. Glutathione S-transferases (GSTs) are a very important family of enzymes that catalyze the detoxification of a range of tobacco carcinogens, such as benzo(a)pyrene, ploycyclic aromatic hydrocarbons, monohalomethanes, etc. (Ketterer, 1988; Hayes and Pulford, 1995). Several studies have been conducted to check the association of genetic polymorphism in GSTs and oral cancer; polymorphism in *GSTM1* has been reported to be associated with oral cancer (Katoh et al., 1999). Alcohol has been regarded as a strong risk factor for oral cancer. Alcohol dehydrogenase type 3 (ADH3) metabolizes ethanol to acetaldehyde, a carcinogen. The ABH3(1-1) genotype is linked with an increased risk of ethanol-related oral cancer (Harty et al., 1997).

2.5.2 HUMAN PAPILLOMAVIRUS (HPV)

Infection with HPV (especially HPV 16 type), which is transmitted sexually, has been linked to a subset of oral cancer (D'Souza et al., 2007). The role of HPV as a sexually transmitted virus and its association with the etiology of oral cancer have been established in several studies (Herrero et al., 2003; Smith et al., 1998). HPV was detected in squamous cell papilloma, condylomas, focal epithelial hyperplasia, and malignant oral lesions (Chang et al., 1990; Jalal et al., 1992; Syrjänen, 1992; Anderson et al., 1994; Ostwald et al., 1994; Chiba et al., 1996; Paz et al., 1997; Wen et al., 1997). The detection rate of the genetic component from HPV in oral cancer biopsies is as large as 46 to 78% (Woods et al., 1993; Miller and White, 1996; Franceschi et al., 1996). It has been observed that only a small fraction of HPV-infected lesions rarely proceed to malignant transformation (zur Hausen, 1996), which suggests the presence of other factors, like carcinogens present in tobacco, is necessary for oral carcinogenesis.

2.5.3 DIET

A diet low in vegetables and fruits may play a role in oral cancer development (McCullough and Giovannucci, 2004; Steinmetz and Potter, 1991, 1996; Winn, 1995). Higher risk is also associated with high intake of meat and processed meat products (Levi et al., 1998). An estimated 50% reduction in risk of oral cancer is noted among people who consume an adequate daily amount of fresh fruits and vegetables (Warnakulasuriya, 2009).

2.5.4 OTHER FACTORS

Oral cancer risk increases with age, and it mostly occurs in people over the age of 40. Exposure to the sun has been linked with cancer of the lip. Lip cancer is reported to be increased after kidney transplantation (King et al., 1995). Lip cancer is significantly associated with the use of immunosuppressive agents like azathioprine and cyclosporin (Van Leeuwen et al., 2009; Li et al., 2003).

2.6 PATHOLOGY

Pathogenesis of oral cancer suggests the appearance of various stages in which the epithelium passes through progressive aggravating dysplasias before transitioning into the invasive cancer stage. These pathological changes correspond to change in characteristics of cells; for instance, cells and nuclei look more primitive, like basal cells with enlarged nuclei (called nuclear hyperplasia), with an elevated nuclear-cytoplasmic ratio. Compared to the normal keratinocytes, these cells are placed more closely together. The mitotic activity is elevated in dysplastic epithelium. Precancerous stages are marked by the presence of enlarged, tripolar, or star-shaped mitotic figures. Oral cancers are also characterized by the presence of premature production of keratin below the surface, which can be seen as individually keratinized cells or tight concentric rings of flattened keratinocytes (epithelial pearls). Poorly differentiated carcinomas are usually characterized by cellular necrosis and loss of cellular cohesiveness (acantholysis) (Walker et al., 2003).

2.6.1 ORAL PREMALIGNANCY

A precancerous lesion is defined as morphologically altered tissue in which malignancies due to cancer are more likely to occur than in its apparently normal counterpart (Schepman and van der Waal, 1995). The genesis of oral cancers is often preceded by premalignant lesions, e.g., leukoplakia, erythroplakia, and oral submucosal fibrosis (SMF) (de Villiers et al., 1985; Pindborg, 1980).

2.6.2 LEUKOPLAKIA

As defined by the World Health Organization, leukoplakia is a "white patch or plaque that cannot be characterized clinically or pathologically as any other disease" (Kramer et al., 1978). Histologically, oral leukoplakia can have hyperkeratosis (thickening of the stratum corneum), parakeratosis (increased number of nucleated cells near the surface), or acanthosis (elongation of rete pegs into the submucosa). The leukoplakia are often developed as a hyperkeratotic response to an irritant, and symptomatic; however, around 20% of leukoplakic lesions have signs of dysplasia or carcinoma during their first clinical detection (Axell et al., 1984). In 40% of leukoplakias, spontaneous regression is observed, 47% remain stable, and malignant transformation is observed in 3 to 6% of cases (Mehta and Hammer, 1993).

The location of oral leukoplakia has a significant correlation with the frequency of finding dysplastic or malignant changes at biopsy. The floor of the mouth was the highest-risk site, with 42.9% of leukoplakias showing some degree of epithelial dysplasia, carcinoma in situ, or unsuspected invasive squamous cell carcinoma. The tongue and lip were also identified as high-risk sites, with dysplasia or carcinoma present in 24.2 and 24%, respectively (Waldron and Shafer, 1975). Leukoplakia is best managed by surgical excision. Leukoplakia is not known to be associated with any causative agents other than tobacco.

2.6.3   DYSPLASIA

Dysplasia is a histologic term that describes varying degrees of abnormal epithelial changes, such as an increased nuclear-cytoplasmic ratio, an increased rate of mitotic figures, cellular pleomorphism, and nuclear hyperchromatism. The risk of malignant progression of dysplasia ranges from 10 to 14% (Silverman et al., 1984).

2.6.4   ERYTHROPLAKIA

Erythroplakia is a term used to describe oral mucosa lesions that appear as red, velvety plaques that cannot be ascribed clinically or pathologically to any other known condition. These lesions are associated with a higher risk of cancer development than leukoplakia (Kramer et al., 1978; de Villiers et al., 1985).

2.6.5   LICHEN PLANUS

Lichen planus is another premalignant mucocutaneous condition that can arise within the oral cavity. Its pathogenesis is a cell-mediated immune reaction to basal keratinocytes, and the potential for malignant progression to squamous cell carcinoma is 4% (Vas'Kovaskaia and Abramova, 1981).

2.6.6   ORAL SUBMUCOUS FIBROSIS (OSMF)

Oral submucous fibrosis (OSMF) a precancerous condition prevalent in India and other South Asian countries, and is attributed to the habit of chewing areca nut with or without tobacco (Pindborg et al., 1984). OSMF is a chronic mucosal condition with mucosal rigidity of varying intensity due to fibroelastic transformation of the juxtaepithelial connective tissue layer. OSMF is causally associated with the habit of areca chewing in the Indian subcontinent; however, its causal factors are different in Western countries, where it is strongly associated with HPV (90%) and *Candida albicans* (50%) (de Villiers et al., 1985). Unlike other precancerous lesions, OSMF does not show regression, even with cessation of tobacco chewing habit.

2.6.7   ORAL SQUAMOUS CELL CARCINOMA (OSCC)

Oral squamous cell carcinoma (OSCC) is the most common cancer of the head and neck. Each year it accounts for more than 300,000 cases worldwide, more than 30,000 cases in the United States, and more than 3,000 cases in Canada. The 5-year survival rate for oral SCC has remained at approximately 50% for the past several decades. A key factor in the lack of improvement in prognosis over the years is the fact that a significant proportion of oral SCCs are not diagnosed or treated until they reach an advanced stage. This diagnostic delay may be caused by patients (who may not report unusual oral features) or healthcare workers (who may not investigate observed lesions thoroughly), and it is presumed that such delays are longer for asymptomatic lesions. The prognosis for patients with oral SCC that is treated early is much better, with 5-year survival rates as high as 80%; in addition, the quality of life improves after early treatment, because a cure can be achieved with less complex and less aggressive treatment than is necessary for advanced lesions. Furthermore, many oral SCCs are believed to develop from oral premalignant lesions, and early detection and diagnosis of these premalignant lesions should be possible. Identification of high-risk oral pre-malignant lesions and intervention at premalignant stages could constitute one of the keys to reducing the mortality, morbidity, and cost of treatment associated with SCC. In addition, certain patients are known to be at high risk for head and neck cancer, specifically those who use tobacco or alcohol and those over 45 years of age. Such patients can be screened by physical examination, and early-stage disease, if detected, is curable. This chapter reviews recent advances in techniques for detecting lesions early and predicting their progression or recurrence (Khan and Bisen, 2013).

SCC is most often characterized as exophytic or ulcerative, or a combination of both (Alvi et al., 1996). Exophytic lesions are generally less common, slower grow-ing, and less infiltrative than ulcerative lesions. Ulcerative lesions are more common, and often appear as red or grayish ulcers with heaped-up edges that bleed easily. These ulcerative lesions can be deeply infiltrative. A complete gross pathologic description of the tumor should include a description of tumor size (described in terms of the lesion's surface area) and the depth of invasion (lesion thickness). SCCs are graded histologically from low-grade tumors that show extensive keratinization, infrequent mitosis, and little nuclear pleomorphism to high-grade tumors showing little keratin, much mitosis, and extreme nuclear pleomorphism. There are several variants of SCC, of which the two most common are basaloid squamous cell carci-noma and verrucous carcinoma. Basaloid SCC is an aggressive variant of squamous cell carcinoma, characterized by basaloid cells that are arranged in nests or cords with pseudoglandular spaces and a high mitotic rate (Alvi et al., 1996). Verrucous carcinoma accounts for fewer than 5% of all oral cavity carcinomas. These lesions are characterized by their whitish, warty, bulky cauliflower-like growth, with a broad base. These lesions have a predilection for the buccal mucosa; they have a more favorable prognosis than SCC and are considered a low-grade malignancy. More than 90% of oral cavity cancers have a squamous cell epithelial histogenesis. Kaposi's sarcoma (KS) is the most common oral malignancy associated with the acquired immunodeficiency syndrome.

2.7 STAGING

The staging of oral cancer is based on the TNM (stands for T-Tumor, N-Node, M-Metatasis) system from the American Joint Committee for Cancer Staging and End-Results Reporting (1998). Although the TNM system of staging has its own

TABLE 2.1

TNM Staging System for the Oral Cavity

Stage	Explanation
	Primary Tumor (T)
Tx	Primary tumor cannot be assessed
T0	No evidence of tumor
Tis	Carcinoma in situ
T1	Tumor $\leq$ 2 cm in greatest dimension
T2	Tumor > 2 cm but < 4 cm in greatest dimension
T3	Tumor > 4 cm in greatest dimension
T4	Tumor invades adjacent structures (mandible, tongue musculature, maxillary sinus, skin)
	Nodal Involvement (N)
Nx	Regional lymph nodes cannot be assessed
N0	No regional lymph node metastasis
N1	Metastasis in single ipsilateral lymph node, $\leq$3 cm in greatest dimension
N2a	Metastasis in single ipsilateral lymph node, >3 cm but <6 cm in greatest dimension
N2b	Metastasis in multiple ipsilateral lymph node, none > 6 cm in greatest dimension
N2c	Metastasis in bilateral or contralateral lymph node, none > 6 cm in greatest dimension
N3	Metastasis in a lymph node, >6 cm in greatest dimension
	Distant Metastasis (M)
M0	No distant metastasis
M1	Distant metastasis
	Stage Grouping
Stage I	$T_1N_1M_0$
Stage II	$T_2N_0M_0$
Stage III	$T_3N_0M_0$
	T_1 or T_2
Stage IV	$T_3N_1M_0$
	T_4N_0 or N_1M_0
	Any T, N_2, or N_3M_0
	Any T, any N, M_1

limitations, it is a useful system for treatment planning, prognostication, and comparison of treatment results. Clinical staging is based on physical examination and radiographic findings. Tumor (T) staging is based on the size of the primary tumor. Nodal (N) staging is based on the status and size of lymph nodes. Metastasis (M) staging refers to the absence (M0) or presence (M1) of distant metastasis (Table 2.1).

The survival of patients with oral and oropharyngeal cancer is strongly related to the stage of disease at diagnosis. According to 1999–2006 SEER data from the National Cancer Institute, the 5-year relative survival rate of patients with localized disease is 82.5%, and this survival rate drops to 32.2% in line with stage progression (Table 2.2).

TABLE 2.2

Stage Distribution and 5-Year Relative Survival by Stage at Diagnosis for 1999–2006, All Races, Both Sexes

Stage at Diagnosis	Stage Distribution (%)	5-Year Relative Survival (%)
Localized (confined to primary site)	33	82.5
Regional (spread to regional lymph nodes)	46	54.7
Distant (cancer has metastasized)	14	32.2
Unknown (unstaged)	7	53.2

2.7.1 MOLECULAR BASIS OF ORAL CARCINOGENESIS

The progression of oral cancer is a multistep and multifactor process requiring the accumulation of multiple genetic alterations (like inactivation of tumor suppressor genes (TSGs) or activation of oncogenes by mechanisms like mutation, hypermethylation, deletion, loss of heterozygosity, transcriptional repression, etc.), influenced by a patient's genetic predisposition (presence of risk allele), as well as by environmental influences, including tobacco, alcohol, chronic inflammation, and viral infection.

Chromosomal aberrations are observed in cancer that induce normal cellular mechanisms toward malignancy by silencing or repressing the expression of TSGs or activation of oncogenes. It has been reported that loss of alleles 3p14 and 9p21 occurs early on in the development of oral squamous cell carcinoma tumors (Uzawa et al., 2001). A high frequency of loss of heterozygosity (LOH) at chromosomal locations 13q and 17p has been reported in premalignant oral lesions and early carcinomas (El-Naggar et al., 1995). Aberrations in chromosome 9 are quite often in early tumor development; allelic losses at 9p21 have been reported in premalignant oral lesions and early carcinomas (Ohta et al., 2009). Cell proliferation regulator genes like p16 and p14 are located in 9p21. Allelic losses at 5p21 to 22, 22q13, 4q, 11q, 18q, and 21q are often found in association with advanced tumor stages and poorly differentiated carcinomas (Moles et al., 2008; Khan and Bisen, 2013).

Oral cancer cells attain malignancy by various adaptive mechanisms, with the objective of attaining limitless proliferative potential to attain tumorigenic growth. Oral cancer cells attain acquired self-sufficiency in growth signals for cell proliferation. erbB-1 and erbB-2 are 20.2% overexpressed in oral carcinomas, which are signs of the carcinogenic process, and such aberrations are very common in histologically nondysplasic premalignant oral lesions (Werkmeister et al., 2000). Overexpression of the epidermal growth factor receptor (EGFR) gene has been reported in several human cancers, including oral squamous cell carcinoma (Whyte et al., 2002; Khan and Bisen, 2013). Various antigrowth signals are active in normal cells to check abnormal growth, and under cancerous conditions such regulation becomes aberrant, which is evident by overexpression of retinoblastoma protein (pRb), cyclin D1, and CCND1 (Koontongkaew et al., 2000; Zhou et al., 2009; Sathyam et al., 2008; Marsit et al., 2008).

The rate of error during DNA replication has been identified as a critical event in tumorigenesis (Nelson and Mason, 1972; Loeb et al., 1974; Cairns, 1975). The infidelity of DNA polymerase was initially implicated for the error during DNA replication; however, of late the role of the DNA repair mechanism genetic instability leading to tumorigenesis has also gained much acceptance (Ishwad et al., 1995). DNA damage is one of the common events in the development of various cancers. Normal cells have multiple mechanisms to repair such DNA damage, which deters cells to enter cancerous growth. Some of the most prominent DNA repair mechanisms are the nucleotide excision repair (NER) system, base excision repair (BER) system, and O6 methylguanine DNA methyltransferase (MGMT) system. In an experiment it was observed that activity of NER is significantly delayed in HPV-immortalized human oral keratinocytes after exposure to UV irradiation (Rey et al., 1999).

Programmed cell death or apoptosis is another mechanism to bypass cell regulation. In oral cancer survival is achieved by aberrant changes for pro-apoptotic genes like p53 (Whyte et al., 2002), Bax (Schliephake, 2003; Cruz et al., 2002), Fas (Schliephake, 2003; Muraki et al., 2000), and survival genes like Bcl2 (Schliephake, 2003; Whyte et al., 2002). HSP27 and HSP70 are associated with mutations of the p53 gene, and HSP70 has been detected in OSCC (Schliephake, 2003). Immortalization is yet another mechanism adapted by cancer cells to achieve limitless growth. Telomerase activity is not undetectable in normal somatic cells; however, it can be evaluated in biopsied tissue from oral cancer (Liao et al., 2000). Factors promoting the growth of new blood vessels (angiogenesis) are imperative for tumorigenesis. Some of the pro-angiogenic genes implicated in OSCC are VEGF/VEGF-R, NOS_2, PD-ECGF, FGF-2, PGF-3, HIF1α, and cyclo-oxygenase 2 (COX-2) (Schliephake, 2003; Wakulich et al., 2002; Sudbo, 2004). Invasion and metastasis are mechanisms by which cancer cells find nutrients necessary for growth by invading to adjacent tissues. Some of the invasion and metastasis factors implicated in oral cancer are matrix metalloproteinases (MMPs), cathepsins, integrins, cadherins and catenins, desmoplakin/placoglobin, and Ets-1 (Schliephake, 2003; Hamidi et al., 2000; Bankfalvi et al., 2002).

The overexpression of human papillomavirus (HPV) E6/E7 genes was detected in tumorigenic oral cancer cells (Chen et al., 1997), which suggests the role of viral oncogenes in oral tumorigenesis. The role of p53 as a tumor suppressor gene is undisputable. Viral oncogene E6 interacts with E6-associated protein (E6-AP) and forms an E6/E6-AP complex, which binds and degrades p53, and thus helps tumorigenesis (Yin et al., 1992). The gene expression is also regulated by epigenetic mechanisms like methylation, acetylation, etc., and such regulation is affected in carcinogenesis. DNMT3B is involved in epigenetic control along with DNMT1. SNPs in DNMT3B have been postulated to play a causative role in several cancers, including OSCC (van Heerden et al., 2001). The expression of Ras association family genes (RASSF) in general and RASSF2 in particular was found to be altered by methylation in OSCC (Imai et al., 2008).

2.7.2 Precancerous Lesions of the Oral Cavity

Most oral cancers are preceded by premalignant lesions, e.g., leukoplakia, erythroplakia, and submucosal fibrosis (SMF), which can be detected early and treated (de Villiers et al., 1985; Pindborg, 1980). SMF is a collagen disorder that is characterized by extreme sensitivity to temperature/spices, whitening of mucosa, progressive trismus, and bleeding. Usual presentation is marble-like blanching of the mucosa, submucosal, palpable fibrotic bands, and white and raised patches with areas of ulceration or erythema. It is usually associated with a habit of areca chewing in tropical countries like India, but in the West 90% have association with HPV and 50% with *Candida albicans* (de Villiers et al., 1985). It is commonly seen in the Indian subcontinent, and 50 to 70% develops into cancer within a decade. Erythroplakia is a chronic red mucosal macule, 80% of which may harbor microinvasive carcinoma. Without therapy 60 to 90% of erythroplakia may turn into cancer in 5 to 10 years (de Villiers et al., 1985).

Leukoplakia is a whitish patch or plaque that cannot be charecterized clinically or pathologically as any other disease and which is not associated with any physical or chemical causative agent except the use of tobacco. Most leukoplakias are a hyperkeratotic response to an irritant and are symptomatic, but about 20% of leukoplakic lesions show evidence of dysplasia or carcinoma at first clinical recognition (Axell et al., 1984). However, some anatomic sites (floor of mouth and ventral tongue) have rates of dysplasia carcinoma as high as 45%. There is no reliable correlation between clinical appearance and the histopathologic presence of dysplastic changes except that the possibility of epithelial dysplasia increases in leukoplakic lesions with interspersed red areas. In one large study, lesions with an erythroplakic component had a 23.4% malignant transformation rate, compared with a 6.5% rate for lesions that were homogenous.

2.7.3 METASTASIS AND SURVIVAL

One of the major causes of high mortality due to oral cancer is its late-stage detection; by the time it is detected, cancer cells usually invade to adjacent regions (metastasis), which makes treatment very complicated, often with cancer recurrence. The migration of cancer cells to regional lymph nodes in the neck, which can happen soon after development of oral carcinoma, is made possible by rich lymphatic drainage from the oral cavity. Metastasis is very critically linked with the patient's survival; patients with carcinomas that have spread to regional lymph nodes have a poor survival rate compared with patients with localized carcinomas. The neoplasms that have spread to lymph nodes are advanced-stage lesions that require aggressive treatment. In some instances cancer cells spread via blood vessels, causing metastatic tumors to arise in distant organ systems (e.g., lungs), which makes a patient's survival essentially zero.

Metastasis is the most important prognostic indicator for patients with OSCC (Regezi and Sciubba, 1989). The process of metastasis consists of sequential and selective steps, including proliferation, induction of angiogenesis, detachment, motility, invasion into the circulation, aggregation and survival in the circulation, cell arrest in distant capillary beds, and extravasation into the organ parenchyma (Fidler, 1990). Cancer cells metastasize by breaking away from their tumor and invading the circulatory or lymphatic system, to be carried away to new locations. The body has several mechanisms to check such cell breakage and migration; however, cancer cells have evolved an elaborate mechanism to overcome such safeguards. To begin the process of metastasis, a malignant cell must first break away from the cancerous tumor. Cells in normal tissue are adhered with neighboring cells and with protein mesh present in the intercellular space. This protein mesh is known as an extracellular matrix (ECM). Metastasis by malignant cells is initiated by their breaking away from the cells around them, and from the extracellular matrix. Cell-extracellular matrix interactions are important for the survival and proliferation of normal epithelial cells. ECM molecules and their receptors, the integrins, are essential in proper tissue development, wound healing, tissue maintenance, and oncogenesis (Albelda and Buck, 1990). Altered expression of ECM molecules like collagens, fibronectin, laminin, and tenascin assists in metastasis. The human trophoblast cell surface antigen (TROP2) has been associated with shortened survival (Fong et al., 2008). The epithelial adhesion molecule (EpCAM) has been associated with tumor size and invasiveness (Yanamoto et al., 2007). Matrix metalloproteinase (MMP-2 and MMP-9) has been associated with the invasive potential of tumors and levels of alcohol, leading to inference that alcohol might play a role in oral carcinogenesis through the stimulation of these genes (Moles et al., 2008).

2.8   INDIVIDUAL CANCERS OF THE ORAL CAVITY

2.8.1   Lip Cancer

Lip cancer mostly affects the lower lip, and it predominantly affects males (Menck et al., 1991) (Figure 2.1). Some of the most possible causal factors are tobacco and UV exposure (Lindqvist and Teppo, 1978; Wynder et al., 1983). Lymph node metastasis

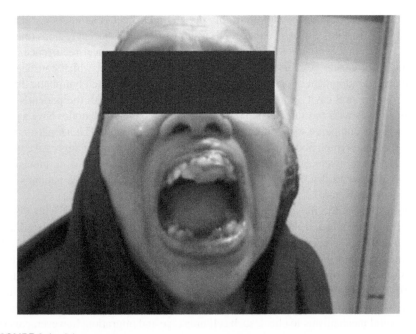

FIGURE 2.1   Lip cancer.

occurs with less frequency, around 5 to 10%. Smaller lesions can be resected with a V excision followed by primary closure of the resulting defect, whereas a larger lesion requires a transposition flap, e.g., Abbe Eastlander flap, Kerapandzic flap, Abbes flap, Nasolabial flap, or Gillies flap (Byers et al., 1981). The 5-year survival rate ranges from 95% (for T1 lesions) to 50% (for advanced metastatic form) (Guillamondegui et al., 1980).

2.8.2 ALVEOLAR AND RETROMOLAR TRIGONE CANCER

Around 10% of oral cancers belong to this class (Menck et al., 1991). Alveolar and retromolar trigone cancers are considered aggressive because of their propensity to metastasize to lymph nodes, mostly levels I and II. The 5-year survival rate ranges from 85–90% (for T1 lesions) to 20% (for advanced metastatic form) (Teichgraeber et al., 1985). Surgical methods usually require extensive resection with reconstruction.

2.8.3 FLOOR OF MOUTH CANCER

Around 10 to 15% of the oral cancer is developed at the floor of mouth (Menck et al., 1991). The 5-year survival rate ranges from 85–90% (for T1 lesions) to 32% (for advanced metastatic form) (Figure 2.2). Some of the features, like perineural invasion, depth of primary tumor invasion, and poor tumor differentiation, are some of that factors that indicate poor prognosis (Bloom and Spiro, 1980).

2.8.4 BUCCAL MUCOSA CANCER

Buccal mucosa is one of the common sites of tumorigenesis among tobacco-eating people. Some of the most overt symptoms are pain, ulcer, bleeding, and trismus. Surgery is the method of choice to treat localized tumor; small lesions are excised intraorally, but larger ones require omohyoid dissection (SOHD) and composite resections (Figure 2.3). Radiotherapy along with surgery is required to manage more advanced tumors. The 5-year survival rate ranges from 77% (for T1 lesions) to 18% (for advanced metastatic form) (Spiro et al., 1974).

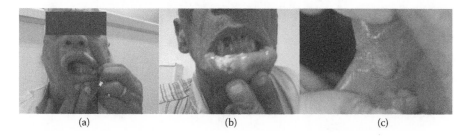

(a)   (b)   (c)

FIGURE 2.2   (a) Early oral cancer. (b) Oral cancer. (c) Early oral cavity cancer.

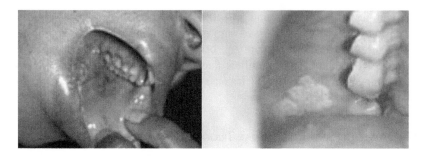

FIGURE 2.3   Erythroleukoplakia and leukoplakia.

2.8.5 Tongue Cancer

The incidence rate of tongue cancer is very high, and it comes next only to cancer of the lip (Menck et al., 1991). It affects younger people relatively more than those from an advanced age group (Evans et al., 1982). It is characterized by the presence of primary symptoms like pain, difficulty in deglutition, and phonation (Figure 2.4). It usually turns aggressive because of the high likelihood for lymph node metastasis (15 to 75%). The common treatment options are surgery, radiotherapy, and chemotherapy.

2.8.6 Hard Palate Cancer

Hard palate cancer constitutes around 5% of all oral cancer cases (Menck et al., 1991). Histologically, around 50% of hard palate cancers are squamous carcinoma, and the remaining 50% comprise adenocarcinoma and adenoid cystic carcinoma. The instances of lymph node metastasis are around 6 to 29%, which is the aggressive form of cancer and usually requires adjuvant radiotherapy as the treatment option, along with surgery. Surgical options for treating early stages involve infrastructure maxillectomy or near total palatectomy if necessary, with immediate prosthetic obturator.

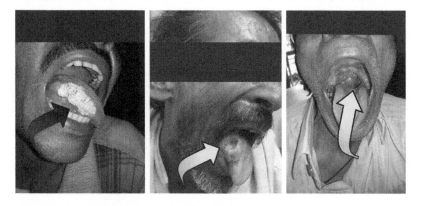

FIGURE 2.4   Oral and tongue cancer.

2.9 CONCLUSION

The mortalities due to oral cancer are rising worldwide, and the rate of increase is even higher in developing and underdeveloped nations. The underlying reasons for oral cancer incidences are primarily extrinsic, like tobacco, alcohol, and HPV infection. From the disease management point of view it looks very plausible to control deaths from oral cancer because of the relative ease of dealing with extrinsic factors compared to manipulating the genetic makeup of a susceptible individual. Tobacco is one of the strongest cancer-causing factors in various cancers, including those of the oral cavity. With effective regulations and policies, tobacco-related products in markets across the world can be controlled, and we can move toward systematic phasing out of these products. There have been several attempts to expose the ill effects of tobacco use, for instance, the statutory warning on cigarette packs, banning of smoking in public places, etc. These attempts can be best appreciated as an awareness drive; however, strong commitment from policy makers is missing from all these attempts. We can hope that a tremendous economic strain caused by a cancer burden worldwide would act as a catalyst for effective policies to control tobacco. Of late, HPV infection is getting notoriously associated with the increasing number of deaths due to cervical cancer, oral cancer, and oropharyngeal cancer. HPV infection can be prevented up to a certain extent through immunization, safe sexual practice, etc. Awareness about these causal factors and commitment at the individual level will make a positive difference in the prevention of oral cancer.

The oral cavity is one of the most readily accessible sites for physical examination, and therefore should help in the early detection of the origin of any malignancy. But it is surprising that most oral cancer cases are detected at relatively advanced stages, which is the main reason for the poor survival rate among cancer patients. The reasons for poor detection could be the ill-defined signs or guidelines for dental physicians to deduce the possible presence of oral cancer and the practice of not going for regular dental checkups, especially in less developed countries. It is a well-known fact that changes at the molecular level start way before the development of physical changes in tissue or organ, and such molecular level changes can be used for designing an effective detection system. Advancements in the field of high-throughput technologies like *omics* and next-generation sequencing have equipped us to accurately identify markers that can help in profiling patients who are at a greater risk of developing oral cancer.

The current treatment plan of oral cancer involves the convergence of various medical practices like surgery, chemotherapy, cosmetic surgery (for facial reconstruction), radiation therapy, pathology, etc. The contribution of chemotherapy in managing various cancers cannot be ruled out; however, it comes with the baggage of unavoidable side effects, which quite often harm vital organs of patients. Targeted therapies are gaining acceptance in the treatment of cancers of the breast, lungs, etc., and are now even prescribed for oral cancer. The efficacy of targeted therapy depends on the presence of conditions at the molecular level (e.g., presence of certain mutations, expression of certain genotypes, etc.); therefore, proper profiling of patients is required to achieve favorable outcomes.

REFERENCES

Albelda, S.M., and C.A. Buck. 1990. Integrins and other cell adhesion molecules. *FASEB J* 4: 2868–2880.

Alvi, A., E.N. Myers, and J.T. Johnson. 1996. Cancer of the oral cavity. In *Cancer of the head and neck*, ed. E.N. Myers and J.Y. Suen, 321–360. 3rd ed. Mosby, Philadelphia.

American Joint Committee for Cancer Staging and End-Results Reporting. 1998. American Joint Committee on Cancer, Chicago.

Anderson, J.A., J.C. Irish, C.M. McLachlin, et al. 1994. H-ras oncogene mutation and human papillomavirus infection in oral carcinomas. *Arch Otolaryngol Head Neck Surg* 120: 755–760.

Atkinson, L., R. Purohit, and Y.P. Reay. 1982. Cancer reporting in Papua New Guinea: 1958–70 and 1971–78. *Natl Cancer Inst Monogr* 62: 65–71.

Axell, T., P. Holmstrup, I.R.H. Kramer, et al. 1984. International seminar on oral leucoplakia and associated lesions related tobacco habits. *Commun Dent Oral Epidemiol* 12: 145–154.

Baan, R., K. Straif, Y. Grosse, et al. 2007. Carcinogenicity of alcoholic beverages. *Lancet Oncol* 8: 292–293.

Baan, R.A., P.T.M. Vander Berg, M.-J.S.T. Steenwinkel, et al. 1988. *Methods for detecting DNA damaging agents in humans: Application in cancer epidemiology and prevention*, 146–151. IARC, Lyon, France.

Bankfalvi, A., M. Krabort, A. Vegh, et al. 2002. Deranged expression of the E-cadherin/β-catenin complex and the epidermal growth factor receptor in the clinical evolution and progression of oral squamous cell carcinomas. *J Oral Pathol Med* 31: 450–457.

Bloom, N.D., and R.H. Spiro. 1980. Carcinoma of the cheek mucosa: A retrospective analysis. *Am J Surg* 140: 556–559.

Byers, R.M., R. Newman, N. Russel, et al. 1981. Results of treatment for squamous carcinoma of the lower gum. *Cancer* 47: 2236–2238.

Cairns, J. 1975. Mutation selection and the natural history of cancer. *Nature* 255: 197–200.

Caplan, D.J., and I. Hertz-Picciotto. 1998. Racial differences in survival of oral and pharyngeal cancer patients in North Carolina. *J Public Health Dent* 58: 36–43.

Chang, F., S. Syrjänen, J. Nuutien, et al. 1990. Detection of human papillomavirus (HPV) DNA in oral squamous cell carcinomas by in situ hybridization and polymerase chain reaction. *Arch Dermatol Res* 282: 493–497.

Chen, J.K., R.V. Katz, and D.J. Krutchkoff. 1990. Intraoral squamous cell carcinoma: Epidemiologic patterns in Connecticut from 1935 to 1985. *Cancer* 66: 1288–1296.

Chen, Z., K.A. Storthz, and E.J. Shillitoe. 1997. Mutations in the long control region of human papillomavirus DNA in oral cancer cells and their functional consequences. *Cancer Res* 57: 1614–1619.

Chiba, I., M. Shindoh, M. Yasuda, et al. 1996. Mutations in the p53 gene and human papillomavirus infection as significant prognostic factors in squamous cell carcinomas of the oral cavity. *Oncogene* 12: 1663–1668.

Cruz, I., P. Snijders, V. Van Houten, et al. 2002. Specific p53 immunostaining patterns are associated with smoking habits in patients with oral squamous cell carcinomas. *J Clin Pathol* 55: 834–840.

Day, G.L., W.J. Blot, D.F. Austin, et al. 1993. Racial differences in risk of oral and pharyngeal cancer: Alcohol, tobacco, and other determinants. *J Natl Cancer Inst* 85: 1416–1417.

De Stefani, E., N. Munoz, J. Esteve, et al. 1990. Mate drinking, alcohol, tobacco, diet, and esophageal cancer in Uruguay. *Cancer Res* 50: 426–431.

de Villiers, E.M., H. Weidauer, H. Otta, et al. 1985. Papilloma virus DNA in human tongue carcinomas. *Int J Cancer* 36: 575–578.

D'Souza, G., A.R. Kreimer, R. Viscidi, et al. 2007. Case-control study of human papillomavirus and oropharyngeal cancer. *N Engl J Med* 356: 1944–1956.

El-Naggar, A.K., K. Hurr, J.G. Batsakis, et al. 1995. Sequential loss of heterozygosity at microsatellite motifs in preinvasive and invasive head and neck squamous carcinoma. *Cancer Res* 55: 2656–2659.

Evans, S.J., J.D. Langdon, A.D. Rapidis, et al. 1982. Prognostic significance of STNMP and velocity of tumour growth in oral cancer. *Cancer* 49: 773–776.

Fahmy, M.S., A. Sadeghi, and S. Behmard. 1983. Epidemiologic study of oral cancer in Fars Province, Iran. *Community Dent Oral Epidemiol* 11: 50–58.

Fidler, I.J. 1990. Critical factors in the biology of human cancer metastasis: Twenty-eighth GHA Clowes Memorial Award Lecture. *Cancer Res* 50: 6130–6138.

Fleming, M., M. Shear, and M. Altini. 1982. Intraoral squamous cell carcinoma in South Africa. *J Dent Assoc S Afr* 37: 541–544.

Fong, D., G. Spizzo, J.M. Gostner, et al. 2008. TROP2: A novel prognostic marker in squamous cell carcinoma of the oral cavity. *Mod Pathol* 21: 186–191.

Franceschi, S., N. Munoz, X.F. Bosch, et al. 1996. Human papillomavirus and cancers of the upper aerodigestive tract: A review of epidemiological and experimental evidence. *Cancer Epidemiol Biomarkers Prev* 5: 567–575.

Goldberg, H.I., S.A. Lockwood, S.W. Wyatt, et al. 1994. Trends and differentials in mortality from cancers of the oral cavity and pharynx in the United States, 1973–1987. *Cancer* 74: 565–572.

Guillamondegui, O.M., B. Oliver, and R. Hayden. 1980. Cancer of the anterior floor of mouth: Selective choice of treatment and analysis of failures. *Am J Surg* 140: 560–562.

Hamidi, S., T. Salo, T. Kainulainen, et al. 2000. Expression of vß6 integrin in oral leukoplakia. *Br J Cancer* 82: 1433–1440.

Harty, L.C., N.E. Caporaso, R.B. Hayes, et al. 1997. Alcohol dehydrogenase 3 genotype and risk of oral cavity and pharyngeal cancers. *J Natl Cancer Inst (Bethesda)* 89: 1698–1705.

Hayes, J.D., and D.J. Pulford. 1995. The glutathione S-transferases supergene family: Regulation of GST and contribution of the isoenzyme to cancer chemoprevention and drug resistance. *Crit Rev Biochem Mol Biol* 30: 445–600.

Herrero, R., X. Castellsague, M. Pawlita, et al. 2003. Human papillomavirus and oral cancer: The International Agency for Research on Cancer Multicenter Study. *J Natl Cancer Inst (Bethesda)* 95: 1772–1783.

Hoffmann, D., K.D. Brunnemann, J.D. Adams, et al. 1982. N-Nitrosamines in tobacco carcinogenesis. In *Nitrosamines and human cancer*, ed. P.N. Magee, 211–225. Banbury Report 12. Cold Spring Harbor Laboratory, Cold Spring Harbor, NY.

Hoffmann, D., M.V. Djordjevic, and K.D. Brunnemann. 1991. New brands of oral snuff. *Food Chem Toxicol* 29: 65–68.

Huberman, E., L. Sachs, S.K. Yang, et al. 1976. Identification of mutagenic metabolites of benzo(a)pyrene in mammalian cells. *Proc Natl Acad Sci USA* 73: 607–611.

Imai, T., M. Toyota, H. Suzuki, et al. 2008. Epigenetic inactivation of RASSF2 in oral squamous cell carcinoma. *Cancer Sci* 99: 958–966.

Ishwad, C.S., R.E. Ferrell, K.M. Rossie, et al. 1995. Microsatellite instability in oral cancer. *Int J Cancer* 64: 332–335.

Jalal, H., C.M. Sanders, S.S. Prime, et al. 1992. Detection of human papillomavirus type 16 DNA in oral squames from normal young adults. *J Oral Pathol Med* 21: 465–470.

Katoh, T., S. Kaneko, K. Kohshi, et al. 1999. Genetic polymorhisms of tobacco- and alcohol-related metabolizing enzymes and oral cavity cancer. *Int J Cancer* 83: 606–609.

Ketterer, B. 1988. Protective role of glutathione transferases in mutagenesis and carcinogenesis. *Mutat Res* 202: 343–361.

Khan, Z., and P.S. Bisen. 2013. Oncoapoptotic signaling and deregulated target genes in cancers: special reference to oral cancer. *Biochim Biophy Acta Reviews on Cancer*, dxdoi.org/10.1016/j.bbcan.2013.04.002.

King, G.N., C.M. Heal, M.T. Glover, et al. 1995. Increased prevalence of dysplastic and malignant lip lesions in renal transplant recipients. *N Engl J Med* 332: 1052–1057.

Koontongkaew, S., A. Chareonkitkajorn, A. Chavitan, et al. 2000. Alterations in p53, pBb, cyclin D1 and cdk4 in human oral and pharyngeal squamous cell carcinomas. *Oral Oncol* 36: 334–339.

Kramer, I.R.H., D. El-Lablan, and K.W. Lee. 1978. The clinical features and risk of malignant transformation in sublingual keratosis. *Br Dent J* 144: 171–180.

Levi, F., C. Pasche, C. La Vecchia, et al. 1998. Food groups and risk of oral and pharyngeal cancer. *Int J Cancer* 31: 705–709.

Li, A.C.Y., S. Warnakulasuriya, and R.P.H. Thompson. 2003. Neoplasia of the tongue in a patient with Crohn's disease treated with azathioprine: Case report. *Eur J Gastroenterol Hepatol* 15: 185–187.

Liao, J., T. Mitsuyasu, K. Yamane, et al. 2000. Telomerase activity in oral and maxillofacial tumors. *Oral Oncol* 36: 347–352.

Lindqvist, C., and L. Teppo. 1978. Epidemiological evaluation of sunlight as a risk factor of lip cancer. *Br J Cancer* 37: 983–989.

Llewellyn, C.D., N.W. Johnson, and K.A. Warnakulasuriya. 2001. Risk factors for squamous cell carcinoma of the oral cavity in young people—A comprehensive literature review. *Oral Oncol* 37: 401–418.

Loeb, L.A., C.F. Springgate, and N. Battula. 1974. Errors in DNA replication as a basis of malignant changes. *Cancer Res* 34: 2311–2321.

Marsit, C.J., C.C. Black, M.R. Posner, et al. 2008. A genotype phenotype examination of cyclin D1 on risk and outcome of squamous cell carcinoma of the head and neck. *Clin Cancer Res* 14: 2371–2377.

Martin, L.M., J.E. Bouquot, P.A. Wingo, et al. 1996. Cancer prevention in the dental practice: Oral cancer screening and tobacco cessation advice. *J Public Health Dent* 56: 336–340.

Matters, C.D., and D. Loncar. 2006. Projections of global mortality and burden of disease from 2002 to 2030. *PLoS Med* 3: e442.

McCullough, M.L., and E.L. Giovannucci. 2004. Diet and cancer prevention. *Oncogene* 23: 6349–6364.

Mehta, F.S. 1982. An intervention study of oral cancer and precancer in rural Indian populations: A preliminary report. *WHO Bull* 60: 441–446.

Mehta, F.S., P.C. Gupta, D.K. Daftary, J.J. Pindborg, and S.K. Choksi. 1972. An epidemiologic study of oral cancer and precancerous conditions among 101,761 villagers in Maharastra, India. *Int J Cancer* 10: 134–141.

Mehta, F.S., and J.E. Hammer. 1993. *Tobacco-related oral mucosal lesions and conditions in India.* Basic Dental Research Unit, TIFR Publication, Mumbai.

Menck, H.R., L. Garfinkel, and G.D. Dodd. 1991. Preliminary report of the national cancer database. *CA Cancer J Clin* 41: 7–18.

Merchant, A., S.S.M. Hosain, M. Hosain, et al. 2000. Epidemiology and cancer prevention. *Int J Cancer* 86: 128–131.

Miller, C.S., and D.K. White. 1996. Human papillomavirus expression in oral mucosa, premalignant conditions, and squamous cell carcinoma: A retrospective review of the literature. *Oral Surg Oral Med Oral Pathol Oral Radiol Endod* 82: 57–68.

Moles, M.A.G., J.A.G. Montoya, and I.R. Avila. 2008. Base moleculares de la carizacion de cavidad oral. *Av Odontosetomatol* 24: 55–60.

Muraki, Y., A. Tateshi, C. Seta, et al. 2000. Fas antigen expression and outcome of oral squamous cell carcinoma. *Int J Oral Maxillofac Surg* 29: 360–365.

Nelson, R.L., and H.S. Mason. 1972. An explicit hypothesis for chemical carcinogenesis. *J Theor Biol* 37: 197–200.

Neville, B.W., D.D. Damm, C.M. Allen, et al. 2002. *Oral and maxillofacial pathology,* 337–369. 2nd ed. Saunders, Philadelphia.

Ohta, S., H. Uemura, Y. Matsui, et al. 2009. Alterations of p16 and p14ARG genes and 9p21 locus in oral squamous cell carcinoma. *Oral Surg Oral Med Oral Pathol Oral Radiol Endod* 107: 81–91.

Ostwald, C., P. Muller, M. Barten, et al. 1994. Human papillomavirus DNA in oral squamous cell carcinomas and normal mucosa. *J Oral Pathol Med* 23: 220–225.

Park, N.H., J.P. Sapp, and E.G. Herbosa. 1986. Oral cancer induced in hamsters with herpes simplex virus infection and simulated snuff-dipping. *Oral Surgery* 62: 164–168.

Paymaster, J.C. 1962. Some observations on oral and pharyngeal carcinomas in the state of Bombay. *Cancer* 15: 578–583.

Paz, I.B., N. Cook, T. Odom-Maryon, et al. 1997. Human papillomavirus (HPV) in head and neck cancer. An association of HPV 16 with squamous cell carcinoma of Waldeyer's tonsillar ring. *Cancer* 79: 595–604.

Pindborg, J.J. 1980. Pathology of oral leukoplakia. *Am J Dermatopathol* 2: 277–278.

Pindborg, J.J., P.R. Murti, R.B. Bhonsle, et al. 1984. Oral submucous fibrosis as pre-cancerous condition. *Scand J Dent Res* 92: 224–229.

Preston-Martin, S. and P. Correa. 1989. Epidemiological evidence for the role of nitroso compound in human cancer. *Cancer Surveys* 8: 459–473.

Pulera, N., S. Petruzzelli, A. Celi, et al. 1997. Presence and persistence of serum anti-benzo[a] pyrene diolepoxide-DNA adduct antibodies in smokers: Effects of smoking reduction and cessation. *Int J Cancer* 70: 145–149.

Regezi, J.A., and Sciubba J.J. 1989. Ulcerative conditions. In *Oral pathology: Clinical-pathologic correlations*, ed. J.A. Regezi and J.J. Sciubba, 70–83. W.B. Saunders, Philadelphia.

Rey, O., S. Lee, and N.H. Park. 1999. Impaired nucleotide excision repair in UV-irradiated human oral keratinocytes immortalized with type 16 human papillomavirus genome. *Oncogene* 18: 6997–7001.

Ries, L.A.G., B.F. Hankey, B.A. Miller, et al. 1991. Cancer statistics review 1973–1988. NIH Publication No. 91-2789. National Cancer Institute, Bethesda, MD.

Sanghvi, L.D., D.K. Jain, and S. Krishnamurthy. 1986. National Cancer Registry. Annual report 1983. Indian Council of Medical Research, New Delhi.

Sathyam, K.M., K.R. Nalinakumari, T. Abraham, et al. 2008. CCND1 ploymorhisms (A870G and C1722G) modulate its protein expression and survival in oral carcinoma. *Oral Oncol* 44: 689–697.

Schantz, S.P., and G.P. Yu. 2002. Head and neck cancer incidence trends in young Americans, 1973–1997, with a special analysis for tongue cancer. *Arch Otolaryngol Head Neck Surg* 128: 268–274.

Schepman, K.P., and I. van der Waal. 1995. A proposal for classification and staging for oral leukoplakia: A preliminary study. *Oral Oncol Eur J Cancer* 31B: 396–398.

Schliephake, H. 2003. Prognostic relevance of molecular markers of oral cancer—A review. *J Oral Maxillofac Surg* 32: 233–245.

Shklar, G., K. Niukian, M. Hassan, et al. 1985. Effects of smokeless tobacco and snuff on oral mucosa of experimental animals. *J Oral Maxillofacial Surg* 43: 80–86.

Silverman, S., Jr. 1998 Epidemiology. In *Oral cancer*, ed. S. Silverman Jr., 1–6. 4th ed. BC Decker, Hamilton, Ontario, Canada.

Silverman, S., Jr. 2001. Demographics and occurrence of oral and pharyngeal cancers: The outcomes, the trends, the challenge. *J Am Dent Assoc* 132: 7S–11S.

Silverman, S., M. Gorsky, and F. Lozada. 1984. Oral leukoplakia and malignant transformation: A follow-up study. *Cancer* 53: 563–568.

Smith, E.M., H.T. Hoffman, K.S. Summersgill, et al. 1998. Human papillomavirus and risk of oral cancer. *Laryngoscope* 108: 1098–1103.

Spiro, R.H., A.E. Alfonso, H.W. Farr, et al. 1974. Cervical node metastasis from epidermoid carcinoma of the oral cavity and oropharynx: A critical assessment of current staging. *Am J Surg* 128: 562–567.

Steinmetz, K.A., and J.D. Potter. 1991. Vegetables, fruit, and cancer. I. Epidemiology. *Cancer Causes Control* 2: 325–357.

Steinmetz, K.A., and J.D. Potter. 1996. Vegetables, fruit, and cancer prevention: A review. *J Am Diet Assoc* 96: 1027–1039.

Stich, H.F., B.B. Parida, and K.D. Brunnemann. 1992. Localized formation of micronuclei in the oral mucosa and tobacco-specific nitrosamines in the saliva of "reverse" smokers, Khaini-tobacco chewers and gudakhu users. *Int J Cancer* 50: 172–176.

Sudbo, J. 2004. Novel management of oral cancer: A paradigm of predictive oncology. *Clin Med Res* 2: 233–242.

Swango, P.A. 1996. Cancers of the oral cavity and pharynx in the United States: An epidemiologic overview. *J Public Health Dent* 56: 309–318.

Syrjänen, S. 1992. Viral infections in oral mucosa. *Scand J Dental Res* 100: 17–31.

Teichgraeber, J., J. Bowman, and H. Goepfert. 1985. New test series for the functional evaluation of oral cavity cancer. *Head Neck Surg* 8: 9–20.

Uzawa, N., D. Akanuma, A. Negishi, et al. 2001. Homozygous deletions on the short arm of chromosome 3 in human oral squamous cell carcinomas. *Oral Oncol* 37: 351–356.

van Heerden, W.F., T.J. Swart, B. Robson, et al. 2001. FHIT RNA and protein expression in human oral squamous cell carcinomas. *Oral Oncol* 37: 498–504.

Van Leeuwen, M.T., A.E. Grulich, S.P. McDonald, et al. 2009. Immunosuppression and other risk factor for lip cancer after kidney transplantation. *Cancer Epidemiol Biomarkers Prev* 18: 561–567.

Vas'Kovaskaia, G.P., and E.I. Abramova. 1981. Cancer development from lichen planus on the oral and labial mucosa. *Stomatologiia (Mosk)* 40: 46–48.

Wakulich, C., L. Jackson-Boeters, T.D. Daley, et al. 2002. Immunohistochemical localization of growth factors fibroblast growth factor-1 and fibroblast growth factor-2 and receptors fibroblast growth factor receptor-2 and fibroblast growth factor-3 in normal oral epithelium, epithelial dysplasias, and squamous cell carcinoma. *Oral Surg Oral Med Oral Pathol Oral Radiol Endod* 93: 573–579.

Waldron, J.E., and W.G. Shafer. 1975. Leukoplakia revisited: A clinicopathologic study of 3256 oral leukoplakias. *Cancer* 36: 1386–1392.

Walker, D.M., G. Boey, and L.A. McDonald. 2003. The pathology of oral cancer. *Pathology* 35: 376–383.

Warnakulasuriya, S. 2009. Food, nutrition and oral cancer. In *Food constituents and oral health*, ed. M. Wilson, 273–295. Woodhead Publishing Ltd. Cambridge.

Wen, S., T. Tsuji, X. Li, et al. 1997. Detection and analysis of human papillomavirus 16 and 18 homologous DNA sequences in oral lesions. *Anticancer Res* 17: 307–312.

Werkmeister, R., B. Brandt, and V. Joos. 2000. Clinical relevance of erbB-1 and -2 oncogenes in oral carcinomas. *Oral Oncol* 194: 303–313.

WHO Fact Sheet. 2011. Oral cancer fact sheet. http://www.michigan.gov/documents/mdch/Oral_Cancer_Fact_Sheet_2011__364756_7.pdf.

Whyte, D.A., C.E. Bronton, and E.J. Shillitoe. 2002. The unexplained survival of cells in oral cancer: What is the role of p53? *J Oral Pathol Med* 31: 25–33.

Winn, D. 1995. Diet and nutrition in the etiology of oral cancer. *Am J Clin Nutr* 61(Suppl): 437S–445S.

Winn, D.M. 1984. Tobacco chewing and snuff dipping: An association with human cancer. In *N-Nitroso compounds: Occurrence, biological effects and relevance to human cancer*, ed. I.K. O'Neill, R.C. von Borstel, C.T. Miller, J. Long, and H. Bartsch, 837–849. IARC Science Publication 57. IARC, Lyon, France.

Woods, K.V., E.J. Shillitoe, M.R. Spitz, et al. 1993. Analysis of human papillomavirus DNA in oral squamous cell carcinomas. *J Oral Pathol Med* 22: 101–108.

Wynder, E.L., G. Kabat, and S. Rosenberg. 1983. Oral cancer and mouth wash use. *J Natl Cancer Inst* 70: 255–260.

Yanamoto, S., G. Kawasaki, I. Yoshitomi, et al. 2007. Clinicopathologic significance of EpCAM expression in squamous cell carcinoma of the tongue and its possibility as a potential target for tongue cancer gene therapy. *Oral Oncol* 43: 897–877.

Yin, Y., M.A. Tainsky, F.Z. Bischoff, et al. 1992. Wild-type p53 restores cell cycle control and inhibits gene amplification in cells with mutant p53 alleles. *Cell* 70: 937–948.

Zhou, X., Z. Zhang, X. Yang, et al. 2009. Inhibition of cyclin D1 expression by cyclin D1 shRNAs in human oral squamous cell carcinoma cells is associated with increased cisplatin chemosensitivity. *Int J Cancer* 124: 483–489.

zur Hausen, H. 1996. Papillomavirus infections—A major cause of human cancers. *Biochim Biophys Acta* 1288: F55–F78.

3 Proliferative and Apoptotic Signaling in Oral Cancer

KEY WORDS

Apoptosis
Bcl2
Caspases
Death receptors
EGFR
Oncogenes
Oral cancer
p53
pRb
Proliferation
Ras
Receptor tyrosine kinase
TRAIL

3.1 INTRODUCTION

Apoptosis is a regulated event of cell death, i.e., a morphologically and biochemically distinct mode of cell death that occurs during embryogenesis, carcinogenesis, cancer treatment, or immune reactions, as well as in cell proliferation (Khan et al., 2010a, 2012; Sah et al., 2006). It is an important regulated process that counterbalances the cells produced due to cell division and complements differentiation in the overall tissue homeostatic mechanisms. The realization that apoptosis is a gene-directed program has had profound implications for our understanding of developmental biology and tissue homeostasis, for it implies that cell numbers can be regulated by factors that influence cell survival as well as those that control proliferation and differentiation. Several lines of evidence link apoptosis to proliferation. A number of dominant oncogenes have a dual role: they can induce both proliferation and apoptosis. As somatic cells proliferate, the cell cycle progression is regulated by positive and negative signals. Cell cycle genes such as p53, Rb, and E2F have been shown to participate in both the cell cycle and apoptosis (Cao et al., 1992; Khan and Bisen, 2013). Thus, the balance between apoptosis and proliferation must be strictly maintained to sustain tissue homeostasis.

Studies revealed a high frequency of apoptosis in spontaneously regressing tumors and in tumors treated with cytotoxic anticancer agents. Together, these observations suggested that apoptosis contributed to the high rate of cell loss in malignant tumors and, moreover, could promote tumor progression. Defects in apoptotic pathways occur during cancer development and are the mechanism by which cancer cells survive in the presence of mutations and in the sites of the body that are distant from the primary tumor. The regulation of apoptosis involves the tumor suppressor genes, oncogenes, bcl-2 gene family, receptor and mitochondrial apoptotic factors, and caspases.

3.2   REGULATION OF CELL HOMEOSTASIS

A balance between cell proliferation and cell death maintains cell number homeostasis. A cell receives a multitude of positive and negative proliferative or apoptotic signals. Most studies of this renewal process have focused on the control of cell proliferation. However, it is becoming increasingly apparent that the control of cell death is equally, if not more, important in the regulation of cell numbers, and ultimately susceptibility to neoplastic transformation. The cell, after integrating all the received signals, will proceed to undergo proliferation or death, depending on the balance between the signals. These signals are preceded to the cell in several steps. For example, in the case of peptide hormones, which are thought to act exclusively outside the cell, signaling is completed in three phases: (1) ligand binds to its receptor and activates it, (2) development of cytoplasmic signaling cascade, and (3) signaling to the nucleus and activation of transcriptional activity, leading to the expression of proteins involved in the regulation of the cell cycle. If the receptor activation is induced by negative regulators, then this process inhibits cell cycle progression and completion. Cancer has been described as a disease of excessive cell proliferation, and there is a great deal of information supporting this view. Genes and gene products that are known to contribute to the rapid growth of tissues are found to be associated with many cancers, including oral (Cantley et al., 1991; Hunter, 1991). Oncogenes are specific genetic sequences that are known to be overrepresented in cancers, and are known to be capable of transforming normal cells into cancer cells (Connelly and Stern, 1990). This seems reasonable by the fact that cancers are capable of more rapid growth than their normal counterparts.

Cell number is not just a function of cell proliferation. The fact that, in adults, cell numbers stay reasonably constant implies that no net growth of cells occurs in these subjects. Though growth continues in an adult in most tissues, with the possible exceptions of nerve and muscle cells, if cell numbers are to remain constant, the rate of cell growth must be closely balanced by the rate of cell death. Cells can die through two processes, necrosis and apoptosis. Necrosis is a process during which damaged cells undergo degeneration. Their cell membranes break down, and their intracellular contents are spilled into the extracellular fluid. This is usually accompanied by the body's inflammatory reaction to these usually sequestered proteins and other intracellular materials. Apoptosis, on the other hand, is a physiological process leading to cell death, with an orderly clearance of these cells without the process of inflammation. In most tissues, apoptosis goes on all the time, just as cell growth continues. Cell growth must be balanced by the sloughing of cells that can occur on the surface, the orderly clearance of apoptotic cells, and the occasional limited cell death through necrosis. This balance sheet between cell growth and cell death is needed to maintain normal homeostasis in cell numbers in adults. This balance sheet is out of balance in cancers, due to either excessive proliferation or insufficient cell death. There is a growing body of evidence that inadequate cell death, as well as excessive cell proliferation, contributes to this loss of homeostasis (Khan and Bisen, 2013).

Cell growth and apoptosis are regulated by many intracellular signals, a complexity that means all of these signals must be ultimately integrated into a system providing for homeostasis in cell number. The signals that contribute to these two processes include phosphorylation on specific intracellular transducing elements at specific sites, signals that alter G protein messages, and signals that bind to, and activate, specific genes and other less well-defined signals (Bishop, 1991; Liebow and Kamer, 1992). The phosphorylation process involves two general systems: those that alter phosphorylation on tyrosine residues and those that alter phosphorylation on serine and threonine residues. Both of these processes can be controlled by kinases, which increase phosphorylation, and phosphatases, which decrease phosphorylation. Excesses in the kinases have been more thoroughly examined and have been found to be associated with the transformation of normal tissues into cancers through the contribution of oncogene products (Hunter, 1987). There has been less exploration into the possible role of phosphatases in the transformation process, but that possibility has received more attention recently. These controlling mechanisms are apparently active in the control of apoptosis as well as in the control of mitosis. The following discussion will attempt to elucidate some of pathways by which proliferation and apoptosis are controlled.

3.3 SOMATIC GENE MUTATION

Several risk factors for the development of oral cancer are very well established. However, not all the people exposed to these risk factors develop oral cancer. It suggests that there could be some other genetic factors involved in the xenobiotic metabolism of carcinogens modulating the risk of oral cancer. Evidence suggests that oral cancer is caused by DNA damage. Many oncogenes, tumor suppressor genes, and regulatory genes are genetically deregulated in oral cancer (Field, 1992). These alterations are usually somatic events, although germ-line mutations can predispose a person to heritable or familial cancer. A single genetic change is rarely sufficient for the development of a malignant tumor. Most evidence points to a multistep process of sequential alterations in several, often many, oncogenes, tumor suppressor genes, or microRNA genes in cancer cells.

Like other cancers, a large variety of chromosomal aberrations can be found in oral cancer, including chromosomes 3p, 4q, 6p, 8p, 9p, 11q, 13q (retinoblastoma (Rb)), 14q, 17p (p53), 18q (DCC), and 9p (CDKN2A) (Partridge et al., 1997; Califano et al., 1996). The genetic hypothesis predicts a role for hyperactive oncogenes (growth-promoting genes) in oral carcinogenesis. The impact of these aberrations varies significantly, and their cellular and clinical significance is frequently uncertain. It is generally believed, though, that the number of aberrations increases steadily during cancer progression: oral leukoplakia has fewer chromosomal aberrations than oral cancer (Weber et al., 1998), and lower tumor stage (T1) is associated with fewer aberrations than higher tumor stage (T2) (Okafuji et al., 2000). Some aberrations have been described as early, or common, and may bear considerable prognostic significance for patients with premalignant lesions or early-stage oral cancer. Chromosome 9 is believed to be one of the earliest targets, and allelic losses in the 9p21 region, possibly associated with the genes encoding the p16 and p14 cyclin-dependent kinase inhibitors, are present in premalignant lesions (Jiang et al., 2001) and oral cancer (Rosin et al., 2000; el Naggar et al., 1995). Aberrations that are usually associated with advanced tumor stage or poor differentiation include allelic losses in 5q21 to 22 (Mao et al., 1998), 22q13 (Reis et al., 2002), 4q, 11q, 18q, and 21q (Bockmühl et al., 1998). Gains in 3q (Oga et al., 2001) are also a common finding in advanced oral cancer.

3.4 ONCOGENES AND TUMOR SUPPRESSORS

It is now recognized that there are two classes of cancer genes, oncogenes and tumor suppressor genes. The oncogene class of genes is probably involved in both initiation and progression of the disease, whereas the tumor suppressor genes produce tumors following mutational damage or inactivation. Many of the major oncogenes and tumor suppressor genes that are implicated in other cancer types also contribute to oral cancer. A large number of these genes promote unscheduled, aberrant proliferation, override the G-S, G-M, and M checkpoints of the cell cycle, prevent apoptosis, and enable cellular survival under unfavorable conditions. Normal cells or proto-oncogenes may become activated oncogenes by a carcinogen, irradiation, or possibly a virus. Activation of the oncogene may be in the form of chromosomal translocations, gene amplification, point mutations, or deletions. The action of oncogenes has been demonstrated at the level of growth factors and their receptors, signal transduction systems, and also nuclear proteins. In contrast to proto-oncogenes, the tumor suppressor genes are normal genes that when activated by a mutation, deletion, or virus, cause deregulation of critical pathways controlling growth and differentiation (Harris, 1991).

3.4.1 RAS GENE FAMILY SIGNALING PATHWAYS

Ras is one of the most frequently genetically deregulated oncogenes in oral cancer. All Ras protein family members belong to a class of protein called small GTPase, and are involved in transmitting signals within cells (cellular signal transduction). This gene family consists of H-*ras*, K-*ras*, and N-*ras*, which encode for a protein of 21 kD molecular mass, which possesses guanosine triphosphate (GTP) activity (Barbacid, 1987). Mutations in the *ras* genes are usually found in codons 12, 13, and 61 (Khan and Bisen, 2013).

When Ras is "switched on" by incoming signals it subsequently switches on other proteins, which ultimately turn on genes involved in cell growth, differentiation, and survival. As a result, mutations in *ras* genes can lead to the production of permanently activated Ras proteins. This can cause unintended and overactive signaling inside the cell, even in the absence of incoming signals. Because these signals result in cell growth and division, overactive Ras signaling can ultimately lead to cancer (Goodsell, 1999). Ras is the most common oncogene in human cancer—mutations

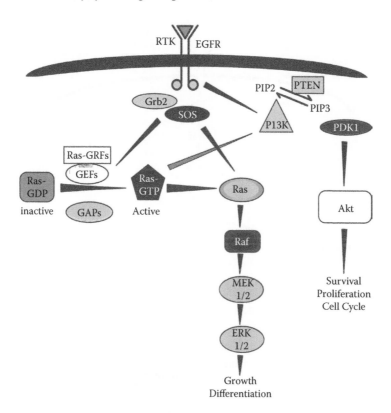

FIGURE 3.1 RAS activation and signaling pathways.

that permanently activate Ras are found in 20 to 25% of all human tumors and up to 90% in certain types of cancer (Downward, 2003). For this reason, Ras inhibitors are being studied as a treatment for cancer, and other diseases with Ras overexpression.

Ras is a G-protein, or a guanosine-nucleotide binding protein. Growth factors activate *ras* through enhanced exchange of guanine nucleotide (Dyson et al., 1989). Activation of Ras by receptor tyrosine kinase (RTK) is mediated by the Grb2/ SOS complex (Avraham et al., 2000) (Figure 3.1). The molecular mechanism of *ras* depends on the whole super family of small G-proteins that function as binary signaling switches with active and inactive states. In the off state, it is bound to the nucleotide guanosine diphosphate (GDP), while in the on state, Ras is bound to guanosine triphosphate (GTP), which has an extra phosphate group compared to GDP. This extra phosphate holds the two switch regions in a loaded-spring con- figuration (specifically the Thr-35 and Gly-60). When released, the switch regions relax, which causes a conformational change into the inactivate state. Hence, activa- tion and deactivation of Ras and other small G-proteins are controlled by cycling between the active GTP-bound and inactive GDP-bound forms (Quilliam et al., 1996; Wittinghofer and Nassar, 1996; Rajalingam et al., 2007). The process of exchanging the bound nucleotide is facilitated by guanine nucleotide exchange factors (GEFs) and GTPase activating proteins (GAPs). GAPs accelerate Ras inactivation by activating

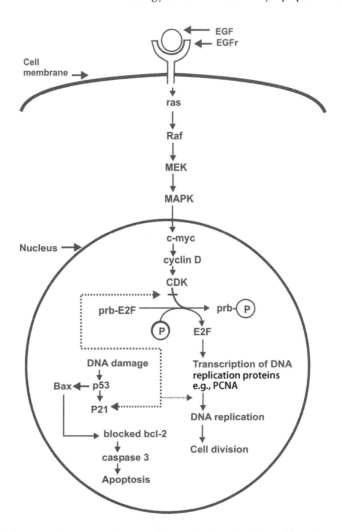

FIGURE 3.2 Normal oral keratinocyte division is stimulated by epidermal growth factor (EGF) binding the epidermal growth factor that activates *ras* (EGF^r). Dotted lines represent the apoptotic cell death pathway following DNA damage.

its GTPase activity. GEFs catalyze a reaction that releases GDP from Ras. Because intracellular GTP is abundant relative to GDP (approximately 10-fold more) (Lodish et al., 2000), GTP predominantly reenters the nucleotide binding pocket of Ras and reloads the spring. Thus, GEFs facilitate Ras activation (Vetter and Wittinghofer, 2001) (Figure 3.2). Well-known GEFs include Son of Sevenless (SOS) and cdc25, which include the RasGEF domain. The balance between GEF and GAP activity determines the guanine nucleotide status of Ras, thereby regulating Ras activity.

The *ras* oncogenes encode cell membrane-associated proteins that are involved in the transduction of extracellular growth, differentiation, and survival signals (Karnoub and Weinberg, 2008). Ras-regulated signal pathways control such

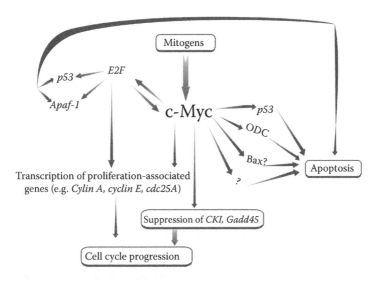

FIGURE 3.3 The role of c-Myc in cell cycle and apoptosis.

processes as proliferation, differentiation, cell adhesion, apoptosis, and cell migration. In a cell where *ras* is mutated, the equilibrium between the GTP and GDP-bound state is impaired. These mutations prevent the intrinsic and GAP-catalyzed hydrolysis of GTP, thereby generating constantly active *ras*, and it accounts for the activation of many downstream effectors whereby the cell undergoes aberrant malfunctioning leading to malignancy.

Growth receptors are known to induce different cellular responses in response to the binding of specific ligands that represent external stimuli. The erbB family of receptors, a subfamily of four closely related receptor tyrosine kinases (RTKs), has received attention in particular (EGFR, also known as erbB1 or Her-1) due to its inherent ability to stimulate the proliferation of epithelial cells (Normanno et al., 2006). Oral cancer is a lesion characterized by dysregulated division of oral keratinocytes. Normally, oral keratinocyte division is stimulated by epidermal growth factor (EGF) binding the epidermal growth factor receptor (EGFR) on the surface of basal keratinocytes (Figure 3.3). Amplification of EGFR is found in a considerable percentage of oral tumors and also in premalignant lesions (Ishitoya et al., 1989; Nagatsuka et al., 2001). Signal transduction from activated RTKs like EGFR depends on a variety of downstream mediators. EGFR-ligand interaction activates the Ras protein on the cytoplasmic side of the EGFR, which then activates two key signal transduction components, including the small GTPase *ras* and the lipid kinase PI3K. Active *ras* could also directly activate PI3K (Bader et al., 2005). The activated *ras* stimulates several different effector pathways; the two most important are the mitogen-activated protein kinase (MAPK) and phosphatidylinositol-3-kinase (PI3K)/Akt pathways. Active *ras* activates the Raf protein and subsequently the other cytoplasmic kinases (MEK, MAPK) in a cascade-like manner (Figures 3.3 and 3.4). MAPK is a mitogen-activated protein (MAP) kinase, MEK is MAP kinase kinase, and Raf is MAP kinase kinase kinase. The kinase cascade transmits the growth signal from

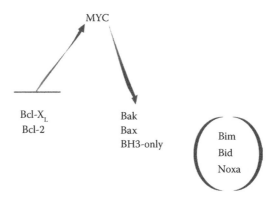

FIGURE 3.4 c-Myc amplifies the mitochondrial apoptotic pathway by inhibiting anti-apoptotic Bcl-2 proteins and activating pro-apoptotic Bcl-2 protein.

the cell membrane to the nucleus, where it activates the expression of genes involved in growth and differentiation. On the other hand, *ras* transduces the PI3K/Akt signaling pathway that inhibits apoptosis (Rodriguez-Viciana et al., 1994). Upon activation of PI3K, the second messenger phosphatidylinositol-3,4,5-triphosphate (PIP3) is generated from phosphatidylinositol-4,5-diphosphate (PIP2). The tumor suppressor phosphatase and tensin homolog (PTEN) antagonizes the PI3K activity by dephosphorylating [PI(3,4,5)P3] to [PI(4,5)P2]. Accumulation of the second messenger [PI(3,4,5)P3] in the membrane pool brings the pleckstrin homology (PH) domain containing proteins including protein serine-threonine kinases Akt (PKB) and phosphatidylinositol-dependent protein kinases 1 and 2 (PDK1 and 2) into proximity and facilitates phosphorylation of Akt by PDK1. This phosphorylation stimulates the catalytic activity of Akt, leading to phosphorylation of its effector proteins that affect cell proliferation, cell death, and cell survival (Murugan et al., 2012; Sjölander et al., 1991; Rodriguez-Viciana et al., 1994).

3.4.2 c-Myc, Growth, and Apoptotic Signaling

c-Myc gene was discovered as the cellular homolog of the retroviral v-Myc oncogene (Bishop, 1982). The proto-oncogene c-Myc has a pivotal function in growth control, differentiation, and apoptosis, and is among the most frequently affected genes in human cancers. Normal expression of the c-Myc gene is tightly regulated, and this can occur at both transcriptional and posttranscriptional levels (Cole, 1996; Eisenman and Thompson, 1986; Alitalo et al., 1983).

Amplification of c-Myc has been reported in a large percentage of human tumors, including cancers of lymphoid, mesenchymal, and epithelial origin (Evan and Littlewood, 1998). Abnormal behavior of the c-Myc gene has also been extensively discussed in oral cancer (Field et al., 1989). The expression of the c-Myc protein is deregulated in human cancers by a number of different mechanisms (Spencer and Groudine, 1991), including chromosomal translocations and amplifications, activation of upstream growth stimulatory signaling cascades, and increased protein stability.

c-Myc is a nuclear phosphoprotein that functions as a transcription factor orchestrating the expression of 10% to more than 15% of all cellular genes. Target genes of c-Myc are involved in many vital cellular activities, including survival, cell growth, protein synthesis, cell adhesion, cytoskeleton, and metabolism. The mechanisms by which c-Myc induces neoplastic transformation and apoptosis are beginning to emerge with the identification of authentic target genes, both direct and indirect. A direct target gene is one whose expression is altered by direct interaction of the c-Myc protein with the gene regulatory elements or with *trans*-acting factors that bind these *cis* elements. An indirect target gene of c-Myc is one whose expression is altered as a consequence of expression of the direct c-Myc target genes and whose expression is connected to c-Myc-dependent phenotypes such as cellular proliferation, transformation, or apoptosis. It activates expression of genes through binding on consensus sequences (enhancer box sequences) and recruiting histone acetyltransferases (HATs). It can also act as a transcriptional repressor. By binding Miz-1 transcription factor and displacing the p300 coactivator, it inhibits expression of Miz-1 target genes. In addition, c-Myc has a direct role in the control of DNA replication (Dominguez-Sola et al., 2007).

c-Myc has a critical role in normal cell cycle progression, especially during the transition from G0 to S phase (Spencer and Groudine, 1991). Myc is activated upon various mitogenic signals via the MAPK/ERK pathway. The kinase cascade transmits the growth signal from the cell membrane to the nucleus where the level of c-Myc protein rises sharply (Figure 3.2). c-Myc is an early response gene; i.e., it responds directly to mitogenic signals to push cells in the G1 phase of the cell cycle (Heikkila et al., 1987; de Alboran et al., 2001). The c-Myc expression is maintained throughout the cell cycle, and some observations also suggest a role for c-Myc in G2 (Hann et al., 1985; Mateyak et al., 1997). c-Myc can exert its effect on cell cycle progression by the transcription of genes with an important role in cell cycle control, i.e., Cdc25A, cyclin D1, cyclin D2, cyclin E, cyclin A, CDK1, CDK2, CDK4, and E2F (Born et al., 1994; Kim et al., 1994; Beier et al., 2000) (Figure 3.5).

Abundant expression of cyclins is a common (36 to 66%) feature of oral cancer (Koontongkaew et al., 2000; Miyamoto et al., 2003; Kushner et al., 1999). Regarding the regulation of G1, the connection between c-Myc and cyclin D1 is complex and may depend on specific stimuli and cell systems (Roussel, 1998). Both c-Myc and cyclin D1 are required for activation through the CSF1 receptor, and their relationship is nonlinear (Roussel et al., 1995). With serum stimulation of fibroblasts, it is expected that c-Myc may activate the subsequent expression of cyclin D1; however, the role of c-Myc in regulating cyclin D1 expression is complex since there are conflicting data in the literature (Daksis et al., 1994; Philipp et al., 1994; Rosenwald et al., 1993). Deregulated c-Myc expression is linked to increased cyclin A and increased cyclin E expression (Daksis et al., 1994; Hanson et al., 1994; Hoang et al., 1994; Jansen-Durr et al., 1993). Recent evidence has been provided that c-Myc is able to transactivate the expression of cyclin E directly, although the mechanism is unclear (Leone et al., 1997; Perez-Roger et al., 1997). c-Myc increases CDK function through several mechanisms. In a study, c-Myc appeared to cooperate with Ras to induce the CDC2 (CDK1) promoter, which does not contain a consensus Myc E box (Born et al., 1994). There are no other data, however, that support the elevation

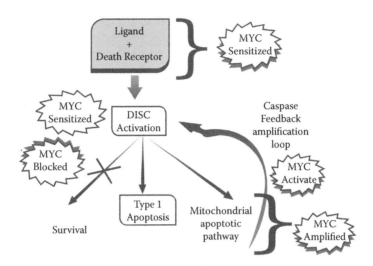

FIGURE 3.5 Myc promotes apoptosis through the death receptor pathways at multiple junctions. Myc influences multiple pathways activated by death receptors at several points.

of CDC2 in response to Myc. More recently, evidence has been provided that the cdc25A gene is a direct target of c-Myc (Galaktionov et al., 1996). The connection between c-Myc and cdc25A has not been confirmed in other studies (Amati et al., 1998), indicating that differences in experimental models might account for the discrepancy. This gene produces a protein phosphatase that activates both CDK2 and CDK4. Thus, a direct link between c-Myc and the cell cycle machinery may exist through its ability to activate the cdc25A and cyclin E genes directly. Another important mechanism of the ability of c-Myc to promote cell growth is suppression of transcription of growth arrest and the DNA damage-inducible gene 45 (Gadd45), Gadd153, and of the CKI genes p15, p21, and p27 (Grandori et al., 2000) (Figure 3.2). c-Myc expression also decreases the levels and interferes with the function of the p27 CDK inhibitor (Leone et al., 1997; Perez-Roger et al., 1997; Rudolph et al., 1996; Vlach et al., 1996). The mechanism by which c-Myc interferes with p27 activity is not known. These activities of c-Myc are all compatible with the ability of c-Myc to promote cell entry into the S phase. The ability of c-Myc to promote cell proliferation suggests that its deregulation contributes to deregulated DNA synthesis and genomic instability (Mai et al., 1996a, 1996b) (Figure 3.6).

Besides its role in the cell cycle, c-Myc also plays a key role in regulating apoptosis. This role was confirmed in different cell types under a wide variety of physiological conditions, and both under- and overexpression of c-Myc can lead to cell death (Conzen et al., 2000; Thompson, 1998). Until now, the molecular events responsible for c-Myc-induced apoptosis are not well understood. Several components important for cell cycle progression, including cyclin A and Cdc25A, have been implicated in apoptotic systems associated with elevated c-Myc. Cyclin A was elevated in rat-1A fibroblasts overexpressing c-Myc and undergoing apoptosis (Hoang et al., 1994). Cdc25A is a well-established transcriptional target of c-Myc and can induce apoptosis in serum-deprived fibroblasts, as does c-Myc (Galaktionov

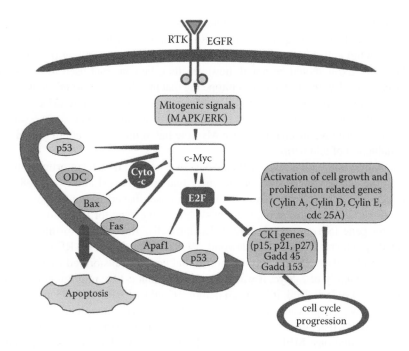

FIGURE 3.6   p53 and pRb/E2F signaling pathways.

et al., 1996). However, not all the Cdc25A substrates are yet known, and their relationship with molecules participating in apoptosis remains to be elucidated (Zornig and Evan, 1996; Thompson, 1998). Another target for transcriptional stimulation by c-Myc is ornithine decarboxylase (ODC), which can cause apoptosis when overexpressed (Packham and Cleveland, 1995). c-Myc-induced apoptosis may involve p53-dependent and -independent pathways. c-Myc transactivates the p53 gene promotor and increases the half-life of p53 (Reisman et al., 1993; Hermeking and Eick, 1994). However, there does not seem to be a universal requirement for p53 in c-Myc-mediated apoptosis (Hsu et al., 1995; Sakamuro et al., 1995). c-Myc-induced apoptosis has been shown to correlate with Fas ligand and Fas receptor expression (Wang et al., 1998). c-Myc-induced apoptosis also seems to be inhibited by Bcl-2 and Mcl-1 (Bissonnette et al., 1992; Reynolds et al., 1994). c-Myc-induced cytochrome c release involves functionally active Bax, and preliminary evidence suggests that c-Myc regulates the transcription of this pro-apoptotic molecule (Duelli and Lazebnik, 2000; Mitchell et al., 2000; Soucie et al., 2001).

Classically, c-Myc was touted to be an immortalizing gene, the ectopic expression of which facilitates the immortalization of primary rodent fibroblasts. This simple view overlooked the initial events following ectopic c-Myc expression and the crisis period that cells must survive to achieve immortality. Since telomerase contributes to the immortality of tumor cells, the ability of increased expression of viral or cellular oncogenes to induce telomerase in normal human mammary epithelial cells and human fibroblasts (IMR-90) was studied (Wang et al., 1998). Among six candidates, c-Myc emerged as a key switch for induction of telomerase activity, as well as an expression of the catalytic subunit of telomerase, termed (telomerase reverse transcriptase) TERT. Intriguingly, whereas TERT increases the life span of human mammary epithelial cells, overexpression of TERT was unable to prolong the life span of IMR-90 cells. It should be noted, however, that the construct used in that study produces a TERT with a C-terminal epitope tag that may have compromised its activity. In contrast to epitope-tagged TERT, c-Myc is able to immortalize IMR-90 cells, even though these cells do not show stabilization of telomeres. These observations suggest an alternative mechanism for c-Myc-mediated immortalization, in addition to the induction of telomerase.

In collaboration with activated Ras, c-Myc was able to transform primary fibroblasts in the classic experiments of Weinberg and coworkers (Land et al., 1983). In this role, c-Myc appears to inactivate cellular responses that are normally required for Ras-mediated growth inhibition, thereby switching the gene for Ras into a growth-promoting gene (Serrano et al., 1997). Reciprocally, Ras is able to inhibit Myc-mediated apoptosis (Evan and Littlewood, 1998). Given that p19 ARF-null murine embryonic fibroblasts (MEFs) are immortal and can be transformed by oncogenic RAS independently of c-Myc, it was hypothesized that c-Myc might regulate ARF (Zindy et al., 1998). Indeed, it has been demonstrated that ARF and p53 are induced by ectopic c-Myc expression in wild-type MEFs, triggering a replicative crisis and apoptosis. MEFs that survive *myc* overexpression and the crisis period sustain ARF loss or p53 mutations. MEFs that lack ARF or p53 showed a decreased apoptotic response to c-Myc overexpression. These observations indicate that ARF participates in a p53-dependent checkpoint that safeguards cells against oncogenic signals, such as overexpression of c-Myc. These new observations indicate that immortalization of primary cells by oncogenes is a complex phenomenon in which normal safeguard apoptotic mechanisms are inactivated, thereby allowing immortalized cells to emerge from a crisis period of massive cell death.

It has been shown that different subregions of the c-Myc N-terminal domain can control distinct biological functions, including apoptosis (Chang et al., 2000; Conzen et al., 2000; Nesbit et al., 2000). However, this issue cannot be directly addressed until the c-Myc target genes essential for apoptosis have been clearly identified. The factors that determine the decision of inducing either cell division or cell death need to be further addressed. In addition, the c-Myc expression is tightly linked to the extracellular milieu, and the function of c-Myc is also probably influenced by the extracellular environment (Oster et al., 2002).

3.4.3   ONCOGENE ACTIVATION

Activation of oncogenes by chromosomal rearrangements, mutations, and gene amplification confers a growth advantage or increased survival of cells carrying such alterations. All three mechanisms cause either an alteration in the oncogene structure or an increase in or deregulation of its expression (Bishop, 1991).

3.4.3.1   Chromosomal Rearrangements

Chromosome inversions and translocations are common cytogenetic abnormalities in cancer cells. In hematopoietic cancers and solid tumors, the translocations and inversions increase or deregulate transcription of the oncogene. In prostate cancer, gene fusion occurs between a gene that carries a promoter that is very active in the target cells and another that carries the oncogenic activity (e.g., ERG1) (Tomlins et al., 2005). In cancers of B- and T-cells, the most common mechanism of activation by translocation resembles c-Myc deregulation, whereas in myeloid cancers and soft tissue sarcomas, gene fusion is more common.

3.4.3.2   Mutations

When an oncogene is activated by a mutation, the structure of the encoded protein is changed in a way that enhances its transforming activity. Many types of mutation occur in oncogenes (Rodenhuis, 1992). Examples are the Ras oncogenes (K-Ras, H-Ras, and N-Ras), which encode proteins with guanosine nucleotide binding activity and intrinsic guanosine triphosphatase activity. When mutated in codon 12, 13, or 61, the Ras genes encode a protein that remains in the active state and continuously transduces signals by linking tyrosine kinases to downstream serine and threonine kinases. These incessant signals induce continuous cell growth. Mutation of oncogenes in the Ras family has been associated with exposure to environmental carcinogens. Activating point mutations of the B-Raf gene occurs in many type of cancers (Davies et al., 2002). Most of the mutations change the valine residue at position 599 to glutamic acid (V599E). This change occurs within the kinase domain of the B-Raf protein, resulting in a constitutively active protein that uncontrollably stimulates the MAP kinase cascade, thereby deregulating genes involved in cell proliferation, differentiation, and survival (Davies et al., 2002; Frattini et al., 2004).

3.4.3.3   Gene Amplification

An example of gene amplification, which usually occurs during tumor progression, is the amplification of the dihydrofolate reductase (DHFR) gene in methotrexate-resistant acute lymphoblastic leukemia (Alt et al., 1978). Amplification of DHFR is accompanied by cytogenetic alterations that mirror amplification of oncogenes (Cowell, 1982; King et al., 1985). The amplified DNA segment segment usually involves several hundred kilobases and can contain many genes. Members of four different oncogene families are often amplified: c-Myc, cyclin D1, EGFR, and Ras. C-Myc is amplified in small cell lung cancer, breast cancer, esophageal cancer, cervical cancer, ovarian cancer, and head and neck cancer, whereas amplification of N-Myc correlates with an advanced tumor stage (Schwab et al., 1983). The t(11; 14) translocation juxtaposes cyclin D1 and immunoglobulin enhancer elements and is characteristic of mantle cell lymphoma (Tsujimoto et al., 1984). Cyclin D1 amplification also occurs in breast, esophageal, hepatocellular, and head and neck cancers. EGFR is amplified in glioblastoma and head and neck cancer.

3.4.4   p53, pRb, AND E2F Signaling

p53 and pRb are widely recognized tumor suppressor proteins that function during the cell cycle and apoptosis. Tumor suppressors usually prevent cells from acquiring malignant characteristics. These genes are usually entrusted with the regulation of discrete checkpoints during the cell cycle progression and with the monitoring of DNA replication and mitosis. Cellular stress and a variety of insults can activate tumor suppressor pathways to arrest the cell cycle. The deregulation of the p53 tumor suppression network is observed in many tumor types, including oral cancer. In fact, the activation of the DNA damage response is one of the earliest findings in the natural history of cancer (Gorgoulis et al., 2005; Bartkova et al., 2005). Immunohistochemical evaluation for p53 is positive in up to 57% of oral tumors (Khan et al., 2009; Ogden et al., 1992; de Vicente et al., 2004). p53 expression may predict poor prognosis in the subset of patients with low-stage, node negative disease (de Vicente et al., 2004) or in those carrying specific TP53 mutations (Yamazaki et al., 2003). Interestingly, tumors with TP53 mutations seem to be more resistant to radiotherapy (Khan et al., 2010b; Shintani et al., 2000; Alsner et al., 2001; Jayasurya et al., 2004), and this information could be vital for the selection of an appropriate treatment.

The p53 gene has been found to be highly conserved in evolution, suggesting an important function in the regulation of cell homeostasis. It is located on chromosome 17 and behaves as a negative regulator; however, it may be converted into a dominant oncogene by a mutation in its gene. Investigations have shown that p53 plays a significant role at the first checkpoint of the cell cycle (Khan et al., 2006, 2009). Its role at the checkpoint situated between the Gl and S phases is exerted through two mechanisms: (1) cell cycle arrest and (2) induction of apoptosis. It has been shown that p53 has a biochemical role as a specific transcription factor and a biological role as a G1 checkpoint control for DNA damage (Farmer et al., 1992; Lane, 1992). The p53 protein is able to enforce cell cycle arrest or apoptosis under replication stress, thus halting the proliferation of potentially malignant cells. The majority of p53 mutations in human tumors have been reported to date in the highly conserved evolutionary domains of exons 5 to 844. Furthermore, the p53 tumor suppressor gene (TSG) is ideally suited to study mutational spectra in association with certain mutagens (Harris, 1991). Chiba et al. (1990) found that 56% of bronchial carcinoma had G-T transversions in the p53 TSG, and Suzuki et al. (1992) have reported similar findings.

Although the complete mechanism by which p53 inhibits cell growth is not understood, evidence implicates p21 in this process. p53 has been found to induce expression of the protein p21[WAFI/CIPI] (El-Deiry et al., 1993). Protein p21[WAFI/CIPI] is a potent inhibitor of cyclin-dependent kinases that are required for the phosphorylation of the Rb protein (Harper et al., 1993). In oral squamous cell carcinoma cell lines, EGF induced inhibition of cell proliferation and increased expression of p21[WAFI/CIPI], consistent with an inhibitory role of this protein in the regulation of oral cancer growth (Jakus and Yeudall, 1996). The dephosphorylated form of the Rb protein prevents the E2F factor from initiating the transcription of genes needed for entrance into the S phase (Buchkovich et al., 1989; Cao et al., 1992). Mutated p53 is associated with many human cancers (Hollstein et al., 1991). This pathway was implicated to be an important mediator of growth inhibition in A431 cells (Fan et al., 1995). This result is consistent with the role of p53 in growth modulation. p53 gene mutations have been detected in over 60% of oral squamous cell carcinomas (Sakai et al., 1992) (Figure 3.3). Several in situ immunohistochemical studies have shown increased p53 protein expression in oral squamous cell carcinoma (Zariwala et al., 1994), and a history of heavy smoking correlates with p53 expression in oral squamous cell carcinoma. Normal oral mucosa and benign oral mucosal lesions are consistently negative for p53 expression, while its presence is highly indicative of malignancy, though it is not found in all oral squamous cell carcinomas (Ogden et al., 1994).

How p53 induces apoptosis is not clear, and there may be more than one mechanism. p53 alters the balance between Bax and Bcl2 (Miyashita et al., 1994). It is evidenced that p53 downregulates Bcl-2 expression and upregulates *bax* expression favoring cell death. Alternatively, p53 increases the expression of at least 14 genes, termed p53-induced genes, of which three are potent generators of reactive oxygen species (Polyak et al., 1997). Whenever a cell undergoes p53-mediated growth arrest or p53-mediated apoptosis, it is probably related to the functional level of p53 (Murphy and Levine, 1998) and $p21^{WAF1/CIP1}$ (Wilson et al., 1998) (Figure 3.3). There is emerging evidence that, analogous to Bcl2, a p53 family of genes is involved in cell cycle control and apoptosis. Two additional members, p51 and p73, have recently been identified (Kaelin, 1999). Both proteins, at least when overproduced, can mimic the ability of p53 to induce apoptosis but, in contrast, appear infrequently mutated in human cancers.

The retinoblastoma protein and its associated molecular network are frequent and early targets in many tumor types. In all retinoblastomas examined, inactivation of both of the Rb alleles has been observed. Rb inactivation also occurs in other cancers, such as breast, lung, prostate, and bladder. Rb inactivation in oral squamous cell carcinoma has not yet been reported, but it has been postulated that Rb inactivation accompanies p53 inactivation in some cases of oral squamous cell carcinoma. Lack of immunohistochemical pRb expression was found in approximately 70% of oral tumors (Pande et al., 1998; Koontongkaew et al., 2000) and 64% of premalignant lesions (Pande et al., 1998). Similarly, in a later study, about half of oral cancer specimens did not express pRb, and 20% of those that did express pRb only contained the inactive, phosphorylated form (Sartor et al., 1999). Most importantly, 84% of premalignant lesions and 90% of oral squamous cell carcinomas show altered expression of at least one of the components of the pRb network (Soni et al., 2005). The cyclin-dependent kinase inhibitors (CDKIs), in particular, are known targets in oral cancer, most likely due to their ability to prevent pRb phosphorylation. The CDKN2A locus that encodes p16INK4A is located in 9p21, which is one of the most vulnerable areas of the genome in oral cancer. Indeed, lack of immunohistochemical p16 expression can be found in up to 83% of oral tumors (Reed et al., 1996; Pande et al., 1998; Wu et al., 1999) and up to 60% of premalignant lesions (Pande et al., 1998). The predominant mode of inactivation is allelic imbalance, but point mutations and promoter methylation also occur with lower frequency (Wu et al., 1999). The alternative CDKN2A transcript, p14ARF, is also commonly suppressed (Shintani et al., 2001), but downregulation of other INK4 family members, like p15INK4B, is less frequent. The prognostic significance of p16INK4A levels is uncertain, although a study has reported favorable prognosis for patients overexpressing p16INK4A (Weinberger et al., 2004).

Rb protein can exist in both phosphorylated and dephosphorylated forms. When in a hypophosphorylated state, the retinoblastoma protein and the other pocket protein family members p107 and p130 bind and inactivate the E2F transcription factors, which are essential for cell cycle progression from G to S. The kinase cascade transmits the growth signal from the cell membrane to the nucleus where the level of oncoproteins such as c-Myc and Ras protein rises sharply. The onco-proteins can bind to DNA and stimulate the transcription of cyclins A, D, and E, which bind and activate the cyclin-dependent kinase (CDK). Active CDK catalyzes the phosphorylation of the retinoblastoma tumor suppressor protein (pRb). The phosphorylated pRb releases the E2F factor from the Rb-E2F complex and allows it to bind DNA and activate the transcription of genes involved in DNA replication and cell cycle progression (Enoch and Norbury, 1995). DNA replication proceeds, followed closely by cell division (Figure 3.3). Cyclin D and most of the DNA rep-lication proteins are degraded and must be newly transcribed with each round of cell division. Many of the proteins that transmit the growth signal from the cell membrane to the nucleus are encoded by oncogenes. As discussed above, oncogene mutation may stimulate excessive keratinocyte proliferation in oral cancer. Any factor capable of dephosphorylating the Rb protein can potentially function as a growth inhibitor.

Besides cell cycle inhibition through E2F suppression, pRb has also been shown to suppress apoptosis. For instance, Rb-deficient embryos show defects in fetal liver hematopoiesis, neurogenesis, and lens development, and extensive apoptosis was observed in these tissues (Morgenbesser et al., 1994; Macleod et al., 1996). The mechanisms by which pRb/E2F influences apoptosis remain unknown. E2F has been shown to induce the expression of the pro-apoptotic factor Apaf-1, and evi-dence suggests a role for E2F in apoptosis following DNA damage (Blattner et al., 1999; Moroni et al., 2001) (Figure 3.2). E2F cannot induce apoptosis when pRb is coexpressed, and pRb possibly has an anti-apoptotic effect through the inhibition of E2F (Fan et al., 1996; Pucci et al., 2000). Previous studies have also shown that, in a hepatoma cell line, apoptosis induced by TGF-3 is associated with the inhibition of both Rb protein expression itself and its phosphorylation. Overexpression of the Rb protein inhibited TGF-3-induced apoptosis (Fan et al., 1996). These results showed that the Rb protein plays a significant role in apoptosis. Overexpression of E2F (non-mutated form, able to bind Rb) in the same cell lines induced the apoptosis that was inhibited by coexpression with the Rb protein. The Rb protein did not affect expres-sion of mutated E2F. These experiments showed that the ability of E2F to induce apoptosis depends on the regulatory stimuli received from the Rb protein. In a head and neck squamous cell carcinoma, introduction of the wild-type p53 gene resulted in significant apoptosis. These results support the regulatory role p53 plays in apop-tosis induction in oral cancers. p53 and pRb/E2F may be directly linked in cell pro-liferation and apoptosis. Activated p53 causes a G1 arrest by inducing p21, followed by an inhibition of cyclin/CDK. In these conditions, pRB is not phosphorylated, and cells do not progress through the cell cycle. In contrast, free E2F directly induces p53 transcription, thus connecting the pRb/E2F pathway to p53-dependent apoptosis (Hiebert et al., 1995) (Figure 3.7). Each of both tumor suppressors (p53 and pRb) may thus be able to compensate for the loss of the other (King and Cidlowski, 1998).

3.4.5 BCL-2 AND BAX

The Bcl-2 protein is a 24 kDa protein that has been localized within the nuclear mem-brane, mitochondrial membrane, and smooth endoplasmic reticulum (Chen-Levy

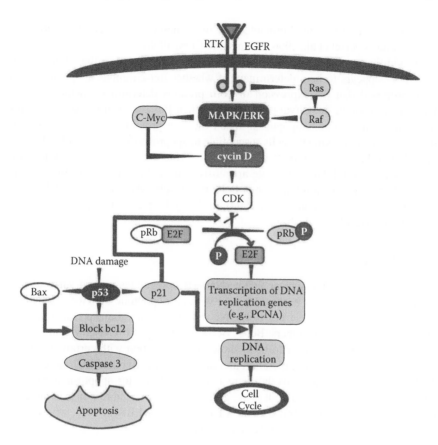

FIGURE 3.7 Regulation of cell growth and apoptosis by c-Myc signaling pathways.

and Cleary, 1990). Bcl-2 mainly has an anti-apoptotic function, but this function can be lost by multisite phosphorylation (Haldar et al., 1995). The regulation of the function of Bcl-2 mainly involves interactions with other proteins of the Bcl-2 protein family, but phosphorylation may also be a crucial event in the regulation of its function (Haldar et al., 1995; Yamamoto et al., 1999). Several signal transduction pathways can be involved in Bcl-2 phosphorylation. Bcl-2 is phosphorylated on serine/threonine residues, and the Bcl-2 kinase(s) is serine/threonine kinase(s). CDK1 is a candidate Bcl-2 kinase and has been shown to phosphorylate Bcl-2 (Furukawa et al., 2000). However, other studies demonstrated that CDK1 did not phosphorylate Bcl-2 (Scatena et al., 1998; Yamamoto et al., 1999). JNK was repeatedly indicated as a potential Bcl-2 kinase (Blagosklonny, 2001). It has been reported to phosphorylate Bcl-2 at four serine/threonine sites (Maundrell et al., 1997). However, it has been suggested that several kinases may be involved in the phosphorylation of Bcl-2 (Blagosklonny, 2001). Importantly, very high levels of Bcl-2 can promote cell death (Shinoura et al., 1999). Bcl-2 can also modulate the cell cycle in a way that is different from the inhibitory effect on apoptosis (Linette et al., 1996). Bcl-2 gene expression can result in an increase of 30 to 60% in the length of the G1 phase, and under suboptimal conditions, Bcl-2 promotes exit into quiescence and retards reentry into the cell cycle (Mazel et al., 1996; Adams and Cory, 1998).

In addition to Bcl-2, another member of the same family—the 21 kDa protein Bax—appears to be a determinant of whether Bcl-2 inhibits apoptosis. It has been proposed that the expression of both proteins determines whether the cell will undergo apoptosis (Oltvai et al., 1993). When Bcl-2 is overexpressed, it forms homodimers and inhibits apoptosis. When the Bax protein is in excess, homodimers are formed, making the cells susceptible to apoptosis (Yin et al., 1995). Factors modulating the expression of Bcl-2 or the Bax protein change the ratio of the two proteins and can inhibit or promote apoptosis (Kinoshita et al., 1995). In addition, the posttranscriptional modification of Bcl-2 by phosphorylation prevents Bcl-2 from opposing apoptosis (Haldar et al., 1994), probably by preventing the formation of homodimers (Haldar et al., 1996).

3.5   NF-κB

The transcription factor nuclear factor (NF)-κB upregulates several survival factors; however, it has also been associated with anti-apoptotic activities (Baichwal and Baeuerle, 1997; Hinz et al., 1999). NF-κB upregulates the expression of anti-apoptotic Bcl-2 family members, Bcl-2, Bcl-Xl, and Bfl-1/A1 (Glasgow et al., 2001). Anti-apoptotic activity has also been shown in conjunction with certain apoptotic stimuli, e.g., TNFα ionizing radiation and daunorubicin (Beg and Baltimore, 1996; Wang et al., 1996). Alternatively, there is also evidence for apoptosis-promoting functions of NF-κB (Kaltschmidt et al., 2000). For instance, in response to anticancer drugs, NF-κB directly transactivates FasL; Khan and Bisen, 2013, whose gene product contributes to cell death (Kuhnel et al., 2000).

3.6   SERINE/THREONINE KINASES

The TGF-3 family of receptors (Fan et al., 1996) activates other apoptotic pathways. These receptors, called type I and type II, are both needed for the TGFβ effect. The binding of TGFβ to its receptor promotes activation of the serine/threonine kinase associated with the type II receptor and phosphorylation of the type I receptor (Massague and Weis-Garcia, 1996). The subsequent steps involved in apoptosis are not known. However, in at least some cells, the Rb protein appears to be involved (Fan et al., 1996). In several systems, apoptosis has been found to be associated with an increase in mRNA for TGFβ (Warri et al., 1993; McCloskey et al., 1996), further implicating this factor in the mediation of the apoptotic signal. TGFβ also induced apoptosis in oral squamous cell carcinoma (Briskin et al., 1996).

3.7   STEROID HORMONES

Examples of factors acting through intracellular receptors are glucocorticoid, estrogen, progesterone, thyroid, and vitamin D hormones. These factors form a complex with their specific receptors and activate the transcription of specific genes. The glucocorticoid-induced apoptosis signal pathway appears to differ with the system studied. Glucocorticoid-induced apoptosis in immature thymocytes is cyclohexi-mide sensitive, while in mature thymocytes, the pathway is not sensitive to cyclo-heximide (Iwata et al., 1995).

Analogs of steroid hormones can stimulate steroid receptors. The antiestrogen factors, like tamoxifen and toremifene, induced apoptosis in breast cancer cell lines (Warri et al., 1993), presumably by blocking estrogen binding to its specific recep-tor. The mechanism by which these factors induced apoptosis included increased expression of genes like TRPM-2 and TGFβ. Estrogen withdrawal induced apop-tosis in these cells as well (Warri et al., 1993). However, the gene expression pat-tern was different from that induced by the antiestrogen factor toremifene. The significance of these results is twofold: First, it demonstrates that antiestrogens can induce apoptosis. Second, it demonstrates that they act by binding to estrogen receptor and stimulating an apoptotic signal independent of their blockade of the estrogen receptor. Several studies attempted to define the signaling pathways for steroid-induced apoptosis. Vitamin D3 (Okazaki et al., 1989) and corticosteroids activate the sphingomyelin pathway by the formation of ceramide, which stimu-lates EGF receptor phosphorylation in A431 cells via sphingosine action (Goldkorn et al., 1991).

3.8 CENTRAL ROLE OF CASPASES

Three major pathways have been elucidated so far, which all result in the activation of caspase-3, a cysteine proteinase that cleaves substrates after aspartic acid (asp) residues. One is the mitochondrial/cytochrome C pathway, largely mediated through Bcl-2 family members, which results in activation of Apaf-1, caspase-9, and then cas-pase-3 (Khan et al., 2006, 2010b, 2012; Sah et al., 2006). The second signals ligation of members of the TNF receptor family (e.g., Fas, TRAIL receptors) and activates caspase-8 and subsequently caspase-3. Finally, granzyme B (a cytolytic T-cell prod-uct) directly cleaves and activates several caspases, resulting in apoptosis.

Apoptosis is a genetically determined process. Cell death during development of the nematode *C. elegans* involves the molecules CED-3 and CED-4, which are required for cell death, and CED-9, which protects cells from death. In mammals, CED-3 homologs constitute a family of cystein proteases with aspartate specific-ity, formerly called the interleukin-1α-converting enzyme (ICE) family and now designated caspases (Alnemri et al., 1996), which are the key effector proteins of apoptosis in mammalian cells (Yuan et al., 1993). The discovery that human Bcl-2 has functional and structural similarity to CED-9 demonstrated that programed cell death in mammalian cells occurs by a highly conserved mechanism as apoptosis in the nematode (Hengartner and Horvitz, 1994; Vaux et al., 1992). The (programmed cell death) PCD cascade can be divided into several stages. Multiple signaling pathways lead from death, triggering extracellular or intracellular agents to a cen-tral control and an execution stage. In this stage, the activation of CED-3/caspases occurs, which leads to the characteristic apoptotic structural lesions accompanying cell death: cytoplasmic and chromatin condensation and DNA fragmentation. Many environmental, pharmacologic, or physiologic stimuli can trigger apoptosis (Bast et al., 2000).

Caspases are zymogens: they exist as inactive polypeptides that can be activated by removal of the regulatory prodomain and assembled into the active heteromeric protease. Currently, the caspase family consists of 13 members. They encompass a death domain (DD), a death effector domain (DED), and a caspase recruitment domain (CARD) (Hofmann et al., 1997). The DD is present in members of the TNF receptor family and is involved in the early events of the signaling pathway. The DED and CARD are critical in the downstream portion of the pathways by recruiting caspases to the plasma membrane before their activation. Recent studies have shown that the apoptotic cascade triggered by cytochrome C and dATP is mediated by binding of caspase-9 to Apaf-1 through CARD-CARD interactions (Li et al., 1997). Caspase-9 becomes activated, and in turn activates and cleaves caspase-3. NMR spectroscopy data provide evidence that basic/acidic surface polarity in the CARD domain is highly conserved and may represent a general mode for CARD-CARD interaction (Chou et al., 1998). Caspase-9 deletion in knockout mice prevents activation of caspase-3 in embryonic brains in vivo, leading to perinatal death with severe brain-specific malformation (Kuida et al., 1998). Caspase-9-deficient thymocytes show resistance to dexamethasone, but not to Fas-mediated apoptosis, implicating functional diversification of caspase cascades, depending on the external stimulus.

Caspases can be grouped into three subfamilies based on their functions (Nicholson and Thornberry, 1997; Talanian et al., 1997; Thornberry et al., 1997). Caspases-8 and -10 each contain two N-terminal located on DEDs (death effector domain) that enable them to associate with death receptors and place these two caspases upstream in the apoptotic activation pathway (Fernandes-Alnemri et al., 1996; Boldin et al., 1996; Muzio et al., 1996). In turn, caspase-3 appears to be a downstream central executioner (Faleiro et al., 1997; MacFarlane et al., 1997; Takahashi et al., 1997) that can directly process procaspases-2, -6, -7, and -9 (Srinivasula et al., 1996; Fernandes-Alnemri et al., 1995). Findings by several groups have revealed that the activation of caspase-3 requires the Ced-4 homolog, Apaf-1, and procaspase-9, as well as dATP and cytochrome C (Liu et al., 1996; Yang et al., 1997a,b; Zou et al., 1997). Hence, caspase-9 is upstream in the pathway and is regulated by Bcl-2 family genes. For murine caspases-11 and -12, no human counterparts have been described so far. Recently, the isolation of human caspase-13 (ERICE) from the ICE subfamily was reported (Humke et al., 1998). The demonstration that activation of ERICE is mediated by caspase-8 has supported a potential downstream role for active ERICE in caspase-8-mediated cell death.

3.9 CASPASES INHIBITORY PROTEINS

The function of caspases, even after activation by cleavage, is subject to inhibition by other physiologic caspase inhibitors, thereby preventing unwanted or accidental proteolysis. Alterations in the expression or function of these proteins may confer resistance of tumor cells to the apoptotic stimuli. Viral proteins including CrmA (inhibitor of caspases-1 and -8) and p35 (which inhibits almost all caspases) were the first described caspase inhibitors (Clem and Miller, 1994). The decoy protein, FLICE-inhibitory protein (FLIP), prevents the binding of FLICE to its cofactor FADD, and thus inhibits caspase-8 FADD-like interleukin-1β-converting enzyme (FLICE), which is required for its activation. The discovery of the CARD domain-containing protein ARC (apoptosis repressor with caspase recruitment domain) suggests the existence of decoy proteins for other caspases. A new family of proteins known as IAPs (for inhibitors of apoptosis proteins) was identified via homology with the baculovirus IAP genes and includes IAP1, IAP2, NAIP, XIAP, and survivin (Clem and Miller, 1994; Duckett et al., 1996; Roy et al., 1995; Rothe et al., 1995; Uren et al., 1996; Hay et al., 1995). Survivin seems to preferentially target the caspase-9. XIAP, c-IAP1, and c-IAP2 prevent the proteolytic processing of procaspases-3, -6, and -7 by preventing the conformational changes of procaspase-9 required for downstream activation (Stennicke et al., 1999; Deveraux et al., 1998; Khan et al., 2006). Additionally, active caspase-3 function is directly inhibited by the binding of cleaved caspase-3 or -7 by XIAP, c-IAP1, c-IAP2, and survivin (Sah et al., 2006). These findings suggest that the ratio of caspases to IAPs is likely to be critical. However, since IAPs function by blocking caspase activation, they may not be able to prevent cell death induced by caspase-independent mechanisms. Loss of IAP-related genes may cause cell death in mammalian cells and certain disorders; that is, NAIP mutations are observed in two-thirds of patients with spinal muscular atrophy (Roy et al., 1995). c-IAP2 and a novel gene MLT are rearranged in the t-(11,18) (Dierlamm et al., 1999) found in MALT lymphomas, potentially conferring a survival advantage to lymphoma cells, and in some cases rendering them immune to Fas-induced apoptosis (Greiner et al., 1998).

A new human gene, survivin, has been described that encodes a structurally unique IAP apoptosis inhibitor that is undetectable in terminally differentiated adult tissues, but prominently expressed in transformed cell lines and in all of the most common human cancers of the lung, colon, pancreas, prostate, and breast, and in high-grade non-Hodgkin's lymphomas (Sah et al., 2006; Khan et al., 2010a, 2012). Survivin is the first apoptosis inhibitor that is selectively expressed in the G2-M cell cycle phase and directly associates with the mitotic spindle microtubules (Ambrosini et al., 1998 Sah et al., 2006). Disruption of survivin-microtubule interactions results in loss of survivin's anti-apoptosis function and increased caspase-3 activity, indicating that survivin may counteract a default induction of apoptosis in the G2/M phase (Li et al., 1998). Therefore, survivin appears to be a novel apoptotic guardian of a cell cycle checkpoint. High levels of survivin expression are associated with poor clinical outcome in oral cancer, neuroblastoma, and colon and gastric cancers (Khan et al., 2009, 2012; Adida et al., 1998a,b; Lu et al., 1998; Kawasaki et al., 1998). Enchantingly, the coding strand of survivin is extensively complementary to that of effector cell protease receptor-1 (EPR-1) (Altieri, 1995), although they are coded from separate genes located at 17q25 (Ambrosini et al., 1998). The finding that downregulation of survivin by overexpression of ERP-1 in vitro increases apoptosis and inhibits growth of transformed cells has supported a potential role for endogenous ERP-1 as a natural antisense (Adida et al., 1998b) and survivin as a potential new target for apoptosis-based therapy.

3.10 DEATH DOMAINS

Cells require both internal and external means of regulating the activation of caspases and the death machinery. Depending on other contextual events, cell surface death receptors can transmit apoptosis signals in response to external stimuli such as death ligands, growth factor withdrawal, or chemotherapeutic agents. Death receptors belong to the tumor necrosis factor (TNF) receptor family and have a characteristic cysteine-rich extracellular domain (Smith et al., 1994) and a homologous cytoplasmic "death domain" (Nagata, 1997) that initiates apoptotic signaling inside the cell (Westendorf et al., 1995). They are set apart from the other receptors by their specific structure. In the cytoplasmic domain, a stretch of 65 amino acids (death domains) has been found, whose activity was necessary for the induction of apoptosis (Baker and Reddy, 1996). These receptors can induce apoptotic cell death within hours after ligand binding and may exert their apoptogenic effects differentially in diverse cell types depending on downstream signaling (Consoli et al., 1998). The mechanism of TNF-mediated signal transduction pathway are poorly understood (Kumar and Harvey, 1995; Yin et al., 1995). Other studies showed that the production of ceramide might play a significant role in some systems. TNF initiated the activation of a 55 kDa receptor and activated sphingomyelinase with the formation of ceramide (Yanaga and Watson, 1992). The treatment of oral cancer cells with TNF-induced significant apoptosis, demonstrates that TNF is a positive regulator of apoptosis in at least some oral cancers (Briskin et al., 1996).

3.11 FAS/FAS LIGAND

Fas ligand (FasL) is a type II transmembrane protein of 40 kDa that belongs to the tumor necrosis factor (TNF) family, predominantly expressed in activated T-cells. Soluble Fas ligand is generated by cleaving membrane-bound FasL at a conserved cleavage site by the matrix metalloproteinase MMP-7. The membrane-bound form of FasL is functional, whereas shedding of soluble FasL inhibits cytotoxicity and may prevent the killing of healthy bystander cells by cytotoxic T-cells (Tanaka et al., 1998). Soluble FasL also plays important roles in immune evasion, by inhibiting Fas-mediated and NKG2D-mediated killing of immune cells, and is able to inhibit Fas and the NKG2D-mediated death of immune cells (Song et al., 2001). The Fas/FasL system is responsible for activation-induced cell death, but also plays an important role in lymphocyte-mediated killing under certain circumstances. In oral squamous cell carcinoma, a high proportion of T-cells in the tumor undergo apoptosis, which correlates with FasL expression on tumor cells. FasL positive microvesicles induced caspase-3 cleavage, cytochrome c release, loss of mitochondrial membrane potential, and reduced TCR-ζ (T-cell receptor-ζ) chain expression in target lymphocytes (Kim et al., 2005). The decreased expression of Fas or increased expression of FasL favors malignant transformation and tumor progression. Downregulation of Fas receptors and killing of activated T-lymphocytes through the constitutive expression of Fas ligand on tumor cells has been suggested as a mechanism for pathologic suppression of immune surveillance (Strand et al., 1996). Such "immune privilege" has been demonstrated in melanomas and colon cancers (Hahne et al., 1996; O'Connell et al.,

1996). Binding of FasL to Fas or cross-linking Fas with agonistic antibodies results in receptor trimerization (Nagata and Golstein, 1997). Adapter proteins (FADD/ MORT1 and RAIDD) bind to DD via their own DDs (Boldin et al., 1995; Chinnaiyan et al., 1995, 1996; Duan and Dixit, 1997). A separate DED (of FADD/MORT1) binds to the prodomain of the caspase-8 (FLICE/MACH) and thereby links of the Fas death-inducing signaling complex (DISC) with proteases (Boldin et al., 1996; Muzio et al., 1996), and thus apoptosis. Another pathway involves the Fas DD (Yang et al., 1997b) binding protein Daxx, which in turn activates the c-Jun NH2-terminal kinase (JNK), the JNK kinase kinase ASK1 (apoptosis signal-regulating kinase 1), and Bcl-2; however, the importance of this pathway is uncertain. Observations from several studies (Muller et al., 1998; Fukazawa et al., 1999; Owen-Schaub et al., 1995) suggest that a functional Fas pathway requires intact p53, and thus provides a potential mechanism for p53-mediated resistance of cancer cells to chemotherapy. A p53 binding sequence has been identified in the Fas promoter (Owen-Schaub et al., 1994), and gene restoration therapy with p53 results in upregulation of Fas (Owen-Schaub et al., 1995).

The CD95 system is an important regulator of T-cell cytotoxicity that is involved in the killing of mature T-cells after immune response and killing of targets by cytotoxic T-cells and natural killer cells. A frame shift mutation that renders cells resistant to Fas-mediated apoptosis has been found in adult T-cell leukemia. This finding has suggested that mutation of the Fas gene may be one of the mechanisms in the progression of ATL (Maeda et al., 1999). Enthusiasm for the clinical use of Fas as a target is dampened by the observation that anti-Fas antibody induces rapid (within hours) death of mice from fulminant hepatic toxicity (Ogasawara et al., 1993; Andreeff et al., 1984). Soluble FasL may be less liver toxic, but induces less apoptosis. This finding may explain why high soluble FasL levels found in many cancers are not associated with toxicity (Schneider et al., 1998), and as a consequence, no trials in humans are underway.

3.12   TRAIL AND ITS RECEPTORS

TRAIL (TNF-related apoptosis-inducing ligand or APO2-L) is a type II transmembrane protein that selectively induces apoptosis in tumor cells but not normal cells (Kemp et al., 2003; Wajant et al., 2002). Because of this differential sensitivity TRAIL is considered an ideal anticancer drug (Ashkenazi, 2002; Yagita et al., 2004). It interacts with four distinct surface receptors, TRAIL-R1 (DR4), -R2 (DR5), -R3 (DcR1), and R4 (DcR2), and with the soluble receptor osteoprotegerin (Kemp et al., 2003, Wajant et al., 2002). DR4 and DR5 receptors bind to and activate caspases through FLICE2 (FADD-like interleukin-1β-converting enzyme 2). Subsequently, nonsignaling decoy receptors (DcR1, DcR2) were identified in normal human tissues, but not in most cancer cell lines examined. Their recognition of TRAIL may prevent TRAIL from binding to functional TRAIL receptors, therefore blocking and not transducing the cell death signal. At this point in time, the definitive role(s) of TRAIL in apoptosis remains to be determined since the presence of "protective" TRAIL receptors does not correspond to resistance or sensitivity to TRAIL-mediated apoptosis in some systems (Griffith et al., 1998; Khan and Bisen, 2013).

The fact that DR4 and DR5 are expressed in many tumors, whereas DcR1 and DcR2 are expressed predominantly in normal tissues, suggests that TRAIL could differentially induce apoptosis in tumor cells, but exceptions to this paradigm already exist.

TRAIL has been evaluated as a possible therapeutic agent and appears to have more promise than Fas. Distinguishing TRAIL from FasL is the observation that TRAIL only seems to induce apoptosis in malignant cell lines and not normal cell lines. In melanoma cell lines, TRAIL induces apoptosis (Thomas and Hersey, 1998; Griffith et al., 1998). Recombinant soluble TRAIL induced significant apoptosis in myeloid and lymphoid cell lines, and decreases in viability were observed in 20% of samples from patients with hematologic malignancies (Snell et al., 1997). Among glioma cell lines, which preferentially express DR4 and 5, but not the decoy receptors, 10/12 cell lines were sensitive to TRAIL (Andreeff et al., 1985). A variety of factors may affect TRAIL sensitivity. There does not appear to be synergism with FasL, and neither ATRA nor MDR1 affects sensitivity (Snell et al., 1997). P53 status also appears unrelated to TRAIL sensitivity; however, high Bcl-2 levels inhibit sensitivity (Andreeff et al., 1985). Among sensitive melanoma cell lines, the levels of DR4/DR5 correlated with sensitivity in one study (Thomas and Hersey, 1998) but not in another (Griffith et al., 1998). Resistance to TRAIL was shown to be secondary to loss of cell surface expression, secondary to either gene loss (4/9 lines) or because it was trapped in the cytoplasm (Golstein, 1997). Resistance has also been correlated with high levels of expression of FLIP, the TRAIL inhibitor, in resistant melanoma cell lines (Griffith et al., 1998); however, a correlation was not observed in all studies. Expression of the inhibitory receptor TRID was also reported (Keane et al., 1999). Stimulation of cells with CD40-CD40L leads to downregulation of TRAIL and upregulation of TNF and Fas to promote B-cell survival (Ribeiro et al., 1998). Combined, these data suggest that the TRAIL is capable of inducing apoptosis in malignancies, including those of hematologic origin, but that multiple mechanisms of resistance likely affect sensitivity to TRAIL (Khan and Bisen, 2013).

Recent preclinical and clinical data have shown the potential utility of TRAIL targeted therapies in advanced cancers, including oral cancer (Yagita et al., 2004; Finnberg and El-Deiry, 2006; Rowinsky, 2005). Currently, there are only limited data available pertaining to the baseline expression levels of TRAIL and its receptors in normal oral mucosa, oral premalignancies (OPMs), and primary or metastatic oral squamous cell carcinoma (OSCC) (Fukuda et al., 2003; Teng et al., 2005). These data are critical to analysis and interpretation of clinical trial data involving recombiant-TRAIL and TRAIL-receptor agonist antibodies in OSCC patients, as well as the understanding of the role of TRAIL and its receptors during oral carcinogenesis.

The functional role of TRAIL, which is constitutively expressed in most normal tissues, is poorly understood. Recently, however, TRAIL has been shown to trigger apoptosis in transformed or dysplastic epithelial cells, suggesting therapeutic potential as a chemopreventive agent against malignant progression of premalignancies (Finnberg and El-Deiry, 2006; Jalving et al., 2006; Lu et al., 2004). Oral premalignancies are considered a progenitor of OSCC and represent an intermediate step in the progression from normal oral mucosa to OSCC (Pindborg et al., 1997). The likelihood for development of OSCC in oral premalignancies is generally proportional to its dysplasia grade (Pindborg et al., 1997). Similar loss of TRAIL expression during the progression from normal epithelia to malignancy has been reported in skin, esophageal, and colon cancers (Popnikolov et al., 2006; Stander and Schwarz, 2005; Strater et al., 2002). The loss of individual TRAIL-R expression in malignant tumors has been attributed to chromosomal gene mutation/deletion (DR4 and DR5) or promotor methylation (DcR1 and DcR2) events (Yagita et al., 2004). However, gene mutation or promotor methylation is not implicated for the malignancy-specific downregulation of TRAIL expression. Recent experimental findings indicate that loss of TRAIL expression during malignant transformation is not mediated by genetic aberration, but by dysregulation of signal transduction pathways common to various cancers. TNFα increases the susceptibility of breast cancer cells to chemotherapy by upregulating TRAIL expression by promoter activation (Xu et al., 2006). Similarly, retinoids and interferons exert their anticancer activity in breast cancer cells by enhancing TRAIL expression in these cells (Clarke et al., 2004). Hyperactivation of the phosphatidylinositol-3-kinase (PI3K)/Akt pathway has been implicated in suppressing TRAIL expression, in colon cancer cells (Wang et al., 2002). Treatment of colon cancer cells with the PI3K inhibitor Wortmannin rescues TRAIL expression in these cells and induces enterocyte-like differentiation (Wang et al., 2002). Interestingly, hyperactivation of the PI3K/Akt pathway via dysregulated EGFR signaling is an important and early event in the pathogenesis of OSCC (Kalyankrishna and Grandis, 2006). Hence, it is plausible that there is a cause-and-effect relationship between dysregulated prosurvival PI3K/Akt signaling and loss of TRAIL expression in oral premalignancies and OSCC. Interestingly, oral carcinogens such as tobacco smoke frequently inactivate the tumor suppressor gene p53, thus depriving its protective role against oral carcinogen-induced DNA damage (Lazarus et al., 1998). TRAIL-induced apoptosis of transformed/dysplastic cells is independent of p53 status (Jalving et al., 2006, El-Deiry, 2001); hence, TRAIL may act as a substitute guardian against malignant transformation by eliminating transformed cells during the initial genesis oral premalignancies. Thus, TRAIL downregulation may allow clonal expansion of transformed cells by protecting them from apoptosis, thereby increasing the risk of malignant progression.

3.13 CONCLUSION

The study of oral cancer is particularly challenging. It appears that, in general, signal transduction in oral cancer is comparable with that described in other cancer models. As described in this chapter, most evidence concerning oral cancers comes from studies evaluating receptors, growth factors, or focusing on the characterization of a specific signal transducer. The regulation of the cell cycle is coupled to cell death and has major significance in cell turnover and tumorigenesis. Several genes are common to cell cycle regulation and apoptosis. The fate of cells is likely to be determined by their interplay. When cells are subjected to adverse (growth) conditions, complex signal transduction networks are initiated. The information received is processed and sent to subcellular organelles. The important decisions of cell death or cell proliferation are likely to be controlled by more than just one signal; most likely this is a mechanism that ensures a proper cellular response. One can imagine catastrophic consequences for the cell, when key players in cell cycle regulation or apoptosis are not coordinated. The last few decades have seen an extraordinary increase in our understanding of proliferation and apoptosis, and its contribution to cancer and cancer therapy, but a sound understanding of underlying cell biology is likely to enable further, much needed progress.

REFERENCES

Adams, J.M., and S. Cory. 1998. The Bcl-2 protein family: Arbiters of cell survival. *Science* 281: 1322.

Adida, C., D. Berrebi, M. Peuchmaur, et al. 1998a. Anti-apoptosis gene, survivin, and prognosis of neuroblastoma. *Lancet* 351: 882–883.

Adida C, P.L. Crotty, J. McGrath, et al. 1998b. Developmentally regulated expression of the novel cancer anti-apoptosis gene survivin in human and mouse differentiation. *Am J Pathol* 152: 43–49.

Alitalo, K., M. Schwab, C.C. Lin, et al. 1983. Homogeneously staining chromosomal regions contain amplified copies of an abundantly expressed cellular oncogene (c-myc) in malignant neuroendocrine cells from a human colon carcinoma. *Proc Natl Acad Sci USA* 80: 1707–1711.

Alnemri, E.S., D.J. Livingston, D.W. Nicholson, et al. 1996. Human ICE/CED-3 protease nomenclature. *Cell* 87: 171.

Alsner, J., S.B. Sørensen, and J. Overgaard. 2001. TP53 mutation is related to poor prognosis after radiotherapy, but not surgery, in squamous cell carcinoma of the head and neck. *Radiother Oncol* 59: 179 –185.

Alt, F.W., R.E. Kellems, J.R. Bertino, et al. 1978. Selective multiplication of dihydrofolate reductase genes in methotrexate-resistant variants of cultured murine cells. *J Biol Chem* 253: 1357–1370.

Altieri, D.C. 1995. Xa receptor EPR-1. *FASEB J* 9: 860–865.

Amati, B., K. Alevizopoulos, and J. Vlach. 1998. Myc and the cell cycle. *Front Biosci* 3: D250–D268.

Ambrosini, G., C. Adida, G. Sirugo, et al. 1998. Induction of apoptosis and inhibition of cell proliferation by survivin gene targeting. *J Biol Chem* 273: 11177–11182.

Andreeff, M., H. Hansen, C. Cirrincione, et al. 1984. Cellular RNA content: A major prognostic factor in adult acute leukemia and non-Hodgkin lymphoma. *Proc Intl Conf Analyt Cytol* 10: 25.

Andreeff, M., G. Wong, B. Koziner, et al. 1985. Non B-non T acute lymphoblastic leukemia (ALL): Evidence for complete B cell differentiation of a quiescent subpopulation and their response to induction therapy. *Proc Am Assoc Cancer Res* 26: 28.

Ashkenazi, A. 2002. Targeting death and decoy receptors of the tumour-necrosis factor superfamily. *Nat Rev Cancer* 2: 420–430.

Avraham H., S.Y. Park, K. Schinkmann, et al. 2000. RAFTK/Pyk2-mediated cellular signalling. *Cell Signal* 12: 123–133.

Bader, A.G., S. Kang, and L. Zhao. 2005. Oncogenic PI3K deregulates transcription and translation. *Nat Rev Cancer* 5: 921–929.

Baichwal, V.R., and P.A. Baeuerle. 1997. Activate NF-kappa B or die? *Curr Biol* 7: R94.

Baker S.I., and E.P. Reddy. 1996. Transducers of life and death: TNF receptor superfamily and associated proteins. *Oncogene* 12: 1–9.

Barbacid, M. 1987. ras genes. *Ann Rev Biochem* 56: 776–827.

Bartkova, J., Z. Horejs, K. Koed, et al. 2005. DNA damage response as a candidate anti-cancer barrier in early human tumorigenesis. *Nature* 434: 864–870.

Bast, R.C., Jr., D.W. Kufe, R.E. Pollock, et al. 2000. *Holland-Frei cancer medicine*. 5th ed. BC Decker, Hamilton, Ontario.

Beg, A.A., and D. Baltimore. 1996. An essential role for NF-kappaB in preventing TNF-alpha-induced cell death. *Science* 274: 782.

Beier, R., A. Burgin, A. Kiermaier, et al. 2000. Induction of cyclin E-cdk2 kinase activity, E2F-dependent transcription and cell growth by Myc are genetically separable events. *EMBO J* 19: 5813.

Bishop, J.M. 1982. Retroviruses and cancer genes. *Adv Cancer Res* 37: 1–32.

Bishop, M.I. 1991. Molecular themes in oncogenesis. *Cell* 64: 235–248.

Bissonnette, R.P., F. Echeverri, A. Mahboubi, et al. 1992. Apoptotic cell death induced by c-myc is inhibited by bcl-2. *Nature* 359: 552–554.

Blagosklonny, M.V. 2001. Unwinding the loop of Bcl-2 phosphorylation. *Leukemia* 15: 869.

Blattner, C., A. Sparks, and D. Lane. 1999. Transcription factor E2F-1 is upregulated in response to DNA damage in a manner analogous to that of p53. *Mol Cell Biol* 19: 3704.

Bockmühl, U., G. Wolf, S. Schmidt, et al. 1998. Genomic alterations associated with malignancy in head and neck cancer. *Head Neck* 20: 145–151.

Boldin, M.P., T.M. Goncharov, Y.V. Goltsev, et al. 1996. Involvement of MACH, a novel MORT1/FADD-interacting protease, in Fas/APO-1- and TNF receptor-induced cell death. *Cell* 85: 803–815.

Boldin, M.P., E.E. Varfolomeev, Z. Pancer, et al. 1995. A novel protein that interacts with the death domain of Fas/APO1 contains a sequence motif related to the death domain. *J Biol Chem* 270: 7795–7798.

Born, T.L., J.A. Frost, A. Schönthal, et al. 1994. C-Myc cooperates with activated Ras to induce the cdc2 promoter. *Mol Cell Biol* 14: 5710–5718.

Bos, J.L. 1989. ras oncogenes in human cancer: A review. *Cancer Res* 49: 4682–4689.

Briskin, K.B., C. Fady, M. Wang, et al. 1996. Apoptotic inhibition of head and neck squamous cell carcinoma cells by tumor necrosis factor {α}. *Arch Otolaryngol Head Neck Surg* 122: 559–563.

Buchkovich, K., L.A. Duffy, E. Harlow, et al. 1989. The retinoblastoma protein is phosphorylated during specific phases of the cell cycle. *Cell* 58: 1097–1105.

Califano, J., P. van der Riet, W. Westra, et al. 1996. Genetic progression model for head and neck cancer: Implications for field cancerization. *Cancer Res* 56: 2488–2492.

Cantley, L.C., K.R. Auger, C. Carpenter, et al. 1991. Oncogenes and signal transduction. *Cell* 64: 281–302.

Cao, L., B. Faha, M. Dembski, et al. 1992. Independent binding of the retinoblastoma protein and p107 to the transcription factor E2F. *Nature* 355: 176–179.

Chang, D.W., G.F. Claassen, S.R. Hann, et al. 2000. The c-Myc transactivation domain is a direct modulator of apoptotic versus proliferative signals. *Mol Cell Biol* 20: 4309.

Chen-Levy, Z., and M.L. Cleary. 1990. Membrane topology of the Bcl-2 proto-oncogenic protein demonstrated in vitro. *Biol Chem* 265: 4929–4933.

Chiba, I., T. Takahashi, M. Nau, et al. 1990. Mutations in the p53 gene are frequent in primary, resected non-small cell lung cancer. *Oncogene* 5: 1603–1610.

Chinnaiyan, A.M., K. O'Rourke, M. Tewari, et al. 1995. FADD, a novel death domain-containing protein, interacts with the death domain of Fas and initiates apoptosis. *Cell* 81: 505–512.

Chinnaiyan, A.M., C.G. Tepper, M.F. Seldin, et al. 1996. FADD/MORT1 is a common mediator of CD95 (Fas/APO1) and tumor necrosis factor receptor-induced apoptosis. *J Biol Chem* 271: 4961–4965.

Chou, J.J., H. Matsuo, H. Duan, et al. 1998. Solution structure of the RAIDD CARD and model for CARD/CARD interaction in caspase-2 and caspase-9 recruitment. *Cell* 94: 171–180.

Clarke, N., A.M. Jimenez-Lara, E. Voltz, et al. 2004. Tumor suppressor IRF-1 mediates retinoid and interferon anticancer signaling to death ligand TRAIL. *EMBO J* 23: 3051–3060.

Clem, R.J., and L.K. Miller. 1994. Control of programmed cell death by the baculovirus genes p35 and iap. *Mol Cell Biol* 14: 5212–5222.

Cole, M.D. 1986. The *myc* oncogenes: Its role in transformation and differentiation. *Ann Rev Genet* 20: 361–384.

Connelly, P.A., and D.F. Stern. 1990. The epidermal growth factor receptor and the product of the neu protooncogene are members of a receptor tyrosine phosphorylation cascade. *Proc Natl Acad Sci USA* 87: 6054–6057.

Consoli, U., I. El-Tounsi, A. Sandoval, et al. 1998. Differential induction of apoptosis by fludarabine monophosphate in leukemic B and normal T cells in chronic lymphocytic leukemia. *Blood* 91: 1742–1748.

Conzen, S.D., K. Gottlob, E.S. Kandel, et al. 2000. Induction of cell cycle progression and acceleration of apoptosis are two separable functions of *c-Myc*: Transrepression correlates with acceleration of apoptosis. *Mol Cell Biol* 20: 6008.

Cowell, J.K. 1982. Double minutes and homogeneously staining regions: Gene amplification in mammalian cells. *Annu Rev Genet* 16: 21–59.

Daksis, J.I., R.Y. Lu, L.M. Facchini, et al. 1994. Myc induces cyclin D1 expression in the absence of de novo protein synthesis and links mitogen-stimulated signal transduction to the cell cycle. *Oncogene* 9: 3635–3645.

Davies, H., G.R. Bignell, C. Cox, et al. 2002. Mutations of the BRAF gene in human cancer. *Nature* 417: 949–954.

de Alboran, I.M., R.C. O'Hagan, F. Gartner, et al. 2001. Analysis of C-MYC function in normal cells via conditional gene-targeted mutation. *Immunity* 14: 45.

de Vicente, J.C., L.M.J. Gutiérrez, A.H. Zapatero, et al. 2004. Prognostic significance of p53 expression in oral squamous cell carcinoma without neck node metastases. *Head Neck* 26: 22–30.

Deveraux, Q.L., N. Roy, H.R. Stennicke, et al. 1998. IAPs block apoptotic events induced by caspase-8 and cytochrome c by direct inhibition of distinct caspases. *EMBO J* 17: 2215–2223.

Dierlamm, J., M. Baens, I. Wlodarska, et al. 1999. The apoptosis inhibitor gene API2 and a novel 18q gene, MLT, are recurrently rearranged in the t(11;18)(q21;q21) associated with mucosa-associated lymphoid tissue lymphomas. *Blood* 93: 3601–3609.

Dominguez-Sola, D., C.Y. Ying, C. Grandori, et al. 2007. Non-transcriptional control of DNA replication by c-Myc. *Nature (London)* 448: 445–451.

Downward, J. 2003. Targeting RAS signalling pathways in cancer therapy. *Nat Rev Cancer* 2003; 3 (1): 11–22.

Duan, H., and V.M. Dixit. 1997. RAIDD is a new "death" adaptor molecule. *Nature (London)* 385: 86–89.

Duckett, C.S., V.E. Nava, R.W. Gedrich, et al. 1996. A conserved family of cellular genes related to the baculovirus IAP gene and encoding apoptosis inhibitors. *EMBO J* 15: 2685–2694.

Duelli, D.M., and Y.A. Lazebnik. 2000. Primary cells suppress oncogene-dependent apoptosis. *Nat Cell Biol* 2: 859–862.

Dyson, N., P.M. Howley, K. Munger, et al. 1989. The human papilloma virus-16, E7 oncoprotein is able to bind to the retinoblastoma gene product. *Science* 243: 934–937.

Eisenman, R.W., and C.B. Thompson. 1986. Oncogenes with potential nuclear function: myc, myb, and fos. *Cancer Surveys* 5: 309–406.

El-Deiry, W.S. 2001. Insights into cancer therapeutic design based on p53 and TRAIL receptor signaling. *Cell Death Differ* 8: 1066–1075.

El-Deiry, W.S., T. Tokino, V.E. Velculescu, et al. 1993. WAFI, a potential mediator of p53 tumor suppression. *Cell* 75: 817–825.

el Naggar, A.K., K. Hurr, J.G. Batsakis, et al. 1995. Sequential loss of heterozygosity at microsatellite motifs in preinvasive and invasive head and neck squamous carcinoma. *Cancer Res* 55: 2656–2659.

Enoch, T., and C. Norbury. 1995. Cellular responses to DNA damage: Cell cycle checkpoints, apoptosis and the role of p53 and AMT. *Trends Biochem Sci* 20: 426–430.

Evan, G., and T. Littlewood. 1998. A matter of life and cell death. *Science* 281: 1317–1322.

Faleiro, L., R. Kobayashi, H. Fearnhead, et al. 1997. Multiple species of CPP32 and Mch2 are the major active caspases present in apoptotic cells. *EMBO J* 16: 2271–2281.

Fan, G., X. Ma, B.T. Kren, et al. 1996. The retinoblastoma gene product inhibits TGF-beta1 induced apoptosis in primary rat hepatocytes and human HuH-7 hepatoma cells. *Oncogene* 12: 1909.

Fan, Z, Y. Lu, X. Wu, et al. 1995. Prolonged induction of p2 lCipl/WAFl/CDK2/PCNA complex by epidermal growth factor receptor activation mediates ligand-induced A431 cell growth inhibition I. *Cell Biol* 131: 235–242.

Farmer, G., J. Burgonetti, H. Zhu, et al. 1992. Wild-type p53 activates transcription in vitro. *Nature* 358: 83–86.

Fernandes-Alnemri, T., R.C. Armstrong, J. Krebs, et al. 1996. In vitro activation of CPP32 and Mch3 by Mch4, a novel human apoptotic cysteine protease containing two FADD-like domains. *Proc Natl Acad Sci USA* 93: 7464–7469.

Fernandes-Alnemri, T., A. Takahashi, R. Armstrong, et al. 1995. Mch3, a novel human apoptotic cysteine protease highly related to CPP32. *Cancer Res* 55: 6045–6052.

Field, J.K. 1992. Oncogenes and tumor-suppressor genes in squamous cell carcinoma of the head and neck. *Eur J Cancer B Oral Oncol* 28: 67–76.

Field, J.K., D.A. Spandidos, P.M. Stell, et al. 1989. Elevated expression of the c-myc oncoprotein correlates with poor prognosis in head and neck squamous cell carcinoma. *Oncogene* 4: 1463–1468.

Finnberg, N., and W.S. El-Deiry. 2006. Selective TRAIL-induced apoptosis in dysplastic neoplasia of the colon may lead to new neoadjuvant or adjuvant therapies. *Clin Cancer Res* 12: 4132–4136.

Frattini, M., C. Ferrario, P. Bressan, et al. 2004. Alternative mutations of BRAF, RET and NTRK1 are associated with similar but distinct gene expression patterns in papillary thyroid cancer. *Oncogene* 23: 7436–7440.

Fukazawa, T., T. Fujiwara, Y. Morimoto, et al. 1999. Differential involvement of the CD95 (Fas/APO-1) receptor/ligand system on apoptosis induced by the wild-type p53 gene transfer in human cancer cells. *Oncogene* 18: 2189–2199.

Fukuda, M., A. Hamao, A. Tanaka, et al. 2003. Tumor necrosis factor-related apoptosis-inducing ligand (TRAIL/APO2L) and its receptors expression in human squamous cell carcinoma of the oral cavity. *Oncol Rep* 10: 1113–1119.

Furukawa, Y., S. Iwase, J. Kikuchi, et al. 2000. Phosphorylation of Bcl-2 protein by CDC2 kinase during G2/M phases and its role in cell cycle regulation. *J Biol Chem* 275: 21661–21667.

Galaktionov, K., X. Chen, and D. Beach. 1996. Cdc25 cell-cycle phosphatase as a target of c-myc. *Nature (London)* 382: 511–517.

Glasgow, J.N., J. Qiu, D. Rassin, et al. 2001. Transcriptional regulation of the Bcl-X gene by NF-kappaB is an element of hypoxic responses in the rat brain. *Neurochem Res* 26: 647–659.

Goldkorn, T., K.A. Dressler, I. Muindi, et al. 1991. Ceramide stimulates epidermal growth factor receptor phosphorylation in A431 human epidermoid carcinoma cells: Evidence that ceramide may mediate sphingosine action. *J Biol Chem* 266: 16092–16097.

Golstein, P. 1997. Cell death: TRAIL and its receptors. *Curr Biol* 7: R750–R753.

Goodsell, D.S. 1999. The molecular perspective: The *ras* oncogene. *Oncologist* 4: 263–264.

Gorgoulis, V.G., L.V.F. Vassiliou, P. Karakaidos, et al. 2005. Activation of the DNA damage checkpoint and genomic instability in human precancerous lesions. *Nature (London)* 434: 907–913.

Grandori, C., S.M. Cowley, L.P. James, et al. 2000. The Myc/Max/Mad network and the transcriptional control of cell behavior. *Annu Rev Cell Dev Biol* 16: 653–699.

Greiner, A., H. Seeberger, C. Knorr, et al. 1998. MALT-type B-cell lymphomas escape the sensoring FAS-mediated apoptosis. *Blood* 92: 484a.

Griffith, T.S., W.A. Chin, G.C. Jackson, et al. 1998. Intracellular regulation of TRAIL-induced apoptosis in human melanoma cells. *J Immunol* 161: 2833–2840.

Hahne, M., D. Rimoldi, M. Schroter, et al. 1996. Melanoma cell expression of Fas(Apo-1/CD95) ligand: Implications for tumor immune escape. *Science* 274: 1363–1366.

Haldar, S., L. Chintapalli, and C.M. Croce. 1996. Taxol induced Bcl-2 phosphorylation and death of prostate cancer cells. *Cancer Res* 56: 1253–1255.

Haldar. S., N. Jena, and C.M. Croce. 1995. Inactivation of Bcl-2 by phosphorylation. *Proc Natl Acad Sci USA* 92: 4507.

Haldar, S., N. Jena, C.M. Croce, et al. 1994. Antiapoptosis potential of Bcl-2 oncogene by dephosphorylation. *Biochem Cell Biol* 72: 455–462.

Hann, S.R., C.B. Thompson, and Eisenman RN. 1985. c-myc oncogene protein synthesis is independent of the cell cycle in human and avian cells. *Nature (London)* 314: 366.

Hanson, K.D., M. Shichiri, M.R. Follansbee, et al. 1994. Effects of c-myc expression on cell cycle progression. *Mol Cell Biol* 14: 5748–5755.

Harper, I.W., G.R. Adami, N. Wei, et al. 1993. The p21 Cdk-interacting protein Cipl is a potent inhibitor of GI cyclin-dependent kinases. *Cell* 75: 805–816.

Harris, C.C. 1991. Chemical and physical carcinogenesis: Advances and perspectives for the 1990's. *Cancer Res* 51: 5023–5044.

Hay, B.A., D.A. Wassarman, and G.M. Rubin. 1995. Drosophila homologs of baculovirus inhibitor of apoptosis proteins function to block cell death. *Cell* 183: 1253–1262.

Heikkila, R., G. Schwab, E. Wickstrom, et al. 1987. A c-myc antisense oligodeoxynucleotide inhibits entry into S phase but not progress from G0 to G1. *Nature (London)* 328: 445–449.

Hengartner, M.O., and H.R. Horvitz. 1994. Programmed cell death in *Caenorhabditis elegans*. *Curr Opin Genet Dev* 4: 581–586.

Hermeking, H., and D. Eick. 1994. Mediation of c-Myc-induced apoptosis by p53. *Science* 265: 2091.

Hiebert, S.W., G. Packham, D.K. Strom, et al. 1995. E2F-1: DP-1 induces p53 and overrides survival factors to trigger apoptosis. *Mol Cell Biol* 15: 6864–6874.

Hinz, M., D. Krappmann, A. Eichten, et al. 1999. NF-kappaB function in growth control: Regulation of cyclin D1 expression and G0/G1-to-S-phase transition. *Mol Cell Biol* 19: 2690–2700.

Hoang, A.T., K.J. Cohen, J.F. Barrett, et al. 1994. Participation of cyclin A in Myc-induced apoptosis. *Proc Natl Acad Sci USA* 91: 6875–6879.

Hofmann, K., P. Bucher, and J. Tschopp. 1997. The CARD domain: A new apoptotic signalling motif. *Trends Biochem Sci* 22: 155–156.

Hollstein, M., D. Sidransky, B. Vogelstein, et al. 1991. p53 mutations in human cancers. *Science* 253: 49–53.

Hsu, B., M.C. Marin, and A.K. el Naggar. 1995. Evidence that c-myc mediated apoptosis does not require wild-type p53 during lymphomagenesis. *Oncogene* 11: 175–179.

Humke, E.W., J. Ni, and V.M. Dixit. 1998. ERICE, a novel FLICE-activatable caspase. *J Biol Chem* 273: 15702–15707.

Hunter, T. 1987. A thousand and one protein kinases. *Cell* 50: 823–829.

Hunter, T. 1991. Cooperation between oncogenes. *Cell* 64: 249–270.

Ishitoya, J., M. Toriyama, N. Oguchi, et al. 1989. Gene amplification and overexpression of EGF receptor in squamous cell carcinomas of the head and neck. *Br J Cancer* 59: 559–562.

Iwata, M., J. Herrington, and R.A. Zager. 1995. Protein synthesis inhibition induces cytoresistance in cultured human proximal tubular (HK-2) cells. *Am J Phys* 286: 1154–1163.

Jakus, J., and W.A. Yeudall. 1996. Growth inhibitory concentrations of EGF induce p21 (WAFI/Cipl) and alter cell cycle control in squamous carcinoma cells. *Oncogene* 12: 2369–2376.

Jalving, M., S. de Jong, J.J. Koornstra, et al. 2006. TRAIL induces apoptosis in human colorectal adenoma cell lines and human colorectal adenomas. *Clin Cancer Res* 12: 4350–4356.

Jansen-Durr, P., A. Meichle, P. Steiner, et al. 1993. Differential modulation of cyclin gene expression by MYC. *Proc Natl Acad Sci USA* 90: 3685–3689.

Jayasurya, R., G. Francis, S. Kannan, et al. 2004. p53, p16 and cyclin D1: Molecular determinants of radiotherapy treatment response in oral carcinoma. *Int J Cancer* 109: 710–716.

Jiang, W.W., H. Fujii, T. Shirai, et al. 2001. Accumulative increase of loss of heterozygosity from leukoplakia to foci of early cancerization in leukoplakia of the oral cavity. *Cancer* 92: 2349–2356.

Kaelin, W.G., Jr. 1999. The emerging p53 gene family. *J Natl Cancer Inst* 91: 594–598.

Kaltschmidt, B., C. Kaltschmidt, T.G. Hofmann, et al. 2000. The pro- or antiapoptotic function of NF-kappaB is determined by the nature of the apoptotic stimulus. *Eur J Biochem* 267: 3828–3835.

Kalyankrishna, S., and J.R. Grandis. 2006. Epidermal growth factor receptor biology in head and neck cancer. *J Clin Oncol* 24: 2666–2672.

Karnoub, A.E., and R.A. Weinberg. 2008. Ras oncogenes: Split personalities. *Nat Rev Mol Cell Biol* 9: 517–531.

Kawasaki, H., D.C. Altieri, C.D. Lu, et al. 1998. Inhibition of apoptosis by survivin predicts shorter survival rates in colorectal cancer. *Cancer Res* 58: 5071–5074.

Keane, M.M., S.A. Ettenberg, M.M. Nau, et al. 1999. Chemotherapy augments TRAIL-induced apoptosis in breast cell lines. *Cancer Res* 59: 734–741.

Kemp, T.J., J.S. Kim, S.A. Crist, et al. 2003. Induction of necrotic tumor cell death by TRAIL/ Apo-2L. *Apoptosis* 8: 587–599.

Khan, Z., and Bisen, P.S. 2013. Oncoapoptotic signaling and deregulated target genes in cancers: Special reference to oral cancer. *Biochim Biophy Acta Reviews on Cancer*, dx.doi. org/10.1016/j.bbcan.2013.04.002.

Khan, Z., P. Bhadouria, R. Gupta, et al. 2006. Tumor control by manipulation of the human anti-apoptotic survivin gene. *Curr Cancer Ther Rev* 2: 73–79.

Khan, Z., N. Khan, R.P. Tiwari, et al. 2010a. Down regulation of survivin by oxaliplatin diminishes radioresistance of head and neck squamous carcinoma cells. *Radiother Oncol* 96: 267–273.

Khan, Z., N. Khan, A.K. Varma, et al. 2010b. Oxaliplatin-mediated inhibition of survivin increases sensitivity of head and neck squamous cell carcinoma cell lines to paclitaxel. *Curr Cancer Drug Targets* 10: 660–669.

Khan, Z., R.P. Tiwari, N. Khan, et al. 2012. Induction of apoptosis and sensitization of head and neck squamous carcinoma cells to cisplatin by targeting survivin gene expression. *Curr Gene Ther* 12: 444–453.

Khan, Z., R.P. Tiwari, R. Mulherkar, et al. 2009. Detection of survivin and p53 in human oral cancer: Correlation with clinicopathological findings. *Head Neck* 31: 1039–1048.

Kim, J.W., E. Wieckowski, D.D. Taylor, et al. 2005. Fas ligand-positive membranous vesicles isolated from sera of patients with oral cancer induce apoptosis of activated T lymphocytes. *Clin Cancer Res* 11: 1010–1020.

Kim, Y.H., M. Buchholz, F.J. Chrest, et al. 1994. Up-regulation of c-myc induces the gene expression of the murine homologues of p34cdc2 and cyclin-dependent kinase-2 in T lymphocytes. *J Immunol* 152: 4328–4335.

King, C.R., M.H. Kraus, and S.A. Aaronson. 1985. Amplification of a novel v-erbB-related gene in a human mammary carcinoma. *Science* 229: 974–976.

King, K.L., and J.A. Cidlowski. 1998. Cell cycle regulation and apoptosis. *Annu Rev Physiol* 60: 601–617.

Kinoshita, T., T. Yokota, K. Arai, et al. 1995. Regulation of Bcl-2 expression by oncogenic Ras protein in hematopoietic cells. *Oncogene* 10: 2207–2212.

Koontongkaew, S., L. Chareonkitkajorn, A. Chanvitan, et al. 2000. Alterations of p53, pRb, cyclin D(1) and cdk4 in human oral and pharyngeal squamous cell carcinomas. *Oral Oncol* 36: 334–339.

Kuhnel, F., L. Zender, Y. Paul, et al. 2000. NFkappaB mediates apoptosis through transcriptional activation of Fas (CD95) in adenoviral hepatitis. *J Biol Chem* 275: 6421–6427.

Kuida, K., T.F. Haydar, C.Y. Kuan, et al. 1998. Reduced apoptosis and cytochrome c-mediated caspase activation in mice lacking caspase 9. *Cell* 94: 325–337.

Kumar, S., and N.L. Harvey. 1995. Role of multiple cellular proteases in the execution of programmed cell death. *FEBS Lett* 375: 169–173.

Kushner, J., G. Bradley, B. Young, et al. 1999. Aberrant expression of cyclin A and cyclin B1 proteins in oral carcinoma. *J Oral Pathol Med* 28: 77–81.

Land, H., L.F. Parada, and R.A. Weinberg. 1983. Tumorigenic conversion of primary embryo fibroblasts requires at least two cooperating oncogenes. *Nature (London)* 304: 596–602.

Lane, D.P. 1992. p53, guardian of the genome. *Nature (London)* 358: 15–16.

Lazarus, P., S.N. Sheikh, Q. Ren, et al. 1998. p53, but not p16 mutations in oral squamous cell carcinomas are associated with specific CYP1A1 and GSTM1 polymorphic genotypes and patient tobacco use. *Carcinogenesis* 19: 509–514.

Leone, G., J. DeGregori, R. Sears, et al. 1997. Myc and Ras collaborate in inducing accumulation of active cyclin E/Cdk2 and E2F. *Nature (London)* 387: 422–426.

Li, F., G. Ambrosini, E.Y. Chu, et al. 1998. Control of apoptosis and mitotic spindle checkpoint by survivin. *Nature (London)* 396: 580–584.

Li, P., D. Nijhawan, I. Budihardjo, et al. 1997. Cytochrome c and dATP-dependent formation of Apaf-1/caspase-9 complex initiates an apoptotic protease cascade. *Cell* 91: 479–489.

Liebow, C., and A.R. Kamer. 1992. Receptor phosphatases and cancer: Models for the therapeutic efficacy of somatostatin and LHRH analogues. *Cancer J* 5: 200–207.

Linette, G.P., Y. Li, K. Roth, et al. 1996. Cross talk between cell death and cell cycle progression: Bcl-2 regulates NFAT-mediated activation. *Proc Natl Acad Sci USA* 93: 9545–9552.

Liu, X., C.N. Kim, J. Yang, et al. 1996. Induction of apoptotic program in cell-free extracts: Requirement for dATP and cytochrome c. *Cell* 86: 147–157.

Lodish, H., A. Berk, S.L. Zipursky, et al. 2000. *Molecular cell biology*. 4th ed. W.H. Freeman, San Francisco.

Lu, C.D., D.C. Altieri, and N. Tanigawa. 1998. Expression of a novel antiapoptosis gene, survivin, correlated with tumor cell apoptosis and p53 accumulation in gastric carcinomas. *Cancer Res* 58: 1808–1812.

Lu, X., J.L. Arbiser, J. West, et al. 2004. Tumor necrosis factor-related apoptosis-inducing ligand can induce apoptosis in subsets of premalignant cells. *Am J Pathol* 165: 1613–1620.

MacFarlane, M., K. Cain, X.M. Sun, et al. 1997. Processing/activation of at least four interleukin-1 beta converting enzyme-like proteases occurs during the execution phase of apoptosis in human monocytic tumor cells. *J Cell Biol* 137: 469–479.

Macleod, K.F., Y. Hu, and T. Jacks. 1996. Loss of Rb activates both p53-dependent and independent cell death pathways in the developing mouse nervous system. *EMBO J* 15: 6178–6188.

Maeda, T., Y. Yamada, R. Moriuchi, et al. 1999. Fas gene mutation in the progression of adult T cell leukemia. *J Exp Med* 189: 1063–1071.

Mai, S., M. Fluri, D. Siwarski, et al. 1996a. Genomic instability in MycER-activated Rat1A-MycER cells. *Chromosome Res* 4: 365–371.

Mai, S., J. Hanley-Hyde, and M. Fluri. 1996b. c-Myc overexpression associated DHFR gene amplification in hamster, rat, mouse and human cell lines. *Oncogene* 12: 277–288.

Mao, E.J., S.M. Schwartz, J.R. Daling, et al. 1998. Loss of heterozygosity at 5q21–22 (adeno-matous polyposis coli gene region) in oral squamous cell carcinoma is common and correlated with advanced disease. *J Oral Pathol Med* 27: 297–302.

Massague, L., and F. Weis-Garcia. 1996. Serine/threoninekinase receptors: Mediators of trans-forming growth factor beta family signals. *Cancer Surv* 27: 41–64.

Mateyak, M.K., A.J. Obaya, S. Adachi, et al. 1997. Phenotypes of c-Myc-deficient rat fibro-blasts isolated by targeted homologous recombination. *Cell Growth Differ* 8: 1039–1048.

Maundrell, K., B. Antonsson, E. Magnenat, et al. 1997. Bcl-2 undergoes phosphorylation by c-Jun N-terminal kinase/stress-activated protein kinases in the presence of the constitu-tively active GTP-binding protein Rac1. *J Biol Chem* 272: 25238–25242.

Mazel, S., D. Burtrum, and H.T. Petrie. 1996. Regulation of cell division cycle progression by bcl-2 expression: A potential mechanism for inhibition of programmed cell death. *J Exp Med* 183: 2219–2226.

McCloskey, D.E., D.K. Armstrong, C. Jackisch, et al. 1996. Programmed cell death in human breast cancer cells. *Rec Prog Horm Res* 51: 493–508.

Mitchell, K.O., M.S. Ricci, T. Miyashita, et al. 2000. Bax is a transcriptional target and media-tor of c-myc-induced apoptosis. *Cancer Res* 60: 6318–6325.

Miyamoto, R., N. Uzawa, S. Nagaoka, et al. 2003. Prognostic significance of cyclin D1 ampli-fication and overexpression in oral squamous cell carcinomas. *Oral Oncol* 39: 610–618.

Miyashita, T., S. Krajewski, M. Krajewska, et al. 1994. Tumor suppressor p53 is a regulator of bcl-2 and bax gene expression in vitro and in vivo. *Oncogene* 9: 1799–1805.

Morgenbesser, S.D., B.O. Williams, T. Jacks, et al. 1994. p53-dependent apoptosis produced by Rb-deficiency in the developing mouse lens. *Nature (London)* 371: 72–74.

Moroni, M.C., E.S. Hickman, E.L. Denchi, et al. 2001. Apaf-1 is a transcriptional target for E2F and p53. *Nat Cell Biol* 3: 552–558.

Muller, M., S. Wilder, D. Bannasch, et al. 1998. p53 activates the CD95 (APO-1/Fas) gene in response to DNA damage by anticancer drugs. *J Exp Med* 188: 2033–2045.

Murphy, M., and A.J. Levine. 1998. The role of p53 in apoptosis. In *Apoptosis genes*, ed. J.W. Wilson, C. Booth, and C.S. Potten, 5–36. Kluwer Academic Publishers, Boston.

Murugan, A.K., A.K. Munirajan, and N. Tsuchida. 2012. Ras oncogenes in oral cancer: The past 20 years. *Oral Oncol* 48: 383–392.

Muzio, M., A.M. Chinnaiyan, F.C. Kischkel, et al. 1996. FLICE, a novel FADDl-homologous ICE/CED-3-like protease, is recruited to the CD95 (Fas/APO-1) death-inducing signal-ing complex. *Cell* 85: 817–827.

Nagata, S. 1997. Apoptosis by death factor. *Cell* 88: 355–365.

Nagata, S., and P. Golstein. 1995. The Fas death factor. *Science* 267: 1449–1456.

Nagatsuka, H., Y. Ishiwari, H. Tsujigiwa, et al. 2001. Quantitation of epidermal growth factor receptor gene amplification by competitive polymerase chain reaction in pre-malignant and malignant oral epithelial lesions. *Oral Oncol* 37: 599–604.

Nesbit, C.E., J.M. Tersak, L.E. Grove, et al. 2000. Genetic dissection of c-myc apoptotic path-ways. *Oncogene* 19: 3200–3212.

Nicholson, D.W., and N.A. Thornberry. 1997. Caspases: Killer proteases. *Trends Biochem Sci* 22: 299–306.

Normanno, N., A.D. Luca, C. Bianco, et al. 2006. Epidermal growth factor receptor (EGFR) signalling in cancer. *Gene* 366: 2–16.

O'Connell, J., G.C. O'Sullivan, J.K. Collins, et al. 1996. The Fas counterattack: Fas-mediated T cell killing by colon cancer cells expressing Fas ligand. *J Exp Med* 184: 1075–1082.

Oga, A., G. Kong, K. Tae, et al. 2001. Comparative genomic hybridization analysis reveals 3q gain resulting in genetic alteration in 3q in advanced oral squamous cell carcinoma. *Cancer Genet Cytogenet* 127: 24–29.

Ogasawara, J., R. Watanabe-Fukunaga, M. Adachi, et al. 1993. Lethal effect of the anti-fas antibody in mice. *Nature (London)* 364: 806–809.

Ogden, G.R., I.G. Cowpe, D.M. Chisholm, et al. 1994. p53 immunostaining as a marker for oral cancer in diagnostic cytopathology-preliminary report. *Cytopathology* 5: 47–53.

Ogden, G.R., R.A. Kiddie, D.P. Lunny, et al. 1992. Assessment of p53 protein expression in normal, benign, and malignant oral mucosa. *J Pathol* 166: 389–394.

Okafuji, M., M. Ita, A. Oga, et al. 2000. The relationship of genetic aberrations detected by comparative genomic hybridization to DNA ploidy and tumor size in human oral squamous cell carcinomas. *J Oral Pathol Med* 29: 226–231.

Okazaki, T., R.M. Bell, and Y.A. Hannun. 1989. Sphingomyelin turnover induced by vitamin D3 in HL-60 cells: Role in the cell differentiation. *J Biol Chem* 264: 19076–19080.

Oltvai, Z.N., C.L. Milliman, and S.I. Korsmeyer. 1993. Bcl-2 heterodimerizes in vivo with a conserved homolog, Bax, that accelerates programmed cell death. *Cell* 74: 609–619.

Oster, S.K., C.S.W. Ho, E.L. Soucie, et al. 2002. The myc oncogene: Marvelously complex. *Adv Cancer Res* 84: 81–154.

Owen-Schaub, L.B., R. Radinsky, E. Kruzel, et al. 1994. Anti-Fas on nonhematopoietic tumors: Levels of Fas/APO-1 and bcl-2 are not predictive of biological responsiveness. *Cancer Res* 54: 1580–1586.

Owen-Schaub, L.B., W. Zhang, J.C. Cusack, et al. 1995. Wild-type human p53 and a temperature-sensitive mutant induce Fas/APO-1 expression. *Mol Cell Biol* 15: 3032–3040.

Packham, G., and J.L. Cleveland. 1995. The role of ornithine decarboxylase in c-Myc-induced apoptosis. *Curr Top Microbiol Immunol* 194: 283–290.

Pande, P., M. Mathur, N.K. Shukla, et al. 1998. pRb and p16 protein alterations in human oral tumorigenesis. *Oral Oncol* 34: 396–403.

Partridge, M., G. Emilion, S. Pateromichelakis, et al. 1997. Field cancerisation of the oral cavity: Comparison of the spectrum of molecular alterations in cases presenting with both dysplastic and malignant lesions. *Eur J Cancer B Oral Oncol* 33B: 332–337.

Perez-Roger, I., D.L.C. Solomon, and A. Sewing. 1997. Myc activation of cyclin E/Cdk2 kinase involves induction of cyclin E gene transcription and inhibition of p27Kip1 binding to newly formed complexes. *Oncogene* 14: 2373–2381.

Philipp, A., A. Schneider, I. Väsrik, et al. 1994. Repression of cyclin D1: A novel function of MYC. *Mol Cell Biol* 14: 4032–4043.

Pindborg, J.J., P.A. Reichart, C.J. Smith, et al. 1997. Histological typing of cancer and precancer of the oral mucosa WHO. 2nd ed. Springer, New York.

Polyak, K., Y. Xia, J.L. Zweier, et al. 1997. A model for p53-induced apoptosis. *Nature (London)* 389: 300–305.

Popnikolov, N.K., Z. Gatalica, P.A. Adegboyega, et al. 2006. Downregulation of TNF-related apoptosis-inducing ligand (TRAIL)/Apo2L in Barrett's esophagus with dysplasia and adenocarcinoma. *Appl Immunohistochem Mol Morphol* 14: 161–165.

Pucci, B., M. Kasten, and A. Giordano. 2000. Cell cycle and apoptosis. *Neoplasia* 2: 291–299.

Quilliam, L.A., M.M. Hisaka, S. Zhong, et al. 1996. Involvement of the switch 2 domain of RAS in its interaction with guanine nucleotide exchange factors. *J Biol Chem* 271: 11076–11082.

Rajalingam, K., R. Schreck, U.R. Rapp, et al. 2007. Ras oncogenes and their downstream targets. *Biochim Biophys Acta* 1773: 1177–1195.

Reed, A.L., J. Califano, P. Cairns, et al. 1996. High frequency of p16 (CDKN2/MTS-1/INK4A) inactivation in head and neck squamous cell carcinoma. *Cancer Res* 56: 3630–3633.

Reis, P.P., S.R. Rogatto, L.P. Kowalski, et al. 2002. Quantitative real-time PCR identifies a critical region of deletion on 22q13 related to prognosis in oral cancer. *Oncogene* 21: 6480–6487.

Reisman, D., N.B. Elkind, and B. Roy. 1993. c-Myc trans-activates the p53 promoter through a required downstream CACGTG motif. *Cell Growth Differ* 4: 57–65.

Reynolds, J.E., T. Yang, L. Qian, et al. 1994. Mcl-1, a member of the Bcl-2 family, delays apoptosis induced by c-Myc overexpression in Chinese hamster ovary cells. *Cancer Res* 54: 6348–6352.

Ribeiro, P., N. Renard, K. Warzocha, et al. 1998. CD40 regulation of death domains containing receptors and their ligands on lymphoma B cells. *Br J Haematol* 103: 684–689.

Rodenhuis, S. 1992. ras and human tumors. *Semin Cancer Biol* 3: 241–247.

Rodriguez-Viciana, P., P.H. Warne, R. Dhand, et al. 1994. Phosphatidylinositol-3-OH kinase as a direct target of Ras. *Nature (London)* 370: 527–532.

Rosenwald, I.B., A. Lazariskaratzas, N. Sonenberg, et al. 1993. Elevated levels of cyclin D1 protein in response to increased expression of eukaryotic initiation factor 4E. *Mol Cell Biol* 13: 7358–7363.

Rosin, M.P., X. Cheng, C. Poh, et al. 2000. Use of allelic loss to predict malignant risk for low-grade oral epithelial dysplasia. *Clin Cancer Res* 6: 357–362.

Rothe, M., M.G. Pan, W.J. Henzel, et al. 1995. The TNFR2-TRAF signaling complex contains two novel proteins related to baculoviral inhibitor of apoptosis proteins. *Cell* 83: 1243–1252.

Roussel, M.F. 1998. Key effectors of signal transduction and G1 progression. *Adv Cancer Res* 74: 1–24.

Roussel, M.F., A.M. Theodoras, M. Pagano, et al. 1995. Rescue of defective mitogenic signaling by D-type cyclins. *Proc Natl Acad Sci USA* 92: 6837–6841.

Rowinsky, E.K. 2005. Targeted induction of apoptosis in cancer management: The emerging role of tumor necrosis factor-related apoptosis-inducing ligand receptor activating agents. *J Clin Oncol* 23: 9394–9407.

Roy, N., M.S. Mahadevan, M. McLean, et al. 1995. The gene for neuronal apoptosis inhibitory protein is partially deleted in individuals with spinal muscular atrophy. *Cell* 80: 167–178.

Rudolph, B.R., J. Saffrich, B. Zwicker, et al. 1996. Activation of cyclin-dependent kinases by Myc mediates induction of cyclin A, but not apoptosis. *EMBO J* 15: 3065–3076.

Sah, N.K., Z. Khan, G.J. Khan, et al. 2006. Structural, functional and therapeutic biology of survivin. *Cancer Lett* 244: 164–171.

Sakai, E., K. Rikimaru, M. Ueda, et al. 1992. The p53 tumor-suppressor gene and ras oncogene mutations in oral squamous cell carcinoma. *Int J Cancer* 52: 867–872.

Sakamuro, D., V. Eviner, K.J. Elliott, et al. 1995. c-Myc induces apoptosis in epithelial cells by both p53-dependent and p53-independent mechanisms. *Oncogene* 11: 2411–2418.

Sartor, M., H. Steingrimsdottir, F. Elamin, et al. 1999. Role of p16/MTS1, cyclin D1 and RB in primary oral cancer and oral cancer cell lines. *Br J Cancer* 80: 79–86.

Scatena, C.D., Z.A. Stewart, D. Mays, et al. 1998. Mitotic phosphorylation of Bcl-2 during normal cell cycle progression and Taxol-induced growth arrest. *J Biol Chem* 273: 30777–30784.

Schneider, P., N. Holler, J.L. Bodmer, et al. 1998. Conversion of membrane-bound Fas(CD95) ligand to its soluble form is associated with downregulation of its proapoptotic activity and loss of liver toxicity. *J Exp Med* 187: 1205–1213.

Schwab, M., K. Alitalo, K.H. Klempnauer, et al. 1983. Amplified DNA with limited homology to myc cellular oncogene is shared by human neuroblastoma cell lines and a neuroblastoma tumour. *Nature (London)* 305: 245–248.

Serrano, M.A.W., M.E. Lin, D. McCurrach, et al. 1997. Oncogenic ras provokes premature cell senescence associated with accumulation of p53 and p16INK4a. *Cell* 88: 593–602.

Shinoura, N., Y. Yoshida, M. Nishimura, et al. 1999. Expression level of Bcl-2 determines anti- or proapoptotic function. *Cancer Res* 59: 4119–4128.

Shintani, S., M. Mihara, Y. Nakahara, et al. 2000. Apoptosis and p53 are associated with effect of preoperative radiation in oral squamous cell carcinomas. *Cancer Lett* 154: 71–77.

Shintani, S., Y. Nakahara, M. Mihara, et al. 2001. Inactivation of the p14(ARF), p15(INK4B) and p16(INK4A) genes is a frequent event in human oral squamous cell carcinomas. *Oral Oncol* 37: 498–504.

Sjölander, A., K. Yamamoto, B.E. Huber, et al. 1991. Association of p21ras with phosphatidylinositol 3-kinase. *Proc Natl Acad Sci USA* 88: 7908–7912.

Smith, C.A., T. Farrah, and R.G. Goodwin. 1994. The TNF receptor superfamily of cellular and viral proteins: Activation, costimulation, and death. *Cell* 76: 959–962.

Snell, V., K. Clodi, S. Zhao, et al. 1997. Activity of TNF-related apoptosis-inducing ligand (TRAIL) in haematological malignancies. *Br J Haematol* 99: 618–624.

Song, E., J. Chen, N. Ouyang, et al. 2001. Soluble Fas ligand released by colon adenocarcinoma cells induces host lymphocyte apoptosis: An active mode of immune evasion in colon cancer. *Br J Cancer* 85: 1047–1054.

Soni, S., J. Kaur, A. Kumar, et al. 2005. Alterations of rb pathway components are frequent events in patients with oral epithelial dysplasia and predict clinical outcome in patients with squamous cell carcinoma. *Oncology* 68: 314–325.

Soucie, E.L., M.G. Annis, J. Sedivy, et al. 2001. Myc potentiates apoptosis by stimulating Bax activity at the mitochondria. *Mol Cell Biol* 21: 4725–4736.

Spencer, C.A., and M. Groudine. 1991. Control of c-myc regulation in normal and neoplastic cells. *Adv Cancer Res* 56: 1–48.

Srinivasula, S.M., T. Fernandes-Alnemri, J. Zangrilli, et al. 1996. The CED-3/interleukin-1 beta converting enzyme-like homolog Mch6 and the lamin-cleaving enzyme Mch2 alpha are substrates for the apoptotic mediator CPP32. *J Biol Chem* 271: 27099–27106.

Stander, S., and T. Schwarz. 2005. Tumor necrosis factor-related apoptosis-inducing ligand (TRAIL) is expressed in normal skin and cutaneous inflammatory diseases, but not in chronically UV-exposed skin and non-melanoma skin cancer. *Am J Dermatopathol* 27: 116–121.

Stennicke, H.R., Q.L. Deveraux, E.W. Humke, et al. 1999. Caspase-9 can be activated without proteolytic processing. *J Biol Chem* 274: 8359–8362.

Strand, S., W.J. Hofmann, H. Hug, et al. 1996. Lymphocyte apoptosis induced by CD95 (APO-1/Fas) ligand-expressing tumor cells—A mechanism of immune evasion? *Nature Med* 2: 1361–1366.

Strater, J., U. Hinz, H. Walczak, et al. 2002. Expression of TRAIL and TRAIL receptors in colon carcinoma: TRAIL-R1 is an independent prognostic parameter. *Clin Cancer Res* 8: 3734–3740.

Suzuki, H., T. Takhsi, T. Kuroishi, et al. 1992. p53 mutations in non-small cell lung cancer in Japan: Association between mutations and smoking. *Cancer Res* 52: 724–736.

Takahashi, A., H. Hirata, S. Yonehara, et al. 1997. Affinity labeling displays the stepwise activation of ICE-related proteases by Fas, staurosporine, and Crm-A-sensitive caspase 8. *Oncogene* 14: 2741–2752.

Talanian, R.V., C. Quinlan, S. Trautz, et al. 1997. Substrate specificities of caspase family proteases. *J Biol Chem* 272: 9677–9682.

Tanaka, M., T. Itai, M. Adachi, et al. 1998. Down regulation of Fas ligand by shedding. *Nat Med* 4: 31–36.

Teng, M.S., M.S. Brandwein-Gensler, M.S. Teixeira, et al. 2005. A study of TRAIL receptors in squamous cell carcinoma of the head and neck. *Arch Otolaryngol Head Neck Surg* 131: 407–412.

Thomas, W.D., and P. Hersey. 1998. TNF-related apoptosis-inducing ligand (TRAIL) induces apoptosis in Fas ligand-resistant melanoma cells and mediates CD4 T cell killing of target cells. *J Immunol* 161: 2195–2200.

Thompson, E.B. 1998. The many roles of c-Myc in apoptosis. *Annu Rev Physiol* 60: 575.

Thornberry, N.A., T.A. Rano, E.P. Peterson, et al. 1997. A combinatorial approach defines specificities of members of the caspase family and granzyme B: Functional relationships established for key mediators of apoptosis. *J Biol Chem* 272: 17907–17911.

Tomlins, S.A., D.R. Rhodes, S. Perner, et al. 2005. Recurrent fusion of TMPRSS2 and ETS transcription factor genes in prostate cancer. *Science* 310: 644–648.

Tsujimoto, Y., J. Yunis, L. Onorato-Showe, et al. 1984. Molecular cloning of the chromosomal breakpoint of B cell lymphomas and leukemias with the t(11;14) chromosome translocation. *Science* 224: 1403–1406.

Uren, A.G., M. Pakusch, C.J. Hawkins, et al. 1996. Cloning and expression of apoptosis inhibitory protein homologs that function to inhibit apoptosis and/or bind tumor necrosis factor receptor-associated factors. *Proc Natl Acad Sci USA* 93: 4974–4978.

Vaux, D.L., I.L. Weissman, and S.K. Kim. 1992. Prevention of programmed cell death in *Caenorhabditis elegans* by human bcl-2. *Science* 258: 1955–1957.

Vetter, I.R., and A. Wittinghofer. 2001. The guanine nucleotide-binding switch in three dimensions. *Science* 294: 1299–1304.

Vlach, J., S. Hennecke, K. Alevizopoulos, et al. 1996. Growth arrest by the cyclin-dependent kinase inhibitor p27Kip1 is abrogated by c-Myc. *EMBO J* 15: 6595–6604.

Wajant, H., K. Pfizenmaier, and P. Scheurich. 2002. TNF-related apoptosis inducing ligand (TRAIL) and its receptors in tumor surveillance and cancer therapy. *Apoptosis* 7: 449–459.

Wang, C.Y., M.W. Mayo, and A.-S.J. Baldwin. 1996. TNF-β and cancer therapy-induced apoptosis: Potentiation by inhibition of NF-kappaB. *Science* 274: 784–787.

Wang, J.L., Y. Xie, S. Allan, et al. 1998. Myc activates telomerase. *Genes Dev* 12: 1769–1774.

Wang, Q., X. Wang, A. Hernandez, et al. 2002. Regulation of TRAIL expression by the phosphatidylinositol 3-kinase/Akt/GSK-3 pathway in human colon cancer cells. *J Biol Chem* 277: 36602–36610.

Warri, A.M., R.L. Huovinen, A.M. Laine, et al. 1993. Apoptosis in Toremifene induced growth inhibition of human breast cancer cells in vivo and in vitro. *J Natl Cancer Inst* 85: 1412–1420.

Weber, R.G., M. Scheer, I.A. Born, et al. 1998. Recurrent chromosomal imbalances detected in biopsy material from oral premalignant and malignant lesions by combined tissue microdissection, universal DNA amplification, and comparative genomic hybridization. *Am J Pathol* 153: 295–303.

Weinberger, P.M., Z. Yu, B.G. Haty, et al. 2004. Prognostic significance of p16 protein levels in oropharyngeal squamous cell cancer. *Clin Cancer Res* 10: 5684–5691.

Westendorf, L.L., I.M. Lammert, and D.F. Jelinek. 1995. Expression and function of Fas (APO-1/CD95) in patient myeloma cells and myeloma cell lines. *Blood* 85: 3566–3576.

Wilson, J.W., D.M. Prichard, J.A. Hickman, et al. 1998. Radiation induced p53 and p21WAF1/CIP1 expression in the murine intestinal epithelium: Apoptosis and cell cycle arrest. *Am J Pathol* 153: 899–909.

Wittinghofer, A., and N. Nassar. 1996. How ras-related proteins talk to their effectors. *Trends Biochem Sci* 21: 488–491.

Wu, C.L., L. Roz, S. McKown, et al. 1999. DNA studies underestimate the major role of CDKN2A inactivation in oral and oropharyngeal squamous cell carcinomas. *Genes Chromosomes Cancer* 25: 16–25.

Xu, J., J.Y. Zhou, and G.S. Wu. 2006. Tumor necrosis factor-related apoptosis-inducing ligand is required for tumor necrosis factor alpha-mediated sensitization of human breast cancer cells to chemotherapy. *Cancer Res* 66: 10092–10099.

Yagita, H., K. Takeda, Y. Hayakawa, et al. 2004. TRAIL and its receptors as targets for cancer therapy. *Cancer Sci* 95: 777–783.

Yamamoto, K., H. Ichijo, and S.J. Korsmeyer. 1999. Bcl-2 is phosphorylated and inactivated by an ASK1/Jun N-terminal protein kinase pathway normally activated at G2/M. *Mol Cell Biol* 19: 8469–8478.

Yamazaki, Y., I. Chiba, A. Hirai, et al. 2003. Specific p53 mutations predict poor prognosis in oral squamous cell carcinoma. *Oral Oncol* 39: 163–169.

Yanaga, F., and S.P. Watson. 1992. Tumor necrosis factor c stimulates sphingomyelinase through the 55 kD receptor in HL-60 cells. *FEBS Lett* 314: 297–300.

Yang, J., X. Liu, K. Bhalla, et al. 1997a. Prevention of apoptosis by Bcl-2: Release of cytochrome c from mitochondria blocked. *Science* 275: 1129–1132.

Yang, X., R. Khosravi-Far, H.Y. Chang, et al. 1997b. A novel Fas-binding protein that activates JNK and apoptosis. *Cell* 89: 1067–1076.

Yin, X.M., Z.N. Oltvai, and S.J. Korsmeyer. 1995. Heterodimerization with Bax is required for Bcl-2 to repress cell death. *Curr Top Microbiol Immunol* 194: 331–338.

Yuan, J., S. Shaham, S. Ledoux, et al. 1993. The *C. elegans* cell death gene ced-3 encodes a protein similar to mammalian interleukin-1 beta-convecting enzyme. *Cell* 75: 641–652.

Zariwala, M., S. Schmid, M. Pfaltz, et al. 1994. p53 gene mutations in oropharyngeal carcinomas: A comparison of solitary and multiple primary tumours and lymph-node metastases. *Int J Cancer* 56: 807–811.

Zindy, F.C.M., D. Eischen, H. Randle, et al. 1998. Myc signaling via the ARF tumor suppressor regulates p53-dependent apoptosis and immortalization. *Genes Dev* 12: 2424–2433.

Zornig, M., and G.I. Evan. 1996. Cell cycle: On target with Myc. *Curr Biol* 6: 1553–1556.

Zou, H., W.J. Henzel, X. Liu, et al. 1997. Apaf-1, a human protein homologous to *C. elegans* CED-4, participates in cytochrome c-dependent activation of caspase-3. *Cell* 90: 405–413.

4 Apoptotic Regulations

KEY WORDS

Apoptosis
Apoptotic signaling
Bcl family
Caspase
Intrinsic and extrinsic pathways
Mitochondrial regulation

4.1 INTRODUCTION

The various components of multicellular organisms involve large-scale reshaping facilitated by well-coordinated cell growth and cell death. Multicellular organisms have an evolved controlled mechanism for cell death, which is known as apoptosis or programmed cell death (PCD). Apoptosis is characterized by a series of biochemical and morphological changes like blebbing, loss of asymmetry and attachment of cell membrane, cell shrinkage, nuclear fragmentation, chromatin condensation, and chromosomal DNA fragmentation. Apoptosis is an integral part of the multicellular organism, having its presence right from developmental stages to adult life. Apoptosis is initiated by a variety of external or internal stimuli and happens by a well-controlled mechanism. The controlled and coordinated steps involved in apoptotic cell death make it different from other types of cell death, like necrosis, which involves uncontrolled cell death resulting in lysis of cells, and invoking inflammatory responses and causing potentially serious health problems. Apoptosis, along with cell proliferation, is responsible for shaping tissues and organs during developmental periods; for instance, apoptosis of cells located in between the toes allows for their separation (Zuzarte-Luis and Hurle, 2002). Apoptosis also plays an important role in the immune system by selective killing ineffective or self-reactive T-cells. T-cells are responsible for destroying damaged or infected cells in the body.

Apoptosis has a critical role in routine execution of various biological functions at the molecular level, and it's not surprising that its dysregulation is implicated in a host of diseases including cancer. Evasion of apoptosis is regarded as one of the hallmarks of cancer. Cell growth and proliferation is a well-regulated phenomenon in normal cells controlled by various growth-promoting and growth-inhibitory mechanisms (Figure 4.1). Cancer cells are characterized by the presence of various mutations that ignore antiproliferative signals and thus help in proliferative growth. Normal cells are marked for apoptotic death on sensing stress signals; however, this mechanism is absent in cancer cells. This unrestrained growth leads to development of a tumor. Apoptosis is one endpoint by which cancer cells are killed by chemotherapeutic agents, and development of mutation in the apoptotic circuit often leads

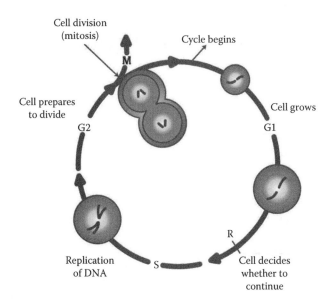

FIGURE 4.1   Different stages of cell cycle.

to development of resistance for these chemotherapeutic agents, thus making cancer management more complicated (Igney and Krammer, 2002).

An optimum level of apoptosis is necessary to maintain the health status of normal cells. Too little (as in cancers, rheumatoid arthritis), as well as too much, apoptosis (neurodegenerative disease, AIDS, and ischemic diseases) has been implicated in the pathogenesis of various diseases (Fadeel et al., 1999). Neurodegenerative diseases like Parkinson's and Alzheimer's have been correlated with progressive loss of neurons due to too much apoptosis.

Apoptosis plays a critical role in developmental stages. Too much apoptosis of the trophoblast cells leads to a pregnancy-related complication known as preeclampsia, in which the maternal environment is not remodeled perfectly to support embryonic development. The development and formation of the organ system involve the production of cells that are produced in excess to support the sculpturing of organs and tissues. The cells, which are produced in excess, have to be removed from the system through means of programmed cell death (Meier et al., 2000). For instance, around half of the neurons produced initially during development of the brain die by the process of apoptosis to form the adult brain (Hutchins and Burger, 1998). Apoptosis can be regarded as a process to maintain a constant number of cells in an organism to carry out various physiological functions (Rathmell and Thompson, 2002). Apoptosis has a ubiquitous presence in various biological functions, including development, differentiation, homeostasis, proliferation, regulation, and immune response. Apoptosis is a widespread phenomenon present across organisms such as metazoans (Tittel and Steller, 2000), nematodes (Liu and Hengartner, 1999), insects (Richardson and Kumar, 2002), and cnidaria (Cikala et al., 1999). Apoptosis also probably plays a role in plant biology (Solomon et al., 1999), and it has also been

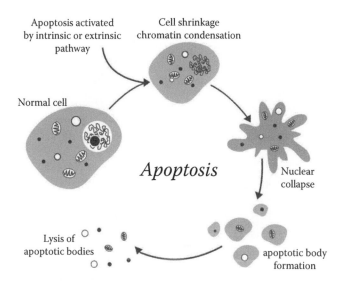

Apoptosis activated
by intrinsic or extrinsic
pathway

Cell shrinkage
chromatin condensation

Normal cell

Apoptosis

Nuclear
collapse

Lysis of
apoptotic bodies

apoptotic body
formation

FIGURE 4.2 Diagrammatical presentation of the phenomenon of apoptosis.

reported in unicellular eukaryotes (like the slime mold *Dictyostelium discoideum*, the ciliate *Tetrahymena thermophila*, and the kinetoplastid parasite *Trypanosoma brucei*), prokaryotes (like myxobacteria, *Bacillus subtilis*, and *Streptomyces*) (Ameisen, 2002) (Figure 4.2).

The cells undergoing apoptosis undergo distinct morphological changes:

1. Shrinkage of cells following the cleavage of lamins and actin filaments in the cytoskeleton.
2. Nuclear condensation due to breakdown of chromatin. At times, nuclei of apoptotic cells take on a horseshoe appearance.
3. The blebbing or budding of the plasma membrane.
4. Formation of apoptotic bodies amenable to be engulfed by macrophages, thus removing debris from the surrounding tissue without causing an inflammatory response.

These morphological changes are the outcome of various molecular and biochemical events occurring in apoptotic cells, like activation of proteolytic enzymes, which are responsible for cleavage of DNA, and other structural proteins (Saraste and Pulkki, 2000).

Apoptosis is a controlled or regulated process by which cells die without causing any damage to their microenvironment, whereas other types of cell deaths, particularly necrosis, are associated with major insults, resulting in swelling, loss of membrane integrity, and rupture of the cells. Necrosis damages surrounding cells by uncontrolled release of the cellular contents into the cellular microenvironment, which induces a strong inflammatory response (Leist and Jaattela, 2001) (Figure 4.3).

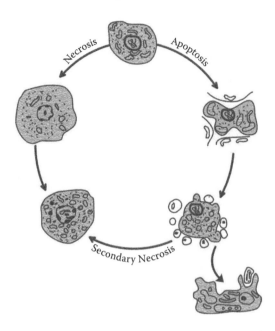

FIGURE 4.3   Cell structural differences between necrosis and apoptosis.

Necrotic cells swell, and finally the plasma membrane ruptures, releasing the cell contents into the surrounding tissue. At a molecular level apoptosis is mediated by various caspases, whereas necrosis is believed to be caused by noncaspase proteases such as calpains or cathepsins (Okada and Mak, 2004). Apoptotic cells shrink, and the plasma membrane invaginates, forming apoptotic bodies, which are engulfed by the phagocytes. The late-stage apoptotic cells undergo secondary necrosis, where the membrane of the apoptotic bodies ruptures. Unlike necrosis, cells undergoing apoptosis exhibit a reduction in size and become granular. Changes to the plasma membrane and cytoskeleton lead to the formation of apoptotic bodies that enclose fragments of the nucleus and cytoplasm. The apoptotic bodies displaying phago-cytic signals on the outer membrane are engulfed and digested by phagocytes and neighboring endothelial cells, thus clearing the cell corpses from the tissue. No inflammatory response is triggered during this process, which is considered the most characteristic feature of apoptosis (Figure 4.4).

4.2   ETYMOLOGY

The principle of apoptosis was first described by German scientist Carl Vogt in 1842. A more precise description of the apoptotic process was proposed by anato-mist Walter Flemming in 1885. Prof. Wyllie and his colleagues at the University of Aberdeen noticed the difference between apoptotic cell death and necrosis when viewed under an electron microscope. The cells, which die through necrosis, tend to swell up and burst abruptly, thus spilling cellular contents around neighboring cells and causing cells to swell. However, when cells die through apoptosis or

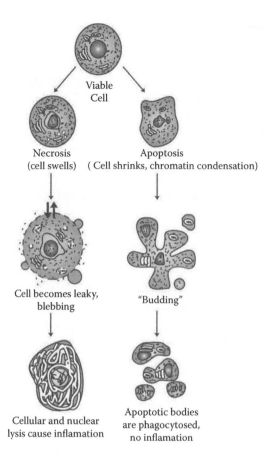

Viable Cell

Necrosis (cell swells)

Apoptosis (Cell shrinks, chromatin condensation)

Cell becomes leaky, blebbing

"Budding"

Cellular and nuclear lysis cause inflamation

Apoptotic bodies are phagocytosed, no inflamation

FIGURE 4.4 Morphological differences between necrosis and apoptosis.

programmed cell death, they are processed in a systematic way and are removed from the system without causing any harm to neighboring cells. Such programmed cell death was given the name *apoptosis*, a word used in a Homeric poem that means falling off, as in leaves from a tree. Prof. Wyllie focused his research on the fine balance between cell death and growth, and disruption of this balance through aberrant regulation of apoptosis.

4.2.1 MAJOR PATHWAYS IN APOPTOSIS: INTRINSIC AND EXTRINSIC PATHWAYS

Apoptosis has been recognized as one of the main types of cell death, and its aberrant behavior leads to the genesis of malignant conditions. The importance of apoptosis has generated a lot of enthusiasm among research communities; consequently, much information is available about the apoptotic process at the molecular level.

4.2.1.1   Key Elements of the Apoptotic Pathways

4.2.1.1.1   Death Receptors

Various cell surface receptors are known to stimulate or induce apoptosis through activation of caspases. A cell surface receptor like CD95 (APO-1, Fas) takes part in peripheral lymphoid compartment and cytotoxic T-lymphocyte (CTL)-mediated targeted cell killing. Fas receptor initiates a signal transduction cascade leading to caspase-dependent programmed cell death, through cross-linking by ligand or agonist antibody.

4.2.1.1.2   Membrane Alterations

Apoptosis is marked by changes at the cell surface and plasma membrane during its early stages. Translocation of phosphatidylserine (PS) is one such change during apoptosis, in which PS is translocated from the inner side of the plasma membrane to the external surface of the cell.

4.2.1.1.3   Protease Cascade

Various caspases (cysteinyl-aspartate-specific proteinases) plays an important role in the execution of apoptosis. Caspase-1 or interleukin-1β-converting enzyme (ICE) is one the best-characterized caspases, which was originally identified as a cysteine protease responsible for the processing of interleukin-1β.

4.2.1.1.4   Mitochondrial Changes

Cell death (apoptosis or necrosis) is marked by disruption of mitochondrial physiology. The membrane permeability of mitochondria is altered during apoptosis, which facilitates the release of activator protease and other factors involved in modulation of the apoptosis process. Cytochrome C is redistributed to the cytosol, because of the discontinuity of the outer mitochondrial membrane, which is followed by subsequent depolarization of the inner mitochondrial membrane. The release of cytochrome C (Apaf-2) promotes activation of the caspase by binding to Apaf-1 and subsequently activates Apaf-3 (caspase-9). Mitochondria also release apoptosis-inducing factor (AIF) with proteolytic activity, which can induce apoptosis (Finucane et al., 1999).

4.2.1.1.5   DNA Fragmentation

The fragmentation of genomic DNA is regarded as a biochemical hallmark of apoptosis. It's an irreversible event that initiates cell death, and sets in even before changes in the plasma membrane permeability. The activation of endogenous Ca^{2+}- and Mg^{2+}-dependent nuclear endonuclease is regarded to cause DNA fragmentation. This enzyme selectively cleaves DNA at sites located between nucleosomal units (linker DNA) generating mono- and oligonucleosomal DNA fragments.

4.2.2   Molecular Events Leading to Apoptosis

Apoptosis pathways can be activated by the host of stimuli ranging from various factors. Apoptosis can be triggered from external factors through activation of receptors on the cell surface like death receptors (DRs) (extrinsic apoptotic pathway) or from internal factors like reactive oxygen species (ROS) (intrinsic apoptotic pathway). Irrespective of origin of apoptotic stimuli (external or internal), apoptosis is finally executed by activation of the common set of effector caspases (Degterev et al., 2003). These caspases are synthesized as inactive proenzymes, which become functional upon cleavage by factors from the apoptotic pathway.

The initiator caspase-8 is activated by interaction of the death receptor with a corresponding death ligand, for instance, interaction of TNF-related apoptosis-inducing ligand (TRAIL) with TRAIL receptors (Lavrik et al., 2005). The activated caspase-3 can engage apoptotic machinery in two ways: (1) it can initiate an apoptotic cascade by cleaving effector caspases like caspase-3, or (2) it can engage the mitochondrial pathway via the cleavage of Bid. The cleaved form of Bid (tBid) enters into mitochondria and induces the release of an apoptogenic substance (such as cytochrome c, apoptosis-inducing factor (AIF), Smac/direct inhibitor of apoptosis (IAP) binding protein with low pI (DIABLO), Omi/HtrA2, or AIF) from mitochondria into the cytosol (Adams and Cory, 2007). Caspase-3 is activated by cytochrome c through the formation of the cytochrome c/Apaf-1/caspase-9-containing apoptosome complex. IAPs (like survivin) are inactivated by Omi/HtrA2 or Smac/DIABLO (Lavrik et al., 2005; Khan and Bisen, 2013). The caspase-independent apoptosis is postulated to be caused by AIF. Apoptosis-inducing cytotoxic drugs act primarily via mediation by a cytochrome c/Apaf-1/caspase-9-dependent pathway linked to mitochondria (Debatin et al., 2002).

4.2.2.1 Intrinsic Pathway

The intrinsic pathway for apoptosis is initiated from factors located within the cell (Figure 2.3). Various molecular events like DNA damage, a defective cell cycle, detachment from the extracellular matrix, hypoxia, loss of cell survival factors, etc., act as triggers for the intrinsic apoptotic pathway. The intrinsic pathway involves the release of a host of pro-apoptotic proteins, which activates caspase enzymes from mitochondria, and which ultimately leads to cell death by apoptosis (Figure 4.5).

The fine balance between proteins of pro- and anti-apoptotic members of the Bcl-2 superfamily plays an important role in regulating the permeability of the mitochondrial membrane, and thus regulating apoptosis. Bcl-2 family members regulate this pathway by modulating the release of key pro-apoptotic polypeptides, including cytochrome c and the second mitochondrial activator of caspases (Smac)/direct inhibitor of apoptosis (IAP) binding protein with low pI (DIABLO), from mitochondria. Several Bcl-2 family members that facilitate mitochondrial permeabilization are transcriptional targets of the p53 tumor suppressor gene, providing a partial explanation for the ability of DNA-damaging agents to induce apoptosis. Other pro-apoptotic Bcl-2 family members are released from cytoskeletal sites upon treatment with paclitaxel or loss of adherence (Figure 4.6).

The anti-apoptotic protein XIAP (X chromosome-linked IAP) binds procaspase-9 and prevents its activation. The gene encoding XIAP is activated by nuclear factor (NF)-κB, contributing to the anti-apoptotic effects of this transcription factor. In this chapter, the various components of the intrinsic pathway are reviewed; alterations in this pathway, in various cancers, are described and evidence given that some of these same anti-apoptotic alterations might contribute to anticancer drug resistance.

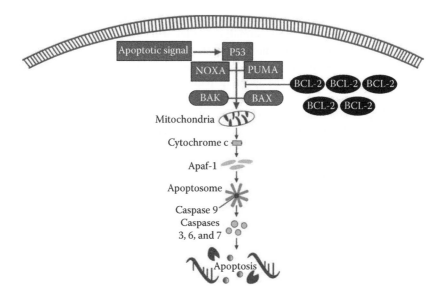

FIGURE 4.5 Elements of intrinsic apoptotic pathway.

The mitochondrial pathway is activated by a variety of extra- and intracellular stresses, including oxidative stress and treatment with cytotoxic drugs. The apoptotic signal leads to the release of cytochrome c from the mitochondrial intermembrane space into the cytosol, where it binds to the apoptotic protease activating factor-1 (Apaf-1), a mammalian CED-4 homolog. Early data suggested early loss of mitochondrial membrane potential and the opening of the mitochondrial permeability pore to be necessary steps for cytochrome c release. However, recent data seem to suggest that both events are not needed for apoptotic cytochrome c release in all instances (Waterhouse et al., 2002; Ly et al., 2003). Binding of cytochrome c to Apaf-1 triggers the formation of the apoptosome, an Apaf-1-containing complex that catalyzes activation of caspases (Figure 4.7).

Apoptosome contains seven Apaf-1, seven cytochrome c, seven (d)ATP, and seven procaspase-9 molecules. Procaspase-9 is the initiator caspase of the apoptosome (Waterhouse et al., 2002). The apoptosome-bound procaspase-9 is activated and can

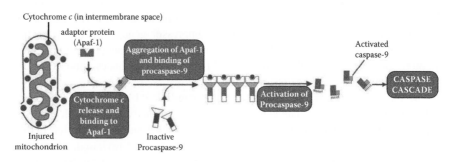

FIGURE 4.6 Process of activation of apoptosis from inside the cell.

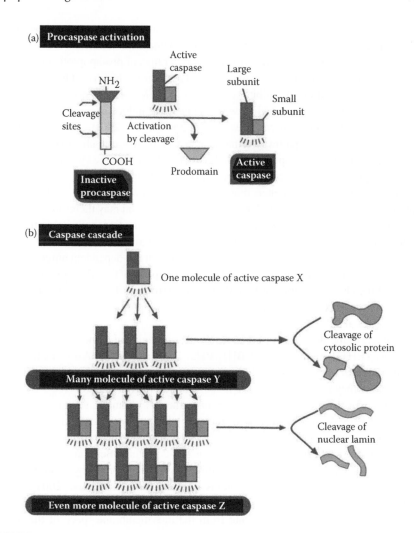

FIGURE 4.7. Process of activation of procaspase and caspase cascades.

then activate an effector caspase (e.g., caspase-3), which then can cleave the cellular substrates needed for the orchestration of apoptosis. The apoptosome structure has been solved by low-resolution kryo-electron microscopy: it forms a "wheel of death," with a sevenfold symmetry. As caspase-9 appears to contain only one active site per tetramer and its activity appears to be three orders of magnitude higher when bound to the apoptosome, one hypothesis is that the seven molecules of procaspase-9 will recruit and activate seven molecules of caspase-9 to the apoptosome to form the caspase-9 holoenzyme (Acehan et al., 2002; Shi, 2002; Salvesen and Renatus, 2002). Brains of Apaf-1, caspase-9, and caspase-3 mutant mice demonstrate forebrain extrusions, reflecting the loss of the capacity to appropriately delete cells, and demonstrating the importance of apoptosis during brain development (Nijhawan et al., 2000).

Proteins of the intrinsic pathway can be upregulated or downregulated by onco-genes (Green and Evan, 2002); e.g., many melanoma expresses very low levels of Apaf-1. Upon activation of the intrinsic pathway, a range of pro-apoptotic molecules, in addition to cytochrome c, are released from the mitochondria. One such molecule is Smac/DIABLO, an inhibitor of cellular IAPs. Mitochondria can also release an apoptosis-inducing factor (AIF), which appears to induce an apoptosis-like cell death that is independent of caspases (Hunot and Flavell, 2001). Two major questions about the intrinsic pathway remain unanswered: Are mitochondria central to the intrinsic pathway of apoptosis (Finkel, 2001)? What forms the cytochrome c (and other pro-apoptotic protein-releasing pores) (Desagher and Martinou, 2000)?

There is increasing evidence that the release of cytochrome c itself may be (directly or indirectly) stimulated by caspase activation, and may therefore constitute an amplification loop rather than a trigger for caspase activation—similar to the situation in the extrinsic pathway. There have been many molecules involved in forming the pore that releases cytochrome c, including Bax, voltage-dependent anion channel (VDAC), the mitochondrial permeability transition pore (PTP), Bax/VDAC, lipids, or something else as yet unidentified. Perhaps the strongest data are available for the involvement of Bax or VDAC in the formation of the pore-releasing cytochrome c (Tsujimoto et al., 2002).

Members of the Bcl-2 family contain signature domains of homology called Bcl-2 homology (BH) domains (termed BH1, BH2, BH3, and BH4) (Ashkenazi, 2008). Pro-apoptotic Bcl-2 proteins are divided into two subgroups based on the number of BH domains they contain. There are those with several BH domains (e.g., Bax and Bak), and then there are those that only have the BH3 domain, such as Bid, Bad, Bim, Bmf, PUMA, and NOXA. BH3-only proteins activate the multi-BH domain pro-apoptotic proteins Bax and Bak, which then allow for permeabilization of the mitochondrial membrane. The anti-apoptotic Bcl-2 proteins Bcl-2 and Bcl-x_L act to prevent permeabilization of the outer mitochondrial membrane by inhibiting the action of the pro-apoptotic Bcl-2 proteins Bax and Bak (Ashkenazi, 2008). Upon membrane permeabilization, cytochrome c and the pro-apoptotic protein Smac/DIABLO are then able to translocate from the intermembrane space of the mitochondria into the cytosol.

The mitochondrial protein Smac/DIABLO directly interacts with the inhibitors of apoptosis (IAPs), which inactivate caspases, and consequently promote apoptosis (Henry-Mowatt et al., 2004; Srinivasula et al., 2000). This is why the intrinsic pathway is sometimes referred to as the mitochondrial pathway. Cytochrome c binds the adaptor apoptotic protease activating factor-1 (Apaf-1) and forms a large multi-protein structure known as the apoptosome. Assembly of the apoptosome is highly regulated and may be driven by nucleotide exchange factors or ATPase-activating proteins (Bao and Shi, 2007). The primary function of the apoptosome seems to be multimerization and allosteric regulation of the catalytic activity of caspase-9 (Bao and Shi, 2007). Initiator caspase-9 is recruited into the apoptosome and activated from within the adaptor protein complex, which in turn activates the downstream effector caspases-3, -6, and -7, thereby causing apoptosis via the intrinsic pathway (Henry-Mowatt et al., 2004).

4.2.2.2 Extrinsic Pathway

Apart from the intrinsic apoptotic pathway, the extrinsic pathway is yet another way through which cells undergo apoptotic death. The extrinsic pathway is triggered by factors (pro-apoptotic ligands) located outside the cell, which activates specific pro-apoptotic receptors on the cell surface. The pro-apoptotic ligands include Apo2L/TRAIL and CD95L/FasL, which bind to their cognate receptors, DR4/DR5 and CD95/Fas, respectively (Debatin and Krammer, 2004). The extrinsic pathway is triggered without association with p53 protein (Figure 4.8), which plays an important role in the intrinsic pathway (Rieger et al., 1998).

The extrinsic pathway is triggered by ligand-receptor interactions; some of this interaction includes tumor necrosis factor (TNF) with tumor necrosis factor receptor-1 (TNFR1), Fas ligand with Fas, and TRAIL with TRAIL receptors. These interactions induce physiological changes like receptor oligomerization, recruitment of death signal adaptor proteins and formation of death-inducing signaling complex (DISC). DISC binds with initiator caspases like caspase-8 and caspase-10, which triggers a caspase cascade, which involves activation of effector caspases like caspase-3, caspase-7, and caspase-9. Ligand binding induces receptor clustering and recruitment of the adaptor protein Fas-associated death domain (FADD) and the initiator caspases-8 or -10 as procaspases, forming a DISC (Boldin et al., 1995). DISC facilitates the procaspase molecule to come into close proximity, and thereby helps in autocatalytic processing and release of active caspase into the cytoplasm, where they activate effector caspases-3, -6, and -7, thereby converging on the intrinsic pathway (Ghobrial et al., 2005). Dimerization may be crucial for caspase-8 activation, and clustering of the receptors and the associated DISC may enhance this activation. DISC formation is modulated by several inhibitory mechanisms, including c-FLICE

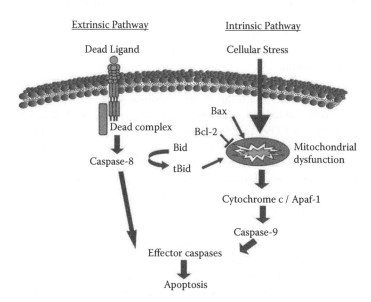

FIGURE 4.8 Differences of intrinsic and extrinsic pathways.

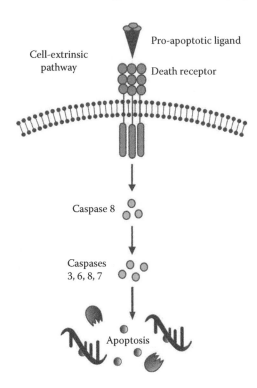

FIGURE 4.9   FADD recruitment and caspase activation via the extrinsic pathway.

inhibitory protein (c-FLIP), which exerts its effects on the DISC by interacting with FADD to block initiator caspase activation, and decoy receptors, which can block ligand binding or directly abrogate pro-apoptotic receptor stimulation. Upon DISC activation, the extrinsic pathway adopts the same effector caspase machinery as the intrinsic pathway (Figure 4.9).

It has been shown that activation of the extrinsic pathway through the binding of CD95L/FasL to CD95/Fas can result in two apoptotic programs, termed type I and type II. Type I cells are able to overcome the need for mitochondrial amplification of the death signal in CD95-mediated apoptosis by producing sufficient amounts of caspase-8 at the DISC, directly cleaving and activating effector caspases and executing cell death (Barnhart et al., 2003). Because type I cells bypass mitochondrial involvement in CD95-mediated apoptosis, expression of Bcl-2 or Bcl-X$_L$ has no inhibitory effect on their apoptotic program. Conversely, type II cells produce minimal amounts of active caspase-8 at the DISC and require the mitochondrial amplification of the CD95 signal. This signal is probably through the pro-apoptotic BH3 domain, which only contains the Bcl-2 family member, Bid. The cleavage of Bid by caspase-8 results in its translocation to the mitochondria, where it initiates the release of mitochondrial factors, which in turn augments cell death. Because type II cells rely on the apoptotic function of mitochondria, expression of Bcl-2/Bcl-xL does confer protection from CD95-mediated apoptosis. An explanation for the differences between type I and type II cells remains unclear, although differential expression of inhibitors of the death receptor signaling cascade, such as c-FLIP or X-linked inhibitor of apoptosis protein (XIAP), has been suggested to play a role (Walter et al., 2008; Khan and Bisen, 2013).

The extrinsic apoptotic pathway is triggered by death receptors belonging to the tumor necrosis factor receptor (TNFR) gene superfamily, which includes TNFR-1, Fas/CD95, and TRAIL receptors DR4 and DR5. The receptors of the TNFR superfamily consist of a cysteine-rich extracellular subdomain, which allows them to recognize specific death ligands, and induce trimerization and activation of the respective death receptor (Naismith and Sprang, 1998). The death signal is further relayed by the cytoplasmic part of the death receptor, which contains a conserved sequence, known as the death domain (DD). Adapter molecules like TRADD or FADD also possess their death domains, by which they are recruited to the DDs of the activated death receptor, and thereby form a death-inducing signaling complex (DISC). The adaptor FADD also contains a death effector domain (DED) along with DD, which through homotypic DED-DED interaction sequesters procaspase-8 to the DISC. The autocatalytic activation and release of caspase-8 is facilitated by local concentration of several procaspase-8 molecules at the DISC. The activated caspase-8 is responsible for processing of downstream effector caspases (Figure 4.10), which cleave specific substrates to cause caspase-dependent cell death in type I cells (Scaffidi et al., 1998).

In type II cells, the death signals generated by the activated receptor are not strong enough to induce direct and caspase-dependent cell death. In type II cells, execution of apoptosis is realized with the help of mitochondrial factors. The link between the extrinsic and intrinsic pathways is provided by the Bcl-2 family member Bid. Bid is cleaved by caspase-8 and is translocated to mitochondria in its truncated form (tBID). tBID acts along with Bax and Bak to induce release of cytochrome c and other pro-apoptotic factors into cytosol (Luo et al., 1998). Cytosolic cytochrome c binds to monomeric Apaf-1, which then, in a dATP-dependent conformational change, oligomerizes to assemble the apoptosome, a complex of a wheel-like structure with sevenfold symmetry, which triggers the activation of the initiator procaspase-9. The activated caspase-9 subsequently initiates a caspase cascade involving downstream effector caspases such as caspase-3, caspase-7, and caspase-6, ultimately resulting in cell death (Slee et al., 1999).

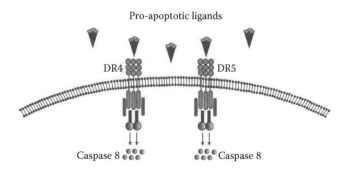

FIGURE 4.10 Elements of extrinsic pathway.

4.2.2.2.1 Role of Death Receptors in the Extrinsic Pathway

The tumor necrosis factor (TNF) receptor gene superfamily consists of more than 20 proteins with wide implications, ranging from cell death regulation to immune regulation to differentiation (Ashkenazi, 2008). The members of the TNF superfamily share similar cysteine-rich extracellular domains and a cytoplasmic domain of about 80 amino acids, which is known as the death domain. The death domain plays a very important role in the transmission of death signals from the cell's surface to the intracellular apoptotic pathway. Some of the well-characterized death receptors are CD95 (APO-1/Fas), TNF receptor-1 (TNFR1), TNF-related apoptosis-inducing ligand-receptor-1 (TRAIL-R1), and TNF-related apoptosis-inducing ligand-receptor-2 (TRAIL-R2) (Khan and Bisen, 2013). These death receptors generate a signal on interacting with corresponding ligands of the TNF superfamily, like CD95 ligand (CD95L), TNFα, lymphotoxin-α, TRAIL, and TWEAK (Walczak and Krammer, 2000).

4.2.2.2.1.1 CD95 The CD95 receptor-ligand system is a key signaling pathway involved in apoptotic regulation in various cell types, including those of the immune system. CD95 is a type I transmembrane receptor of 48 kDa, which is expressed in various tissues of lymphoid or nonlymphoid origin, including tumor cells. Activated T-cells produce CD95 ligand, which is involved in immune regulatory function and triggers autocrine suicide or paracrine death in lymphocytes or other target cells.

The expression of CD95L has also been implicated in immune escape by tumors, wherein a receptor-ligand interaction of CD95 is believed to kill attacking antitumor T-cells. However, this model of tumor counterattack has been challenged because of the absence of any conclusive study supporting this hypothesis (Igney and Krammer, 2002).

4.2.2.2.1.2 TNF-Related Apoptosis-Inducing Ligand and Its Receptors TNF-related apoptosis-inducing ligand/Apo-2L is constitutively expressed in a wide range of tissues; it shares a sequence homology to other members of the TNF superfamily. TRAIL-R1 and TRAIL-R2 are capable of relaying a death signal because of the presence of the cytoplasmic death domain (LeBlanc and Ashkenazi, 2003).

4.2.2.2.1.3 Signaling through CD95 or TRAIL Receptors The ligation of CD95 or TRAIL-R1/TRAIL-R2 by their cognate ligands results in receptor trimerization, clustering of the death domain of receptors, and recruitment of adaptor molecules such as Fas-associated death domain (FADD) through homophilic interaction mediated by the death domain (Walczak and Krammer, 2000). Fas-associated death domain in turn recruits caspase-8 to the activated CD95 receptor to form the CD95 death-inducing signaling complex (DISC). Caspase-8 is activated through self-cleavage as a result of its oligomerization upon DISC. The activated caspase-8 in turn activates other downstream effector caspases like caspase-3. Based on the dependence of mitochondrial factors for full caspase activation, two types of cell types have been identified. In type I cells, the activated caspase-8 is produced in quantities sufficient to induce full activation of downstream caspases required for apoptosis, whereas in the type II cell, caspase-8 requires mitochondrial amplification for execution of apoptosis (Scaffidi et al., 1998).

4.2.2.2.2   Role of Apoptotic Pathways in Anticancer Chemotherapy

4.2.2.2.2.1   Disruption of the Extrinsic Pathway in Cancers   Cancer cells are known to develop various survival strategies/mechanisms that help them to survive and thrive. One such mechanism is to evade apoptosis induced by the extrinsic pathway, by augmentation of anti-apoptotic molecules or by silencing of pro-apoptotic proteins. The expression of death receptors is downregulated in drug-resistant leukemia or neuroblastoma cells. Colon cancer cells control apoptosis by modulating transport of agonistic TRAIL receptors (TRAIL-R1 and TRAIL-R2) from the endoplasmic reticulum to the cell surface (Jin et al., 2004). Mutation is yet another mechanism by which cancer cells suppress expression of apoptotic factors like CD95 (Debatin et al., 2003). Overexpression of decoy receptors like DcR3 and TRAIL-R3 interferes with death receptor-ligand interaction by competitively binding CD95L and TRAIL, respectively (Sheikh et al., 1999). Silencing of the agonistic TRAIL receptors TRAIL-R1 and TRAIL-R2 associated with mutation (Dechant et al., 2004) or loss of heterozygosity (LOH) in chromosome 8p in tumors accounts for TRAIL resistance (LeBlanc and Ashkenazi, 2003).

Epigenetic silencing by CpG island hypermethylation of gene promoters of CD95 or TRAIL receptors causes impaired expression of death receptors in neuroblastoma or colon carcinoma cells. Chromatin remodeling by histone deacetylation may also block the transcription process by preventing access of transcription factors to DNA (Marks et al., 2001). The role of epigenetic silencing in tumorigenesis was verified in studies in which expression of CD95 was restored by histone deacetylase inhibitors, resulting in impairment of tumor growth and restoration of chemosensitivity. Epigenetic changes in CD95 expression have been implicated in chemoresistance and immune escape (Maecker et al., 2002).

The negative regulation of the death signal relayed by cytoplasmic domains of death receptors through association with proteins like FLIP or PED/PEA-15 has been reported (Stassi et al., 2005). Splice variants of FLIP, FLIP$_L$, and FLIP$_S$ are homologous to caspase-8 and caspase-10, but lack its catalytic site; therefore, recruitment of these FLIP splice variants to DISC blocks downstream caspase activation (Krueger et al., 2001). The overexpression of FLIP has been reported in various tumor cells and has been associated with resistance to TRAIL-, CD95- and chemotherapy-induced apoptosis in osteosarcoma and colorectal cancer cells (Kinoshita et al., 2000); however, sensitivity of some other cancer cells, like T-cell leukemia cells, is unaffected by its overexpression (Kataoka et al., 1998).

4.2.2.2.3   Signaling through the Extrinsic Pathway in Cancer Therapy

Apoptosis is regarded as a common end process by which chemotherapeutic drugs kill cancer cells. The extrinsic pathway has been shown to play an important role in drug-induced apoptosis in cancer therapy. The anticancer treatment triggers expression of CD95, which in turn stimulates the receptor pathway in an autocrine or paracrine manner by binding to its receptor CD95 (Fulda et al., 1997, 1998). The treatment of colon carcinoma cells with 5-fluorouracil is supposed to be mediated via CD95 receptor-ligand system. Drug treatment was reported to be associated with the increase of CD95L and surface expression of CD95, especially in cells harboring wild-type p53 (Muller et al., 1997). The correlation between CD95 and p53 can be corroborated by the presence of p53-responsive elements (PREs) in the first intron of the CD95 gene, as well as three putative p53 binding sites within the CD95 promoter, which showed limited homology with the p53 consensus binding site (Muller et al., 1998). Anticancer agents have been reported to activate the CD95 pathway by modulating expression and recruitment of pro- or anti-apoptotic components of the CD95 DISC. The colon carcinoma cells treated with doxorubicin, cisplatin, or mitomycin C were found with increased expression of FADD and procaspase-8 (Micheau et al., 1999). The role of death receptor signaling in drug-mediated apoptosis has been challenged (Villunger et al., 1997) in some studies in which drug-mediated apoptosis was achieved even in the presence of agents that negated components of death receptor signaling; however, the premise of such experiments were questioned by other groups (Hyer et al., 2000).

4.2.3.3 Schematic Representation of Some Major Apoptotic Signaling Pathways

Apoptosis can be triggered by various stimuli from inside (cellular stress caused by oncogenes, irradiation, or drugs) or outside (mediated by death receptors) the cell. Death receptors upon interaction with corresponding ligands are responsible for DISC-mediated activation of initiator caspase-8. Activated caspase-8 initiates a caspase cascade, which involves processing and activation of effector caspases-3, -6, and -7. Based on the strength of caspase cascade signals to induce apoptosis, physiologically cells can be divided into two groups: type I cells and type II cells. In type I cells, the direct caspase cascade is sufficient to elicit apoptosis on its own, whereas in type II cells, the death signals needs amplification through mitochondrial apoptotic signaling via proteolytic activation of Bid by caspase-8 (Figure 4.11).

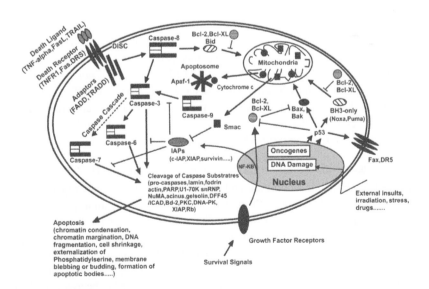

FIGURE 4.11 Schematic representation of some major apoptotic pathways.

Mitochondria contribute to apoptotic signaling through the release of cytochrome c, which forms apoptosome (cytochrome c, Apaf-1, and dATP), involved in activation of effector caspases like caspase-9 and caspase-3. Mitochondria are also responsible for release of Smac, a pro-apoptotic factor responsible for inhibiting IAPs from blocking caspase activity. IAPs are overexpressed in the presence of survival signals generated by growth factor receptors, e.g., by activation of the transcription factor NF-κB. Anti-apoptotic Bcl-2 family members such as Bcl-2 and Bcl-XL counteract action of pro-apoptotic proteins like Bid, Bax, and Bad. The intrinsic apoptotic pathway is often initiated by DNA damage induced by drugs, irradiation, or other cellular stress. p53 is activated by DNA damage, and its activation promotes expression of pro-apoptotic Bcl-2 members and suppresses anti-apoptotic Bcl-2 and Bcl-XL.

4.2.3.4   Relationship between Intrinsic and Extrinsic Apoptotic Pathways

Various studies have pointed out the existence of extensive cross talk between intrinsic and extrinsic apoptotic pathways. There exists a dynamic balance between pro- and anti-apoptotic proteins in the cellular environment, responsible for the survival and death of the cell. Various therapeutic interventions are focused on devising strategies to augment pro-apoptotic proteins and suppress anti-apoptotic proteins to kill cancerous cells by leveraging the apoptotic pathway (Khan et al., 2012). The apoptotic pathway is one of the most sophisticated pathways discovered in the cell to date. Its activity is tightly regulated and monitored by the cell. Recent advances and understanding of the apoptotic pathway have led to better and more innovative treatments against cancer and other diseases. However, the detailed mechanism of the apoptotic pathway is yet to be unraveled.

4.2.3.5   Signal Transduction in Apoptosis

Apoptotic signaling (intrinsic or extrinsic) converges on a common machinery of cell destruction that is activated by a family of cysteine proteases (caspases) that cleave proteins at aspartate residues. The apoptotic cell is dismantled and degraded progressively by proteolysis of vital cellular constituents, DNA degradation, and phagocytosis by neighboring cells (Figure 4.12). Much of our current knowledge about molecular details of apoptosis is derived from genetic studies in *C. elegans* and their comparison with results from genetic and biochemical experiments in mammalian cells.

All developmentally programmed deaths of somatic cells in *C. elegans* require three proteins: the caspase Ced-3, the adaptor protein Ced-4, and Egl-1, a BH3-only pro-apoptotic member of the Bcl-2 family. The Bcl-2 homolog Ced-9 is needed for cell survival (Hengartner and Horvitz, 1994). Protein-protein interactions between Ced-3, Ced-4, Ced-9, and Egl-1 proteins have made a direct link between the caspases, the effector arm of the cell death pathway, and the Bcl-2 protein family. The connecting element between the two types of proteins appears to be Ced-4, and its mammalian homolog, Apaf-1, is thought to fulfill a similar function (Chinnaiyan et al., 1997). Transgenic experiments in *C. elegans* (Shaham and Horvitz, 1996) and the discovery that binding of Ced-9 to Ced-4 is essential for its survival function (Wu et al., 1997) implied that Ced-4 acts upstream of or in parallel to the Ced-3 caspase, and that Ced-9 acts as an inhibitor of Ced-4. Ced-4 can bind simultaneously to the caspase Ced-3 and to Ced-9.

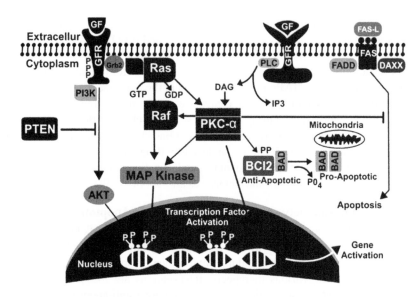

FIGURE 4.12 Systematic presentation of signal transduction in apoptosis.

This mechanism appears to be conserved in mammals, because the human Ced-4 homolog Apaf-1 can bind to and activate human caspase-9 zymogens, and this can be inhibited by Bcl-xL (Hu et al., 1998). Experiments with transgenic and gene knockout mice have shown that different initiator caspases, together with their specific adaptors and regulators, are required for control and execution of different death stimuli. In thymocytes and embryonic fibroblasts, caspase-9 and its adaptor, Apaf-1, are needed for DNA damage-, corticosteroid-, and staurosporine-induced cell killing, but they are dispensable for CD95 (Fas/APO-1)- and TNFRI-transduced apoptosis (Hakem et al., 1998).

In contrast, caspase-8 and its adaptor, FADD, are needed for CD95- and TNFRI-transduced apoptosis, but they are dispensable for the other pathways to cell death (Varfolomeev et al., 1998). Bcl-2 and its homologs are potent inhibitors of cell death caused by growth factor deprivation, DNA damage, or treatment with corticosteroids or staurosporine, all of which require Apaf-1 and caspase-9. Conversely, in at least some cell types, particularly lymphocytes, Bcl-2 and its homologs are poor antidotes to apoptosis transduced through CD95 or TNFRI (Strasser et al., 1995). Consistent with these observations, Bcl-2 and its homologs have been shown to interact with Apaf-1 and prevent it from activating caspase-9.

However, FADD-induced activation of caspase-8 and apoptosis are not blocked by these anti-apoptotic molecules. It therefore appears that mammals have two distinct mechanisms for activating effector caspases. One is initiated by stress-induced signals inside the cell, requires Apaf-1 and caspase-9, and is regulated by the Bcl-2 protein family. The other is activated by CD95 and related receptors, requires FADD and caspase-8, and cannot be blocked by Bcl-2 or its homologs. It has been reported that Bcl-2 can inhibit hepatocyte apoptosis caused by injection of antibodies to CD95, and that Bcl-2-insensitive and Bcl-2-inhibitable pathways leading from CD95 to apoptosis can coexist within the single cell. However, all of these experiments were performed with antibodies to CD95, and it is now known that their action is not identical to that of FasL. It is therefore imperative to reevaluate CD95 signaling in experiments by using the physiological ligand (Strasser and O'Connor, 1998).

The role of adaptor proteins in regulating pro- and anti-apoptotic signaling pathways induced by activation of death receptors is very critical. Adaptor proteins, such as FADD (Fas-associated death domain) and TRADD (TNF receptor-associated death domain), are recruited to ligand-activated, oligomerized death receptors to mediate apoptotic signaling pathways. Apoptotic adaptors associate through domains homologous with the death domain (a cytoplasmic portion) of death receptors. After association with a ligand-bound receptor, the adaptor proteins recruit procaspase-8 or procaspase-10 with the help of their death effector domains (DEDs) to form the death-inducing signaling complex (DISC), which initiates a caspase signaling cascade and consequently induces apoptosis. Other adaptor proteins initiate apoptotic signaling through distinct mechanisms. CRADD/RAIDD (caspase and RIP adaptor with a death domain) and PIDD (p53-induced protein with a death domain) are adaptor proteins that associate with procaspase-2 to form the PIDDosome following DNA damage, and lead to the cleavage and activation of caspase-2. RIP1 (receptor interacting protein-1) and DAXX (Fas death domain-associated-xx) are adaptor proteins with a death domain, which are recruited to TNFR1 or Fas, respectively, upon ligand binding. Apart from promoting apoptosis, ligand binding to DR3 (TWEAK) or TNFR1 (TNFα) can cause the intracellular adaptor proteins to recruit TNF receptor-associated factor-2 (TRAF-2), which subsequently leads to the activation of NF-κB responsible for inducing the expression of anti-apoptotic genes such as FLIP and Bcl-2.

Cell survival and death are well-regulated phenomena that are partly decided by type of ligands that bind to the surface receptors. Death ligands like Fas and TNFα bind to their receptors to promote apoptosis via caspase activation. Anti-apoptotic ligands like growth factors bind to the cell surface receptors to activate intracellular kinases (e.g., PI3K, Akt/PKB, PKA, PKC, MEKs, ERK1/2, p90RSK, and p38 MAPK), which promotes cell survival. The activation of PI3K leads to the phosphorylation and activation of Akt/PKB, which results in the Akt-dependent phosphorylation and inhibition of pro-apoptotic molecules such as Bad, GSK-3, and caspase-9 (Figure 4.13).

Kinase-dependent activation of transcription factors such as CREB and Stat family members (Mad, Max, and c-Myc) promotes cell survival by inducing expression of Bcl-2 and Bcl-xL. Max heterodimers are associated with transcriptional repression and cellular differentiation, while Myc-Max heterodimers are associated with transcriptional activation and cellular proliferation. The Stat family of transcription factors is activated by intracellular kinases such as Jak (1 to 3), TyK2, and Src.

The intrinsic pathway is triggered by a signal emanating within the cell; for instance, DNA damage, loss of contact with the extracellular matrix, and growth factor withdrawal are some of the factors that initiate the intrinsic apoptotic pathway. Such intrinsic factors promote signaling pathways, which leads to caspase activation and loss of mitochondrial membrane integrity. The proteins from the Bcl-2 family play an important role in regulating mitochondrial membrane permeability, and

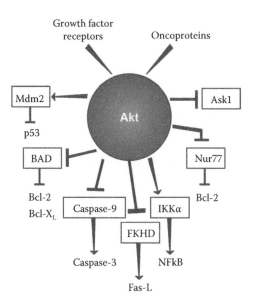

FIGURE 4.13 Suppression of various regulators.

thereby release pro-apoptotic factors such as cytochrome-c. The Bcl-2 protein family can be broadly grouped into the following categories:

1. Anti-apoptotic proteins inhibit apoptosis, e.g., Bcl-2, Bcl-xL, Bcl-w, Mcl-1, Bcl-10, and Bcl2-related protein A1.
2. Pro-apoptotic proteins promote apoptosis, e.g., BAK, Bax, Bcl-rambo, and BOK/Mtd.
3. BH3-only proteins bind and regulate the anti-apoptotic Bcl-2 proteins, e.g., Bad, BID, Bik/Nbk, BIM, BLK, Bmf, and Hrk/DP5.
4. Others like Bag-1, BNIP3, and BNIP3L are Bcl-2-interacting proteins that are also involved in modulating apoptosis.

In normal cells, Bad is phosphorylated and sequestered in the cytoplasm by the adapter protein, 14-3-3. The increase of cytoplasmic levels of free Bad causes anti-apoptotic proteins Bcl-2 and Bcl-xL to bind with Bad, and consequently release Bax and BAK. Bax and BAK initiate release of cytochrome c from the mitochondrial membrane.

The scientific literature is full of examples showing that a large number of chemical or biological compounds can trigger apoptosis, in at least some cell types under certain conditions. This has been interpreted as evidence that a given stimulus is physiologically or clinically important, but this is often not the case. This section concentrates on those apoptotic stimuli that are the most relevant to normal physiology, development, and tissue homeostasis. Cell death is critical for animal development. In mammals, some developmental cell deaths are autonomous, meaning that pro-apoptotic signals from neighboring cells are not needed for the demise of the doomed cell (Coucouvanis et al., 1995).

In fact, it is believed that in many of these deaths, neighboring cells provide survival signals, and that cell deaths are initiated by withdrawal of growth factors or loss of cell attachment (Raff, 1992). These pathways to cell death are often referred to as death by neglect and can, in many instances, be blocked by anti-apoptotic members of the Bcl-2 protein family. Consistent with this, mice lacking Bcl-2 or Bcl-xL exhibited specific defects in organogenesis. Other developmental cell deaths are by design and require apoptosis-inducing signals from neighboring cells. Death receptors and their ligands are prime candidates for this function.

A role for the TNFR family in development can be inferred from the fact that defects in its signal transducers FADD and caspase-8 cause cardiac abnormalities and early embryonic lethality. The absence of obvious developmental defects in those mice lacking members of the TNF or TNFR families so far studied is likely to be explained by the functional overlap between related molecules. As we have discussed above, these two physiologically activated cell death pathways are subject to distinct control. Death by neglect is regulated by the Bcl-2 protein family, and in neuronal tissues, it requires Apaf-1, caspase-9, and caspase-3. Not all death by neglect requires caspase-9 and Apaf-1, however, because the spontaneous death of cultured thymocytes from mice lacking Apaf-1 or caspase-9 appears normal. This may mean that some other initiator caspases can be activated by Apaf-1 in caspase-9-null cells, and it may also imply that mammals have Ced-4-like adaptors other than Apaf-1. Cell loss through apoptosis occurs in many tissues at multiple stages of cell differentiation. Mice expressing *bcl-2* transgenes have been used to determine which of these physiological cell deaths occur by a mechanism that can be blocked by Bcl-2. For example, Bcl-2 expression restored normal development and function of T-lymphocytes in mice lacking the interleukin-7 receptor, but did not block the death of thymocytes bearing autoreactive antigen receptors or those unable to express a pre-T-cell receptor. This indicates that two (or more) distinct pathways to physiological cell death exist—one that can be inhibited by Bcl-2 and one (or more) that cannot.

Transgenic mice have been generated that overexpress Bcl-2 in myeloid cells, neuronal cells (Farlie et al., 1995) hepatocytes, spermatogonia, or a relatively wide range of cell types. Three transgenic animal studies of Bcl-xL were performed in B- and T-lymphoid cells and one in pancreatic cells. As expected, cells expressing these anti-apoptotic transgenes exhibit extended survival in tissue culture and in the whole animal. Gene knockout technology has been used to determine the essential functions of Bcl-2 family proteins. Mice deficient in Bcl-2 show increased cell death during nephrogenesis and premature demise of their mature lymphocytes. These results were expected, given that Bcl-2 is normally expressed at high levels in these tissues (Hockenbery et al., 1991).

Unexpectedly, there was no detectable increase in apoptosis in other tissues in which Bcl-2 is known to be highly expressed, such as the nervous system, intestine, and skin. These results may imply functional redundancy among the anti-apoptotic Bcl-2 family members because Bcl-xL, for instance, is also known to be highly expressed in some of the tissues that express Bcl-2. Mice deficient in *bcl-x* die at embryonic day 13 and show abnormally increased apoptosis in the central nervous system and in erythroid cells from fetal liver (Motoyama et al., 1995). This last observation now has a molecular explanation, because it has been shown that erythropoietin can directly induce expression of Bcl-xL via a signaling pathway involving Stat5. Mice lacking Bcl-w have defective spermatogenesis, and mice lacking A1 have accelerated neutrophil apoptosis.

We believe that differences in the phenotypes of mice lacking anti-apoptotic Bcl-2 family members can be explained by distinct expression patterns and are not the result of differences in biochemical function. Bax, Bim, and Bid are the only pro-apoptotic Bcl-2 family members for which studies with gene knockout analysis have been published. Cell production and cell death appear to be normal in most somatic tissues of *bax*– mice, with the exception of mild T-cell hyperplasia in the thymus and in peripheral lymphoid organs. Surprisingly, Bax-/- male mice exhibit marked hypoplasia in testicular cells and are sterile (Knudson, 1995).

This phenotypic abnormality is similar to that of mice overexpressing Bcl-2 in spermatogonia, and probably reflects the dependence of spermatid production on apoptosis of neighboring cells, rather than any fundamental difference in the apoptosis process in germ cells. The finding that most physiologically induced cell deaths occur normally in Bax-/- mice is consistent with the notion that Bcl-2/Bax heterodimers are dispensable for both apoptosis and cell survival.

Cell production and cell death appear to be normal in untreated Bid-deficient mice, but as mentioned above, their hepatocytes have reduced sensitivity to anti-CD95 antibodies. Bim is required for early mouse development and hemopoietic cell homeostasis. It is essential for certain apoptosis signaling pathways, notably that induced by cytokine withdrawal, and it acts as a barrier against autoimmune disease. It is anticipated that phenotypic analyses of mice lacking other pro-apoptotic Bcl-2 family members (and crosses between such animals) will reveal individual and collective roles of these proteins in cell death control.

4.2.3.6 Mitochondrial Regulation of Apoptosis

Mitochondria play a very important role in the survival of eukaryotic cells, such as oxidative phosphorylation of adenine, electron transport in the respiratory chain, the β-oxidation of fatty acids, and the Krebs cycle. The role of mitochondria in apoptosis and cellular survival is also very crucial. The disruption of mitochondrial transmembrane potential and release of apoptogenic protein caused by opening of the permeability transition (PT) pore are key events in the apoptosis process (Mayer and Oberbauer, 2003; Li et al., 1997).

It was found that the export of cytochrome c from mitochondria during apoptosis proceeds by a two-step process, involving the detachment of the hemoprotein from its binding to cardiolipin in the inner mitochondrial membrane (IMM), followed by its translocation into the cytosol (Ott et al., 2002). It was also observed that microinjection of cytochrome c results in the induction of apoptosis in a number of different cell types (Zhivotovsky et al., 1998). It was found that aberrant mitochondrial iron distribution is responsible for spontaneous cytochrome c release from mitochondria in early erythroid precursor cells in myelodysplastic syndrome (Figure 4.14).

The mitochondria consist of two distinct compartments that are morphologically and functionally different from each other. The outer mitochondrial membrane (OMM) is made up of a phospholipid bilayer with insertions of voltage-dependent anion channels (VDACs). VDACs are permeable to solutes of up to 5 kDa, preferably

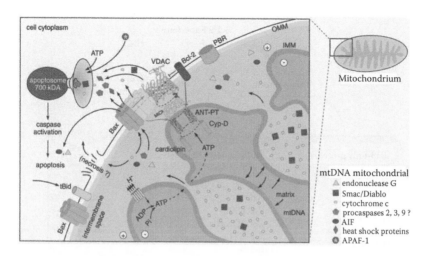

FIGURE 4.14   Mitochondrial regulation of apoptotic cell death.

anionic solutes. The inner mitochondrial membrane (IMM) is permeable only for carbon dioxide, oxygen, and water. The IMM is a highly wrinkled structure that forms lamellar structures known as cristae. The IMM is a site attributed for metabolic activity and also stores about 90% of apoptogenic cytochrome c. These two membranes generate two separate spaces, viz., the intermembrane space and the mitochondrial matrix.

4.2.3.6.1a   Mitochondrial Regulation

Various entities involved in apoptosis are located in mitochondrial substructures. The OMM harbors the VDAC and Bcl-2 members as apoptosis-relevant proteins. The IMM harbors the adenine nucleoside translocator (ANT) as an apoptosis-relevant protein. The intermembrane space harbors the mitochondrial creatine kinase (mtCK) as an apoptosis-relevant protein.

4.2.3.6.2   The Role of VDAC in Apoptosis

The voltage-dependent anion channel (VDAC) or porin acts as a major permeability pathway through the outer mitochondrial membrane. It acts as bidirectional passage for solutes less than 5 kDa. The transport of bigger apoptogenic proteins through VDAC is supposed to be made possible by conformational changes, by pro-apoptotic proteins of the Bcl-2 family (Vander Heiden et al., 2001).

4.2.3.6.3   Bcl-2 Superfamily Members

The Bcl-2 superfamily consists of more than 20 proteins, which play a very important role in regulation of apoptosis. The members of the Bcl-2 superfamily share conserved motifs known as Bcl-2 homology domains (BH1 to BH4). The members of the Bcl-2 superfamily are categorized on the basis of their Bcl-2 homology domains (BH1 to BH4) in their α-helical regions, and also on the basis of their role in the apoptosis process (Figure 4.15). Bcl-2 and its pro-survival homologs, Bcl-xL and

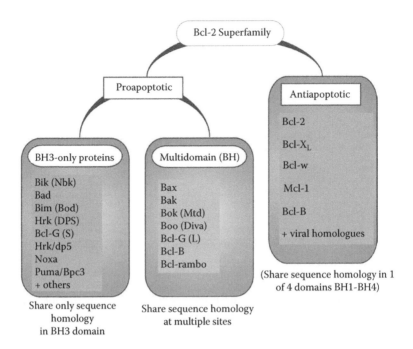

FIGURE 4.15 Members of the Bcl-2 superfamily as key regulators of mitochondrial apoptosis.

Bcl-w, contain all four BH domains; however, other pro-survival members contain at least BH1 and BH2. The members of the Bax subfamily contain BH1, BH2, and BH3, and resemble Bcl-2 fairly closely. The seven mammalian members of the BH3 subfamily possess only the central short (9 to 16 residues) BH3 domain and are unrelated to any known protein (Figure 4.15). The OMM contains anti-apoptotic members such as Bcl-2, Bcl-xL, and Bcl-w, which are responsible for inhibiting apoptosis by preventing opening of VDAC. The pro-apoptotic members of the Bcl-2 superfamily are mainly located in the cytoplasm and can enter OMM on demand. Anti- and pro-apoptotic members can form homo- or heterodimers by binding to their BH domains. The Bax monomers cannot initiate apoptosis induction; however, tBid and Bim have the potential to form pores by activation of VDAC, leading to release of apoptogenic substances into the cytoplasm. The permeability transition (PT) pore is considered the main route through which cytochrome c is released. Bcl-2 is hypothesized to inhibit apoptosis by blocking the formation of the PT. The decrease in the level of cellular Bcl-2, as occurs during ischemic injury, can be considered a reason for increasing the susceptibility of cells for apoptosis (Schwarz et al., 2002).

The members of the Bcl-2 superfamily are under transcriptional regulation of certain cytokines and regulatory proteins like p53. Additionally, Bcl-2 members are regulated by posttranslational mechanisms: in IL-3 stimulated hemopoietic cells the BH3 subfamily member Bad, for example, is phosphorylated and the product is sequestered in the cytosol by 14-3-3 proteins, preventing its inhibition of Bcl-XL. In the case of pro-survival members, phosphorylation may both augment and suppress the activity: a candidate region for phosphorylation and activity regulation is the unstructured loop of about 60 amino acids in Bcl-XL and Bcl-2 (the loop sequence is not conserved between those two Bcl-2 members).

Loop deletion mutants of Bcl-XL and Bcl-2 were demonstrated to have an enhanced anti-apoptotic function; this means those mutant proteins confer a level of resistance to apoptosis significantly greater than the resistance imparted by full-length Bcl-XL or Bcl-2. Interestingly, heterodimerization with Bax is unaffected in the case of loop deletion. But the loop domain influences the phosphorylation status of Bcl-2: wild-type Bcl-2 with loop domain is susceptible toward phosphorylation, while loop-deleted Bcl-2 is not phosphorylated. Either there are phosphorylation sites (serines and threonines) within the loop itself or the loop is a recognition site or a kinase that phosphorylates the protein at different sites.

4.2.3.6.4 The ANT

The mitochondrial permeation plays an important role in the apoptotic process. The adenine nucleotide translocator (ANT) is a dual functional protein that facilitates the exchange of ATP and ADP between mitochondria and cytosol and also participates in apoptotic cell death (Belzacq et al., 2002). The ANT subfamily consists of four isoforms, whose expression is highly regulated, and its dysregulation has been implicated in cancer (Brenner et al., 2011). The isoforms ANT1 and ANT3 are known to induce mitochondrial apoptosis, whereas isoforms ANT2 and ANT4 are reported to be anti-apoptotic (Gallerne et al., 2010). ANT is an integral part of the mitochondrial permeability transition (PT) pore along with cyclophilin D (Cyp-D), a voltage-dependent anion channel (VDAC). The members of the Bcl-2 family are known to interact with ANT to regulate PT; for instance, Bcl-rambo was reported to induce the PT via interaction with ANT (Kim et al., 2012).

4.3 MECHANISMS OF APOPTOSIS IN CANCER

Apoptosis plays a crucial role in carcinogenesis and cancer progression. The current understanding of apoptosis at a molecular level is an outcome of remarkable research done during the last decade. Various research studies have firmly established an implicating role of apoptosis in cancer and have also identified genes involved in apoptosis. Now, we are in a better position to connect the dots and form a picture, albeit incomplete. Apoptosis has also been established as a converging pathway for various cytotoxic drugs, and its dysregulation has been implicated in drug resistance. Failure or a reduced amount of apoptosis is a characteristic feature in various cancers.

The fine balance between apoptotic and survival signals decides the survival status of the cell. Apoptosis acts as a controlled mechanism by which faulty and worn-out cells are removed from the system and replaced by new cells through cell division and growth (Guicciardi and Gores, 2009). In conditions like cancer the apoptotic process is dysregulated, which leads to disruption of the equilibrium between cell growth and cell death. Restoration of the apoptotic machinery in cancer cells is one of the major objectives of therapeutic interventions for cancer management. Restoration of apoptosis has been associated with favorable outcomes, such as increased sensitivity to cytotoxic drugs.

The balance between cell proliferation and apoptosis is influenced by genes that contribute to the development of cancer (oncogenes) and those that encode proteins that normally suppress tumor formation (tumor suppressor genes). Oncogenes are mutated forms of normal cellular genes known as proto-oncogenes, which typically control key cellular functions such as proliferation, cell cycle, and cell death. The mutation of a proto-oncogene is usually an activating mutation, and thereby a mutation in one allele is enough to turn a proto-oncogene into an active oncogene. Functional tumor suppressor genes reduce the probability that a cell will turn into a tumor cell, and their loss allows cellular functions to proceed with little or no control (Vogelstein and Kinzler, 2004). For most tumor suppressor genes, both alleles must be mutated before an effect is manifested.

Mutations are often inherited, but additional somatic mutations can be acquired through inaccurate DNA replication, reactive oxygen species, and other environmental genotoxic stresses. Genetic stability genes are a specific class of tumor suppressor genes that preserve genome integrity. This class of genes does not necessarily drive carcinogenesis, but when altered, they allow for increased mutagenesis rates, which then facilitate carcinogenesis. Virtually all cancer cells contain mutations that enable evasion of apoptosis. Cancer cells bypass apoptosis through a variety of mechanisms involving dynamic interplays between oncogenes and mutated tumor suppressor genes. More than 50% of cancers are marked by a mutation in the tumor suppressor gene p53, a key pro-apoptotic regulator that, when mutated, facilitates tumorigenesis (Hanahan and Weinburg, 2000).

Persistent or elevated signaling from oncogenes such as Myc, Ras, and E1A drives cellular proliferation. However, studies of Myc overexpression suggest that these oncogenes also may affect apoptosis in different ways, depending on certain circumstances. For example, Myc is a powerful inducer of apoptosis under adverse conditions such as cellular stress, DNA damage, or when levels of survival factors are low. Myc may, in fact, enhance tumor cells' sensitivity to apoptotic signaling via the extrinsic pathway, including signaling through DR4 and DR5 (Lutz et al., 1998). The pro-apoptotic function of Myc is abrogated by survival signals like overexpression of Bcl-2 and Bcl-xL, IGF-1, and disruption of the FAS death signaling circuit. Collectively, the data suggest that although a cell's apoptotic program can be initiated by an overexpressed oncogene, other compounding factors can attenuate such effects (Hanahan and Weinburg, 2000).

Other means by which cancer cells overcome apoptosis is through upregulation of the anti-apoptotic PI3 kinase (PI3K)/Akt/PKB survival pathway. This can be achieved through extracellular factors such as IGF-1/2, EGF, or IL-3, by intracellular signals triggered by the oncogene Ras, or by loss of the tumor suppressor gene PTEN, a phospholipid phosphatase that normally downregulates the Akt survival signal. The upregulation of a decoy receptor for the FAS ligand offers yet another mechanism of cancer cell survival (Hanahan and Weinburg, 2000).

4.3.1 MOLECULAR MECHANISMS OF APOPTOSIS WITH PATHOGENICITY

Apoptotic cell death is induced by exogenous and endogenous stimuli such as oxidative stress, genotoxic chemicals, and UV radiation. DNA damage acts as a trigger for

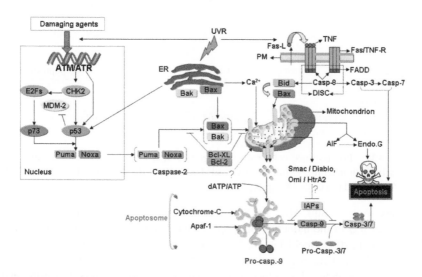

FIGURE 4.16 A generalized pathway of DNA damage-induced p53-mediated apoptosis.

the intrinsic pathway, which is activated via stabilization and activation of p53, which is known to regulate various biomolecules from apoptosis pathways (Figure 4.16). Apart from well-regulated apoptotic death, cells can also die by alternate types of cell death such as necrosis (Debnath et al., 2005; Qiao et al., 2007). Dysregulation of apoptosis has been implicated in various diseases like cancer, autoimmune, and neurodegenerative disorders.

Apoptosis is an imperative component of various processes, including normal cell turnover, proper development and functioning of the immune system, multiplication of mutated chromosomes, hormone-dependent atrophy, normal embryonic development, elimination of indisposed cells, maintenance of cell homeostasis DNA damage, and production of the predominant lesions, such as cyclobutane pyrimidine dimers (CPDs), 6-4 photoproducts (6-4PPs), and certain other lesions (Kumari et al., 2008), as a result of UV radiation (UVR), ionizing radiation (IR), oxidative stress, replication, or recombination errors, as well as from environmental and therapeutic genotoxins. One of the principal triggers of apoptosis is UVR (mostly UV-B; 280 to 315 nm), which is one of the most potent carcinogens that affect normal life processes of all organisms, ranging from bacteria to humans (Friedberg et al., 2006). UV radiation is known to cause DNA damage, which activates and stabilizes nuclear and cytoplasmic p53. Activated p53 regulates a host of genes involved in cell growth and apoptosis (intrinsic and extrinsic apoptotic pathways genes). p53 is functionally conserved in multicellular animals such as *C. elegans* (Cep-1) to human cells, but is absent from unicellular species.

The key components of apoptotic pathways are well conserved in *C. elegans* as Egl-1, Ced-9, Ced-4, and Ced-3 with their corresponding mammalian homologs as BH3, Bcl-2, Apaf-1, and Casp-9, respectively (Figure 4.17). DNA damage is responsible for activation and stabilization of p53, which leads to transcriptional activation of pro-apoptotic members of the Bcl-2 family, such as PUMA, Noxa, Bim, Bid, Bik,

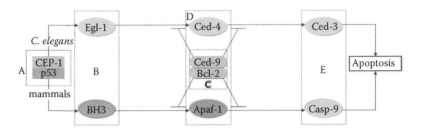

FIGURE 4.17 Programmed cell death in *C. elegans* and mammals showing highly conserved cell death machinery (in box): (A) CEP-1/p53, (B) Egl-1/BH3, (C) Ced-9/Bcl-2, (D) Ced-4/Apaf-1, and (E) Ced-3/Casp-9.

Bak, Bax, Apaf-1, Bmf, Hrk, Pag608, Drs, Fas, and Gadd45 (Tian et al., 2007). The role of p53 in the induction of apoptosis has been questioned in some studies (Kakudo et al., 2005; Speidel et al., 2006). Senescent cells have been reported to be resistant to UV radiation-induced apoptosis (Rochette and Brash, 2008).

UV radiation-induced apoptosis is reported to be caused by p53 transcriptional dependent and independent pathways (Wu et al., 2008). The MDM2-mediated degradation is responsible for maintaining the p53 level under physiological conditions. The transcriptional independent apoptosis regulation by p53 is supposed to be effected by its direct/physical interaction with anti-apoptotic proteins (Bcl-2, Bcl-xL), thus blocking their activity and consequently promoting apoptosis.

Pro-apoptotic Bax/Bak are essential regulators of the mitochondrial or intrinsic pathway of apoptosis. Bax is found in the cytosol of normal healthy cells and translocates to the outer mitochondrial membrane (OMM) during apoptosis, whereas Bak is found on the OMM. The translocation of Bax by UV irradiation was demonstrated to be a Bid-independent event, which is inhibited by overexpression of Bcl-xL and delayed by the p53 inhibitor (Wu et al., 2007, 2008). The conformational changes in UV radiation-induced Bax is induced by p53 and BimL (Chen et al., 2007). The E2Fs' (E2F1 to 3) transcriptional factor can induce apoptosis by the p53-dependent or -independent pathway. In the p53-independent pathway E2Fs' transcription factors directly activate the p53-related protein p73, which induces the BH3-only protein PUMA responsible for mitochondrial translocation and release of cytochrome c and triggering cell death. The transcription factor Dp family member 3 (TFDP3) has been reported to be a negative regulator of E2F1-induced apoptosis (Qiao et al., 2007).

4.3.2 CASPASE (CYSTEINE PROTEASE) ORGANIZATION

Caspases (c: cysteine protease mechanism, aspase: ability to cleave after aspartic acid) are aspartate-directed cysteine proteases that act as initiators and effectors of apoptosis, and their dysregulation is linked with tumor development and several autoimmune diseases (Ghavami et al., 2009; Green et al., 2009). Structurally caspases consist of the functional unit that is a homodimer, with each monomer having a large (~17 to 21 kDa) and a small (~10 to 13 kDa) subunit (Yan and Shi, 2005; Tait and Green, 2008).

The active site of caspases is formed by both large and small subunits. The large subunit harbors active site residues Cys and His, while Asp at the P1 site (P1Asp),

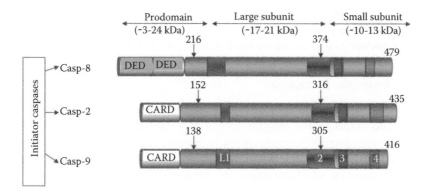

FIGURE 4.18 Graphic illustration showing the organization of mammalian initiator proteases (procaspases).

expressed as the S1 site, is derived from both large and small subunits. Other subsites, like S2, S3, and S4, are formed by small subunits. The active site of caspases is conserved and is made up of five protruding loops (L1 to L4 and L2'). Loops L1 to L4 are present on one monomer, and they together decide sequence specificity of substrates. Loop L2' is present on the nearby monomer and is responsible for stabilization of the active site conformation. Caspases are synthesized as zymogens where an N-terminal prodomain is followed by the sequence programming of a large and subsequently a small subunit. The crystal structure of an active caspase has revealed its tetramer structure (resulting from the combination of two procaspase zymogens) containing two large subunits (homodimer) and two small subunits (heterodimer) folded into a dense canister having a central six-stranded β-sheet flanked by five helices (Figure 4.18).

In the tetramer that is formed by the prodomain-facilitated dimerization of caspase. Two of these cylinders align in a head-to-tail configuration, so that the two active sites are at opposite ends of the molecule (Boatright and Salvesen, 2003). The initiator caspases contain a death domain (DD) such as a caspase activation and recruitment domain (CARD) (e.g., caspases-1, -2, -4, and -9) or a death effector domain (DED) (e.g., caspases-8 and -10), which facilitates caspases, to act together with other molecules that direct their activation. The DED has been found to be involved in interactions with DEDs of signaling adapter proteins such as MORT1/FADD and TRADD. The CARD protein interaction motif is conserved among multiple key apoptotic regulators such as caspases, adaptor molecules RAIDD and Apaf-1, and inhibitors of apoptosis c-IAP1 and c-IAP2 (Ho and Hawkins, 2002). The pathways of apoptosis are extremely complicated, where energy-dependent flow of molecular events takes place.

Another pathway of apoptosis has also been recognized that involves perforin/granzyme A or B (released by cytotoxic T-lymphocytes and natural killer (NK) cells: perforin/granzyme-induced apoptosis). It has been found that granzyme A-induced apoptosis is a caspase-independent pathway through single-stranded DNA damage (Martinvalet et al., 2005). Irrespective of type of apoptotic pathway (intrinsic, extrinsic, and granzyme B induced) active in the cell, the final steps involved in cell death converge to the same point, which involves activation of caspase-3/7 followed by cell shrinkage, chromatin condensation and fragmentation of chromosomal DNA, degradation of nuclear as well as cytoskeleton proteins, etc. (Elmore, 2007).

4.3.3 APOPTOTIC REGULATORS

The survival decisions in cells and their relationship with apoptotic regulatory proteins and mitochondrial integrity have drawn much attention of researchers to elucidating the apoptotic mechanism. Such apoptotic regulators can be categorized broadly into two types: (1) modulators of mitochondrial function and (2) regulators of caspase activation, which are responsible for activation and execution of the apoptotic cascade.

Mitochondrial regulators include Bcl-2 and Bcl-xL, which are responsible for maintaining mitochondrial integrity by prevention of loss of mitochondrial membrane potential or release of pro-apoptotic proteins such as cytochrome c into the cytosol. Other mitochondrial regulators include pro-apoptotic proteins like Bax, BAK, and Bim, which are involved in the regulation of the intrinsic, mitochondrial apoptotic pathway and responsible for release of cytochrome c. The regulators of caspase activation include proteins such as the inhibitors of apoptosis proteins (e.g., XIAP) or FLIP, which are responsible for blocking the activation of caspases, particularly those involved in the engagement of the receptor-related extrinsic apoptotic pathway.

Various signal transduction pathways are also involved in the regulation of apoptosis, mediated via posttranslational modifications in members of the Bcl-2 protein family. For instance, the phosphorylation of Bcl-2 through a JNK-dependent mechanism is believed to cause apoptosis by the taxane class of cytotoxic agents. Researchers are actively pursuing small-molecule-mediated modulation of apoptosis, as an attempt to control various diseases. Two classes of molecules, the Bcl-2 family of regulators and the ICE family of proteases, have emerged from different vertebrate, invertebrate, and viral systems that have been used to elucidate the pathways leading to apoptosis.

The mitochondria have gained prominence as the center of the apoptotic process in recent times. Various studies on cell-free systems of apoptosis have demonstrated a rate-limiting role of mitochondrial products in the activation of caspases and endonucleases in cell extracts. Various studies have suggested a critical role played by the permeability transition pore in inducing or preventing apoptosis. The precise role of mitochondria in the apoptotic process would be important in our understanding its role in carcinogenesis, and thereby the development of novel cytotoxic and cytoprotective drugs.

4.4 ROLE OF APOPTOTIC REGULATORS IN RHEUMATOID ARTHRITIS (RA)

Rheumatoid arthritis (RA) is characterized by low levels of apoptosis, resulting into accumulation of leukocytes at the RA joint. This accumulation of cells promotes inflammation in the joints caused by production of pro-inflammatory chemokines and cytokines, which includes interleukins (IL-1/8), tumor necrosis factor-α (TNFα), and monocyte chemoattractant protein-1 (MCP-1) (Pope, 2002).

The extrinsic apoptosis pathway mediated by death receptors has been studied in detail in isolated synoviocytes from RA patients. The lack of positive staining of TRAIL or TRAIL receptors on isolated synovial fibroblasts or synovial fluid cells was reported (Perlman et al., 2003). However, another research group has reported the presence of expression of the TRAIL-R2 receptor on synovial fibroblasts, and they have also reported that induction of apoptosis by the TRAIL-R2 antibody leads to prevention of bone erosion and cartilage destruction in the RA-SCID model (Miranda-Carus et al., 2004). The loss of proper Fas-FasL function in synoviocytes can be attributed to the decreased death of synoviocytes, resulting in their contribution to inflammation and joint destruction. The expression of Fas and FasL is positive in the RA joint; however, the amount of expected Fas-induced apoptosis is negligible in these tissues. The reason for this anomalous observation can be attributed to the increased level of Fas-inhibitory protein (Flip) identified in both macrophages and synovial fibroblasts of the RA joint and in rats with adjuvant-induced arthritis (Perlman et al., 2001). The increased level of Flip is believed to be under the control of NF-κB, which is activated by TNFα found abundantly in joints. In a same study it was demonstrated that the transcriptional activation of Flip is dependent on NF-κB signaling, and that Flip enhances the transactivation of the NF-κB pathway following stimulation with Fas (Palao et al., 2006). Reactivation of Fas-mediated apoptosis was seen in studies aimed to suppress Flip transcription through the use of histone deactylase inhibitors or antisense to Flip.

While the precise etiology of RA is as of yet unknown, there is a preponderance of evidence to suggest that apoptosis plays an important role in preventing the development of RA. Both the extrinsic and intrinsic apoptosis pathways have been shown to impact the severity of RA in patients and animal models. While it has been well established that the extrinsic and anti-apoptotic intrinsic family members regulate disease, the role of pro-apoptotic Bcl-2 family members has only more recently been elucidated. These new data, which highlight the important role of the BH3-only pro-apoptotic proteins, reveal the delicate balance that is required to maintain proper immune homeostasis within an organism.

4.5   ROLE OF APOPTOTIC REGULATORS
IN PROSTATE CARCINOMA

Members of the Bcl-2 protein family are key apoptosis regulators (Adams and Cory, 1998). Protein Bcl-2 was discovered during studies of the t(14;18) chromosome translocations often found in non-Hodgkin's lymphoma and follicular lymphoma, which are characterized by upregulation of Bcl-2 gene expression and resistance to apoptosis induction. The Bcl-2 family of proteins consists of both inhibitors and promoters of programmed cell death. In mammals, not less than 15 members of the Bcl-2 family have been identified. Others have been found in viruses. Among anti-apoptotic proteins are Bcl-2, Bcl-XL, Mcl-1, Al, and Bcl-w. Pro-apoptotic proteins include

Bax, Bak, Bcl-XS, Bad, Bik, and Bid. Best described are Bcl-2 and Bax (Konopleva et al., 1999).

Each family member has at least one out of four evolutionarily conserved motifs, known as Bcl-2 homology domains (BH1 to BH4). Pro-apoptotic and anti-apoptotic family members can form heterodimers and homodimers, and their relative concentration determines the apoptotic potential of cells (Adams and Cory, 2001). For example, Bcl-2 can form homodimers with different molecules of Bcl-2 or heterodimers with Bax molecules or other family members. The prevalence of Bax determines apoptosis, while the excess of Bcl-2 blocks it. Overall these findings, the mechanisms regulating apoptosis in the case of proteins Bcl-2 and Bax, remain unclear. While dimerization seems to be important, the significance of heterodimerization is still controversial, and recently, studies have shown that both Bcl-2 and Bax can regulate the ability to commence apoptosis independently (Otter et al., 1998; Khan and Bisen, 2013).

Expression of Bcl-2 has been described in many epithelial malignancies, including that of the prostate gland. It has been confirmed that elevated levels of Bcl-2 protein can contribute to the progression of prostate cancer, to the development of metastasis, and to the hormone-insensitive stage. The participation of Bcl-2 in multistep carcinogenesis of the prostate has been shown by in vivo experiments in transgenic mice (Bruckheimer et al., 2000). The role of Bcl-2 in tumorigenesis may be in one of two ways: it can participate in apoptosis suppression and or in stimulating tumor angiogenesis.

In normal prostate tissue, Bcl-2 expression is limited to basal cells of the glandular epithelium. These are resistant to the effects of androgen deprivation. In contrast, secretory epithelial cells, which have no detectable Bcl-2 expression, undergo apoptosis following androgen deprivation. A strong association between Bcl-2 expression and the development of androgen-independent prostate cancer was reported (McDonnell et al., 1992). Since then, several other studies have confirmed that androgen-independent prostatic tumors are typically immunoreactive to Bcl-2 protein. A negative correlation between Bcl-2 expression and clinical prognosis of nonmetastasizing prostate tumors has also been shown in patients undergoing radical prostatectomy (Brewster et al., 1999). The elevated level of Bcl-2 expression corresponds to poor prognosis.

The increase of Bcl-2 expression has been described in patients with locally spread or metastatic forms of prostate cancer treated using hormone ablation therapy. For this reason, Bcl-2 expression is a negative prognostic indicator in these patients. It seems that Bcl-2 enables prostate tumor cells to survive after castration, and that hormone deprivation therapy may select Bcl-2 positive cells that do not react to lowered levels of hormone by undergoing cell death. In vitro studies using prostate cancer cell lines confirm the role of Bcl-2 in inhibition of apoptosis and in the progression to androgen independence. Studies have shown that prostate cancer cells that overexpress Bcl-2 are more resistant to apoptosis induction after hormone ablation in vivo and in vitro.

In addition, Bcl-2 expression in prostate tumor lines is associated with lowered apoptotic response after treatment with different chemotherapeutics, ionizing radiation, and resistance to apoptosis caused by nutritional deprivation. These data confirm that the anti-apoptotic function of Bcl-2 is oncogenic and contributes to the resistance to antiandrogen treatment (Kajiwara et al., 1999). Baltaci et al. (2000) determined the occurrence of Bcl-2 protein expression in low- and high-grade prostatic intraepithelial neoplasia (PIN) lesions, and thus investigated the role of Bcl-2 in tumorigenesis of the prostate. They reached the conclusion that Bcl-2 protein expression is associated with early tumorigenesis of the prostate (Baltaci et al., 2000). In contrast, pro-apoptotic protein Bax was expressed in each tested sample regardless of tumor grade. It has also been shown in cell lines that lowered sensitivity to the effects of cytotoxic chemotherapy is influenced by overexpression of the anti-apoptotic member of the family Bcl-XL protein. Resistance to apoptosis in the absence of pro-apoptotic Bax has also been described (Marcelli et al., 2000).

It has been demonstrated that the absence of Bax may be caused by higher degradation of the protein by the ubiquitin/proteasome pathway. Higher levels of Bax degradation correlate well with lower levels of Bax protein and a higher Gleason score. Studies using antisense oligonucleotides directed against *bcl-2* mRNA in different chemoresistant cells showed a restoration of chemosensitivity to chemical compounds (Berchem et al., 1995).

A similar approach using ribozymes (enzymes connected with antisense nucleotides) against *bcl-2* and *bcl-XL* also promoted apoptosis. These observations show that ectopic Bcl-2/Bcl-XL expression leads to a death-resistant phenotype. This finding also provides new impetus to the development of potential treatment. Antisense strategies aside, therapies influencing Bcl-2/Bcl-XL protein configuration can be used, and in this way cause inhibition of homo- and heterodimerization, for example, by phosphorylation of Bcl-2.

In cell lines derived from prostatic carcinoma, Bcl-2 is phosphorylated by antitumor substances causing a break in production and in the stability of microtubules (taxol, vinkristin, colchicine, and others). Cell lines with no Bcl-2 expression are resistant to taxol and similar substances. Bcl-2 phosphorylation is achieved in the G2/M phase of the cell cycle, where a blockade occurs in polymerization and depolymerization of cell microtubules (Basu and Haldar, 1998). The above-mentioned studies have also shown that posttranslational phosphorylation of protein Bcl-2 by the substances mentioned is induced in the serine position.

4.6   CONCLUSION

Cell death by suicide is called apoptosis. The apoptosis has the following important characters:

1. Cell movement is disabled.
2. Cells shrink.
3. Cells appear as nonregular shapes on their surface.
4. Water is lost.
5. Cell organelles break down.
6. Cells break into small fragments.

The apoptotic process is executed in such a way as to safely dispose of cell corpses and fragments. Apoptosis plays a complementary but opposite role to mitosis in the regulation of an animal cell population; if a cell is unable to undergo apoptosis, due to mutation or biochemical inhibition, it can continue dividing and leading to development of a tumor. Apoptosis allows the cells to be efficiently phagocytosed and their components reused without releasing potentially harmful intracellular substances into the surrounding tissue. Apoptosis is an important process in a wide variety of different biological systems, including normal cell turnover, the immune system, embryonic development, metamorphosis, and hormone-dependent atrophy, and also in chemical-induced cell death. There are three different mechanisms of apoptosis: (1) generated by signals in a cell (DNA damage, stress, etc.), (2) triggered by death factors on the cell surface, and (3) triggered by toxic factors.

Apoptosis caused by internal signals is mediated through factors like Bcl-2, Apaf-1 (apoptotic protease activating factor-1), Bax, cytochrome c, caspase-9, ATP, etc., while Fas, FasL, TNF, and TNF receptor may be related with external signals.

Various studies regarding the role of caspases in apoptosis have suggested the existence of a well-coordinated hierarchy among caspases, wherein a group of initiator caspases (like caspase-8) activates an amplifier caspase like caspase-1, which in turn activates effector caspases like caspase-3/7. Much of the evidence for this, and the concept of a hierarchy of caspases, is based on in vitro data with recombinant enzymes; the limitations of such approaches have already been discussed. It is important, where possible, to determine if such a cascade of caspases occurs in cells undergoing apoptosis. There is a closed relationship between apoptosis and cancer, and many studies have been reported for this relationship. The research results show that several human papillomaviruses could cause cervical cancer, for instance, inactivation of p53 by E6 to evade apoptosis. These actions make the cell more resistant to apoptosis.

Evasion of apoptosis is regarded as one of the most important hallmarks of cancer. Some strategies adapted by cancer cells to evade apoptosis are:

- B-cell leukemias and lymphomas block apoptotic signals.
- Inhibition of expression of Apaf-1, as found in melanoma cells.
- Overexpression of decoy receptors that occupy death receptors like FasL, thus negating interaction with Fas to induce a death signal.
- Some cancer cells express high levels of FasL, and they can kill cytotoxic T-cells that kill them because cytotoxic T-cells also express Fas.

Apoptosis offers potential targets for therapeutic interventions; with an array of drugs already targeting the apoptotic pathway, we should expect a wealth of knowledge to be generated that will act as a guideline for designing cancer treatment strategies in the future. We can expect the emergence of path-breaking treatment that would selectively target multiple key regulators in cancer cells.

REFERENCES

Acehan, D., X. Jiang, D.G. Morgan, et al. 2002. Three dimensional structure of the apoptosome: Implications for assembly, procaspase-9 binding, and activation. *Mol Cell* 9: 423–432.

Adams, J.M., and S. Cory. 1998. The Bcl-2 protein family: Arbiters of cell survival. *Science* 281: 1322–1326.

Adams, J.M., and S. Cory. 2001. Life-or-death decisions by the Bcl-2 protein family. *Trends Biochem Sci* 26: 61–66.

Adams, J.M., and S. Cory. 2007. The Bcl-2 apoptotic switch in cancer development and therapy. *Oncogene* 26: 1324–1337.

Ameisen, J.C. 2002. On the origin, evolution, and nature of programmed cell death: A timeline of four billion years. *Cell Death Differ* 9: 367–393.

Ashkenazi, A. 2008. Directing cancer cells to self-destruct with pro-apoptotic receptor agonists. *Nat Rev Drug Discov* 7: 1001–1012.

Baltaci, S., D. Orhan, G. Ozer, et al. 2000. Bcl-2 proto-oncogene expression in low- and high-grade prostatic intraepithelial neoplasia. *BJU Int* 85: 155–159.

Bao, Q., and Y. Shi. 2007. Apoptosome: A platform for the activation of initiator caspases. *Cell Death Diff* 14: 56–65.

Barnhart, B.C., E.C. Alappat, and M.E. Peter. 2003. The CD95 type I/type II model. *Sem Immunol* 15: 185–193.

Basu, A., and S. Haldar. 1998. Microtubule-damaging drugs triggered bcl2 phosphorylation-requirement of phosphorylation on both serine-70 and serine-87 residues of bcl2 protein. *Int J Oncol* 13: 659–664.

Belzacq, A.S., H.L. Vieira, G. Kroemer, et al. 2002. The adenine nucleotide translocator in apoptosis. *Biochimie* 84(2–3): 167–176.

Berchem, G.J., M. Bosseler, L.Y. Sugars, et al. 1995. Androgens induce resistance to bcl-2-mediated apoptosis in LNCaP prostate cancer cells. *Cancer Res* 55: 735–738.

Boatright, K.M., and G.S. Salvesen. 2003. Mechanisms of caspase activation. *Curr Opin Cell Biol* 15: 725–731.

Boldin, M.P., E.E. Varfolomeev, Z. Pancer, et al. 1995. A novel protein that interacts with the death domain of Fas/APO1 contains a sequence motif related to the death domain. *J Biol Chem* 270: 7795–7798.

Brenner, C., K. Subramaniam, C. Pertuiset, et al. 2011. Adenine nucleotide translocase family: Four isoforms for apoptosis modulation in cancer. *Oncogene* 30: 883–895.

Brewster, S.F., J.D. Oxley, M. Trivella, et al. 1999. Preoperative p53, bcl-2, CD44 and E-cadherin immunohistochemistry as predictors of biochemical relapse after radical prostatectomy. *J Urol* 161: 1238–1243.

Bruckheimer, E.M., S. Brisbay, D.J. Johnson, et al. 2000. Bcl-2 accelerates multistep prostate carcinogenesis in vivo. *Oncogene* 19: 5251–5258.

Chen, M., D. Xing, T. Chen, et al. 2007. BimL involvement in Bax activation during UV irradiation induced apoptosis. *Biochem Biophys Res Commun* 358: 559–565.

Chinnaiyan, A.M., K. O'Rourke, B.R. Lane, et al. 1997. Interaction of CED-4 with CED-3 and CED-9: A molecular framework for cell death. *Science* 275: 1122–1126.

Cikala, M., B. Wilm, E. Hobmayer, et al. 1999. Identification of caspases and apoptosis in the simple metazoan Hydra. *Curr Biol* 9: 959–962.

Coucouvanis, E.C., G.R. Martin, and J.H. Nadeau. 1995. Genetic approaches for studying programmed cell death during development of the laboratory mouse. *Methods Cell Biol* 46: 387–440.

Debatin, K.M., and P.H. Krammer. 2004. Death receptors in chemotherapy and cancer. *Oncogene* 23: 2950–2966.

Debatin, K.M., D. Poncet, and G. Kroemer. 2002. Chemotherapy: Targeting the mitochondrial cell death pathway. *Oncogene* 21: 8786–8803.

Debatin K.M., K. Stahnke, and S. Fulda. 2003. Apoptosis in hematological disorders. *Semin Cancer Biol* 13: 149–158.

Debnath, J., E.H. Baehrecke, and G. Kroemer. 2005. Does autophagy contribute to cell death? *Autophagy* 1: 66–74.

Dechant, M.J., J. Fellenberg, C.G. Scheuerpflug, et al. 2004. Mutation analysis of the apoptotic "death-receptors" and the adaptors TRADD and FADD/MORT-1 in osteosarcoma tumor samples and osteosarcoma cell lines. *Int J Cancer* 109: 661–667.

Degterev, A., M. Boyce, and J. Yuan. 2003. A decade of caspases. *Oncogene* 22: 8543–8367.

Desagher, S., and J.C. Martinou. 2000. Mitochondria as the central control point of apoptosis. *Trends Cell Biol* 10: 369–377.

Elmore, S. 2007. Apoptosis: A review of programmed cell death. *Toxicol Pathol* 35: 495–516.

Fadeel, B., B. Gleiss, K. Högstrand, et al. 1999. Phosphatidylserine exposure during apoptosis is a cell-type-specific event and does not correlate with plasma membrane phospholipid scramblase expression. *Biochem Biophys Res Commun* 266: 504–511.

Farlie, P.G., R. Dringen, S.M. Rees, et al. 1995. bcl-2 transgene expression can protect neurons against developmental and induced cell death. *Proc Natl Acad Sci USA* 92: 4397–4401.

Finkel, E. 2001. The mitochondrion: Is it central to apoptosis? *Science* 292: 624–626.

Finucane, D.M., N.J. Waterhouse, G.P. Amarante-Mendes, et al. 1999. Collapse of the inner mitochondrial transmembrane potential is not required for apoptosis of HL60 cells. *Exp Cell Res* 251: 166–174.

Friedberg, E.C., G.C. Walker, W. Siede, et al. 2006. *DNA repair and mutagenesis*. ASM Press, Washington, DC.

Fulda, S., M. Los, C. Friesen, et al. 1998. Chemosensitivity of solid tumor cells in vitro is related to activation of the CD95 system. *Int J Cancer* 76: 105–114.

Fulda, S., H. Sieverts, C. Friesen, et al. 1997. The CD95 (APO-1/Fas) system mediates drug-induced apoptosis in neuroblastoma cells. *Cancer Res* 57: 3823–3829.

Gallerne, C., Z. Touat, Z.X. Chen, et al. 2010. The fourth isoform of the adenine nucleotide translocator inhibits mitochondrial apoptosis in cancer cells. *Int J Biochem Cell Biol* 42: 623–629.

Ghavami, S., M. Hashemi, S.R. Ande, et al. 2009. Apoptosis and cancer: Mutations within caspase genes. *J Med Genet* 46: 497–510.

Ghobrial, I.M., T.E. Witzig, and A.A. Adjei. 2005. Targeting apoptosis pathways in cancer therapy. *Cancer J Clin* 55: 178–194.

Green, D.R., and G.I. Evan. 2002. A matter of life and death. *Cancer Cell* 1: 19–30.

Green, D.R., T. Ferguson, L. Zitvogel, et al. 2009. Immunogenic and tolerogenic cell death. *Nat Rev Immunol* 9: 353–363.

Guicciardi, M.E., and G.J. Gores. 2009. Life and death by death receptors. *FASEB J* 23: 1625–1637.

Hakem, R., A. Hakem, G.S. Duncan, et al. 1998. Differential requirement for caspase 9 in apoptotic pathways in vivo. *Cell* 94: 339–352.

Hanahan, D., and R.A. Weinberg. 2000. The hallmarks of cancer. *Cell* 100: 57–70.

Hengartner, M.O., and H.R. Horvitz. 1994. Programmed cell death in *Caenorhabditis elegans*. *Curr Opin Genet Dev* 4: 581–586.

Henry-Mowatt, J., C. Dive, J.C. Martinou, et al. 2004. Role of mitochondrial membrane permeabilization in apoptosis and cancer. *Oncogene* 23: 2850–2860.

Ho, P.-K., and C.J. Hawkins. 2002. Mammalian initiator apoptotic caspases. *FEBS J* 272: 5436–5453.

Hockenbery, D.M., M. Zutter, W. Hickey, et al. 1991. BCL2 protein is topographically restricted in tissues characterized by apoptotic cell death. *Proc Natl Acad Sci USA* 88: 6961–6965.

Hu, Y.M., M.A. Benedict, D.Y. Wu, et al. 1998. Bcl-XL interacts with Apaf-1 and inhibits Apaf-1-dependent caspase-9 activation. *Proc Natl Acad Sci USA* 95: 4386–4391.

Hunot, S., and R.A. Flavell. 2001. Apoptosis. Death of a monopoly? *Science* 292: 865–866.

Hutchins, J.B., and S.W. Barger. 1998. Why neurons die: Cell death in the nervous system. *Anat Rec* 253: 79–90.

Hyer, M.L., C. Voelkel-Johnson, S. Rubinchik, et al. 2000. Intracellular Fas ligand expression causes Fas-mediated apoptosis in human prostate cancer cells resistant to monoclonal antibody-induced apoptosis. *Mol Ther* 2: 348–358.

Igney, F.H., and P.H. Krammer. 2002. Immune escape of tumors: Apoptosis resistance and tumor counter attack. *J Leuk Biol* 71: 907–920.

Jin, Z., E.R. McDonald, D.T. Dicker, et al. 2004. Deficient tumor necrosis factor-related apoptosis-inducing ligand (TRAIL) death receptor transport to the cell surface in human colon cancer cells selected for resistance to TRAIL-induced apoptosis. *J Biol Chem* 279: 35829–35839.

Kajiwara, T., T. Takeuchi, T. Ueki, et al. 1999. Effect of Bcl-2 overexpression in human prostate cancer cells in vitro and in vivo. *Int J Urol* 6: 520–525.

Kakudo, Y., H. Shibata, K. Otsuka, et al. 2005. Lack of correlation between p53-dependent transcriptional activity and the ability to induce apoptosis among 179 mutant p53s. *Cancer Res* 65: 2108–2114.

Kataoka, T., M. Schroter, M. Hahne, et al. 1998. FLIP prevents apoptosis induced by death receptors but not by perforin/granzyme B, chemotherapeutic drugs, and gamma irradiation. *J Immunol* 161: 3936–3942.

Khan Z., R.P. Tiwari, N. Khan, et al. 2012. Induction of apoptosis and sensitization of head and neck squamous carcinoma cells to cisplatin by targeting survivin gene expression. *Curr Gene Ther* 12: 444–453.

Khan, Z., and Bisen, P.S. 2013. Oncoapoptotic signaling and deregulated target genes in cancers: Special reference to oral cancer. *Biochim Biophy Acta Reviews on Cancer*, dx.doi. org/10.1016/j.bbcan.2013.04.002.

Kim, J.Y., K.J. So, S. Lee, et al. 2012. Bcl-rambo induces apoptosis via interaction with the adenine nucleotide translocator. *FEBS Lett* 586: 3142–3149.

Kinoshita, H., H. Yoshikawa, K. Shiiki, et al. 2000. Cisplatin (CDDP) sensitizes human osteosarcoma cell to Fas/CD95-mediated apoptosis by down-regulating FLIP-L expression. *Int J Cancer* 88: 986–991.

Knudson, A.G. 1995. Mutation and cancer: A personal odyssey. *Adv Cancer Res* 67: 1–23.

Konopleva, M., S. Zhao, Z. Xie, et al. 1999. Apoptosis: Molecules and mechanisms. *Adv Exp Med Biol* 457: 217–236.

Krueger, A., S. Baumann, P.H. Krammer, et al. 2001. FLICE-inhibitory proteins: Regulators of death receptor-mediated apoptosis. *Mol Cell Biol* 21: 8247–8254.

Kumari, S., R.P. Rastogi, K.L. Singh, et al. 2008. DNA damage: Detection strategies. *EXCLI J* 7: 44–62.

Lavrik, I., A. Golks, and P.H. Krammer. 2005. Death receptor signaling. *J Cell Sci* 118: 265–267.

LeBlanc, H.N., and A. Ashkenazi. 2003. Apo2L/TRAIL and its death and decoy receptors. *Cell Death Differ* 10: 66–75.

Leist, M., and M. Jaattela. 2001. Four deaths and a funeral: From caspases to alternative mechanisms. *Nat Rev Mol Cell Biol* 2: 589–598.

Li, P., D. Nijhawan, I. Budihardjo, et al. 1997. Cytochrome c and dATP-dependent formation of Apaf-1/caspase-9 complex initiates an apoptotic protease cascade. *Cell* 91: 479–489.

Liu, Q.A., and M.O. Hengartner. 1999. The molecular mechanism of programmed cell death in *C. elegans*. *Ann NY Acad Sci* 887: 92–104.

Luo, X., I. Budihardjo, H. Zou, et al. 1998. Bid, a Bcl2 interacting protein, mediates cytochrome c release from mitochondria in response to activation of cell surface death receptors. *Cell* 94: 481–490.

Lutz, W., S. Fulda, I. Jeremias, et al. 1998. MycN and IFNγ cooperate in apoptosis of human neuroblastoma cells. *Oncogene* 17: 339–346.

Ly, J.D., D.R. Grubb, and A. Lawen. 2003. The mitochondrial membrane potential (Dcm) in apoptosis; an update. *Apoptosis* 8: 115–128.

Maecker, H.L., Z. Yun, H.T. Maecker, et al. 2002. Epigenetic changes in tumor Fas levels determine immune escape and response to therapy. *Cancer Cell* 2: 139–148.

Marcelli, M., M. Marani, X. Li, et al. 2000. Heterogeneous apoptotic responses of prostate cancer cell lines identify an association between sensitivity to staurosporine-induced apoptosis, expression of Bcl-2 family members, and caspase activation. *Prostate* 42: 260–273.

Marks, P., R.A. Rifkind, V.M. Richon, et al. 2001. Histone deacetylases and cancer: Causes and therapies. *Nat Rev Cancer* 1: 194–202.

Martinvalet, D., P. Zhu, and J. Lieberman. 2005. Granzyme A induces caspase independent mitochondrial damage, a required first step for apoptosis. *Immunity* 22: 355–370.

Mayer, B., and R. Oberbauer. 2003. Mitochondrial regulation of apoptosis. *Physiology* 18: 89–94.

McDonnell, T.J., P. Troncoso, S.M. Brisbay, et al. 1992. Expression of the protooncogene bcl-2 in the prostate and its association with emergence of androgen-independent prostate cancer. *Cancer Res* 52: 6940–6944.

Meier, P., A. Finch, and G. Evan. 2000. Apoptosis in development. *Nature (London)* 407: 796–801.

Micheau, O., A. Hammann, E. Solary, et al. 1999. STAT-1-independent upregulation of FADD and procaspase-3 and -8 in cancer cells treated with cytotoxic drugs. *Biophys Res Commun* 256: 603–611.

Miranda-Carus, M.E., A. Balsa, M. Benito-Miguel, et al. 2004. Rheumatoid arthritis synovial fluid fibroblasts express TRAIL-R2 (DR5) that is functionally active. *Arthritis Rheum* 50: 2786–2793.

Motoyama, N., F. Wang, K.A. Roth, et al. 1995. Massive cell death of immature hematopoietic cells and neurons in Bcl-x-deficient mice. *Science* 267: 1506–1510.

Muller, M., S. Strand, H. Hug, et al. 1997. Drug-induced apoptosis in hepatoma cells is mediated by the CD95 (APO-1/Fas) receptor/ligand system and involves activation of wild-type p53. *J Clin Invest* 99: 403–413.

Muller, M., S. Wilder, D. Bannasch, et al. 1998. p53 activates the CD95 (APO-1/Fas) gene in response to DNA damage by anticancer drugs. *J Exp Med* 188: 2033–2045.

Naismith, J.H., and S.R. Sprang. 1998. Modularity in the TNF-receptor family. *Trends Biochem Sci* 23: 74–79.

Nijhawan, D., N. Honarpour, and X. Wang. 2000. Apoptosis in neural development and disease. *Annu Rev Neurosci* 23: 73–87.

Okada, H., and T.W. Mak. 2004. Pathways of apoptotic and non-apoptotic death in tumour cells. *Nat Rev Cancer* 4: 592–603.

Ott, M., J.D. Robertson, V. Gogvadze, et al. 2002. Cytochrome c release from mitochondria proceeds by a two-step process. *Proc Natl Acad Sci USA* 99: 1259–1263.

Otter, I., S. Conus, U. Ravn, et al. 1998. The binding properties and biological activities of Bcl-2 and Bax in cells exposed to apoptotic stimuli. *J Biol Chem* 273: 6110–6120.

Palao, G., B. Santiago, M.A. Galindo, et al. 2006. Fas activation of a proinflammatory program in rheumatoid synoviocytes and its regulation by FLIP and caspase 8 signaling. *Arthritis Rheum* 54: 1473–1481.

Perlman, H., H. Liu, C. Georganas, et al. 2001. Differential expression pattern of the anti-apoptotic proteins, Bcl-2 and Flip in experimental arthritis. *Arthritis Rheum* 44: 2899–2908.

Perlman, H., N. Nguyen, H. Liu, et al. 2003. Rheumatoid arthritis synovial fluid macrophages express decreased tumor necrosis factor-related apoptosis-inducing ligand R2 and increased decoy receptor tumor necrosis factor-related apoptosis-inducing ligand R3. *Arthritis Rheum* 48: 3096–3101.

Pope, R.M. 2002. Apoptosis as a therapeutic tool in rheumatoid arthritis. *Nat Rev Immunol* 2: 527–535.

Qiao, A., L. Distefano, C. Tian, et al. 2007. Human TFDP3, a novel DP protein, inhibits DNA binding and transactivation by E2F. *J Biol Chem* 282: 454–466.

Raff, M.C. 1992. Social controls on cell survival and cell death. *Nature* 356: 397–400.

Rathmell, J.C., and C.B. Thompson. 2002. Pathways of apoptosis in lymphocyte development, homeostasis, and disease. *Cell* 109(Suppl): S97–S107.

Richardson H., and S. Kumar. 2002. Death to flies: Drosophila as a model system to study programmed cell death. *J Immunol Methods* 265: 21–38.

Rieger, J., U. Naumann, T. Glaser, et al. 1998. APO2 ligand: A novel lethal weapon against malignant glioma? *FEBS Lett* 427: 124–128.

Rochette, P.J., and D.E. Brash. 2008. Progressive apoptosis resistance prior to senescence and control by the antiapoptotic protein BCL-xL. *Mech Age Dev* 129: 207–214.

Salvesen, G.S., and M. Renatus. 2002. Apoptosome: The seven-spoked death machine. *Dev Cell* 2: 256–257.

Saraste, A., and K. Pulkki. 2000. Morphologic and biochemical hallmarks of apoptosis. *Cardiovasc Res* 45: 528–537.

Scaffidi, C., S. Fulda, A. Srinivasan, et al. 1998. Two CD95 (APO-1/Fas) signaling pathways. *EMBO J* 17: 1675–1687.

Schwarz, C., P. Hauser, R. Steininger, et al. 2002. Failure of BCL-2 up-regulation in proximal tubular epithelial cells of donor kidney biopsy specimens is associated with apoptosis and delayed graft function. *Lab Invest* 82: 941–948.

Shaham, S., and H.R. Horvitz. 1996. An alternatively spliced *C. elegans* ced-4 RNA encodes a novel cell death inhibitor. *Cell* 86: 201–208.

Sheikh, M.S., Y. Huang, E.A. Fernandez-Sales, et al. 1999. The antiapoptotic decoy receptor TRID/TRAIL-R3 is a p53-regulated DNA damage-inducible gene that is overexpressed in primary tumors of the gastrointestinal tract. *Oncogene* 18: 4153–4159.

Shi, Y. 2002. Apoptosome: The cellular engine for the activation of caspase-9. *Structure (Camb)* 10: 285–288.

Slee, E.A., M.T. Harte, R.M. Kluck, et al. 1999. Ordering the cytochrome c-initiated caspase cascade: Hierarchical activation of caspases-2, -3, -6, -7, -8, and -10 in a caspase-9-dependent manner. *J Cell Biol* 144: 281–292.

Solomon, M., B. Belenghi, M. Delledonne, et al. 1999. The involvement of cysteine proteases and protease inhibitor genes in the regulation of programmed cell death in plants. *Plant Cell* 11: 431–444.

Speidel, D., H. Helmbold, and W. Deppert. 2006. Dissection of transcriptional and non-transcriptional p53 activates in the response to genotoxic stress. *Oncogene* 25: 940–953.

Srinivasula, S.M., P. Datta, X.J. Fan, et al. 2000. Molecular determinants of the caspase-promoting activity of Smac/DIABLO and its role in the death receptor pathway. *J Biol Chem* 275: 36152–36157.

Stassi, G., M. Garofalo, M. Zerilli, et al. 2005. PED mediates AKT-dependent chemoresistance in human breast cancer cells. *Cancer Res* 65: 6668–6675.

Strasser, A., A.W. Harris, D.C.S. Huang, et al. 1995. Bcl-2 and Fas/APO-1 regulate distinct pathways to lymphocyte apoptosis. *EMBO J* 14: 6136–6147.

Strasser, A., and L. O'Connor. 1998. Fas ligand—Caught between Scylla and Charybdis. *Nat Med* 4: 21–22.

Tait, S.W., and D.R. Green. 2008. Caspase-independent cell death: Leaving the set without the final cut. *Oncogene* 27: 6452–6461.

Tian, C., D. Lv, H. Qiao, et al. 2007. TFDP3 inhibits E2F1-induced, p53-mediated apoptosis. *Biochem Biophys Res Commun* 361: 20–25.

Tittel, J.N., and H. Steller. 2000. A comparison of programmed cell death between species. *Genome Biol* 1: REVIEWS0003. PubMed: 11178240.

Tsujimoto, Y., N. Nonomura, H. Takayama, et al. 2002. Utility of immunohistochemical detection of prostate-specific Ets for the diagnosis of benign and malignant prostatic epithelial lesions. *Int J Urol* 9: 167–172.

Vander Heiden, M.G., X.X. Li, E. Gottleib, et al. 2001. Bcl-xL promotes the open configuration of the voltage-dependent anion channel and metabolite passage through the outer mitochondrial membrane. *J Biol Chem* 276: 19414–19419.

Varfolomeev, E.E., M. Schuchmann, V. Luria, et al. 1998. Targeted disruption of the mouse caspase 8 gene ablates cell death induction by the TNF receptors, Fas/Apo1, and DR3 and is lethal prenatally. *Immunity* 9: 267–276.

Villunger, A., A. Egle, M. Kos, et al. 1997. Drug-induced apoptosis is associated with enhanced Fas (Apo-1/CD95) ligand expression but occurs independently of Fas (Apo-1/CD95) signaling in human T-acute lymphatic leukemia cells. *Cancer Res* 57: 3331–3334.

Vogelstein, B., and K.W. Kinzler. 2004. Cancer genes and the pathways they control. *Nat Med* 10: 789–799.

Walczak, H., and P.H. Krammer. 2000. The CD95 (APO-1/Fas) and the TRAIL (APO-2L) apoptosis systems. *Exp Cell Res* 256: 58–66.

Walter, D., K. Schmich, S. Vogel, et al. 2008. Switch from type II to I Fas/CD95 death signaling on in vitro culturing of primary hepatocytes. *Hepatology* 48: 1942–1953.

Waterhouse, N.J., J.E. Ricci, and D.R. Green. 2002. And all of a sudden it's over: Mitochondrial outer-membrane permeabilization in apoptosis. *Biochimie* 84: 113–121.

Wu, D., H.D. Wallen, and G. Nunez. 1997. Interaction and regulation of subcellular localization of CED-4 by CED-9. *Science* 275: 1126–1129.

Wu, Y., D. Xing, W.R. Chen, et al. 2007. Bid is not required for Bax translocation during UV-induced apoptosis. *Cell Signalling* 19: 2468–2478.

Wu, Y., D. Xing, L. Liu, et al. 2008. Regulation of Bax activation and apoptotic response to UV irradiation by p53 transcription-dependent and independent pathways. *Cancer Lett* 271: 231–239.

Yan, N., and Y. Shi. 2005. Mechanisms of apoptosis through structural biology. *Annu Rev Cell Dev Biol* 21: 35–56.

Zhivotovsky, B., S. Orrinius, O.T. Brustugun, et al. 1998. Injected cytochrome c induces apoptosis. *Nature* 391: 449–450.

Zuzarte-Luis, V., and J.M. Hurle. 2002. Programmed cell death in the developing limb. *Int J Dev Biol* 46: 871–876.

5 Dynamics of p53 in Oral Cancer

KEY WORDS

Apoptosis
Cell cycle
Mutation
Oncogene
p53
Survivin
Tumor suppressor gene

5.1 INTRODUCTION

p53 protein plays a key regulatory role in carcinogenesis; it has been referred to with various names like "guardian of the genome" (Lane, 1992), "death star" (Vousden, 2000), "good and bad cop" (Sharpless and DePinho, 2002), and "an acrobat in tumorigenesis" (Moll and Schramm, 1998).

p53 was discovered in 1979 as a result of two types of studies involving a virologic and a serologic approach. It was demonstrated in studies of simian virus 40 (SV40)-transformed cells that a 55 kDa protein coprecipitates with the large T-antigen (Clayman et al., 1998; Chang et al., 1979; Kress et al., 1979; Lane and Crawford, 1979; Linzer and Levine, 1979; Melero et al., 1979). Linzer and Levine (1979) found that the 54 kDa protein was overexpressed not only in a wide variety of murine SV40 transformed cells, but also in uninfected embryonic carcinoma cells. It was then postulated that SV40 infection or transformation of mouse cells stimulates the synthesis or stability of a cellular 54 kDa protein. Through serological studies, it was demonstrated that the humoral response of mice to a methylcholanthrene-induced tumor cell line, such as MethA, was directed toward the p53 protein (De Leo et al., 1979). Crawford et al. (1982) first described antibodies against human p53 protein in 9% of breast cancer patient sera. It was later found that such antibodies were present in sera of children with a wide variety of cancers (Caron de Fromentel et al., 1987). The average frequency was 12%, but the figure was 20% in Burkitt lymphoma.

p53 protein is encoded by the TP53 gene, and its name can be attributed to its apparent molecular mass. On SDS-PAGE (sodium dodecyl sulfate polyacrylamide gel electrophoresis) it runs as a 53 kDa protein; however, its actual mass is 43.7 kDa. The difference between apparent and actual molecular mass can be attributed to the high number of proline residues, which slows down its migration in SDS-PAGE (Zeimer et al., 1982). p53 belongs to the protein family consisting of three members:

p53, p63, and p73 (Schmale and Bamberger, 1997; Kaghad et al., 1997). Among them p53 has evolved in higher organisms as a tumor suppressor, whereas the other two are involved in normal developmental biology (Irwin and Kaelin, 2001).

p53 protein is made up of 393 amino acids, which in turn is an ensemble of structural/functional domains:

- N terminus domain:
  - Amino-terminal domain (residues 1–42): It is required for transactivation activity and interacts with various transcription factors, including acetyltransferases and MDM2 (Fields and Jang, 1990; Lin et al., 1994).
  - Proline-rich region (residues 61–94): It plays a role in providing stability against degradation by regulation via MDM2 (Sakamuro et al., 1997).
- Central core domain: It's required for sequence-specific DNA binding. This domain harbors mutation hot spots reported in various cancers, viz., Arg175, Gly245, Arg248, Arg249, Arg273, and Arg282 (Cho et al., 1994).
- C terminus domain: The basic C terminus of p53 also functions as a negative regulatory domain (Vousden and Lu, 2002) and has also been implicated in induction of cell death (Chen et al., 1996).

5.2 FUNCTIONS AND ROLES IN CAUSING DISEASE

p53 occupies a critical place in a network of signaling pathways that are essential for the regulation of cell growth and apoptosis induced by genotoxic and nongenotoxic stresses (Vogelstein et al., 2000; Melino et al., 2002; Vousden and Lu, 2002).

The base level expression of p53 in normal unstressed cells is low, which is regulated by proteins such as MDM2, MULE, PIRH2, or JNK, which promotes its degradation via the ubiquitin-proteosome pathway. Most of these genes are upregulated by p53, which leads to a regulation loop that will keep the p53 level very low in normal cells (Figure 5.1).

After genotoxic or nongenotoxic stresses, activation of p53 is a two-step process. First, the p53 protein level is increased via the inhibition of its interaction with MDM2 and the other negative regulators. Second, a series of modulators (kinases, acetylases) will activate p53 transcriptional activity (Figure 5.2).

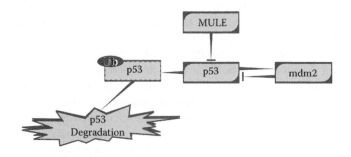

FIGURE 5.1   p53 control switch.

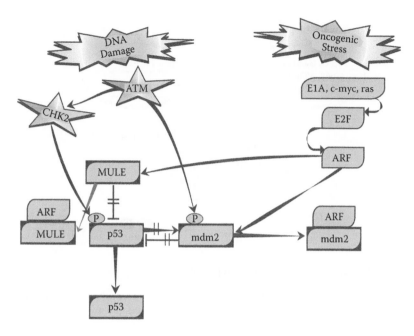

FIGURE 5.2   Activation of p53 in response to cellular stimuli.

The downstream signaling includes a large series of genes that are activated by the transactivating properties of p53. This occurs via specific DNA binding of the p53 protein to a p53 response element (p53 RE) that is found either in the promoter or in the intron of target genes (El-Deiry et al., 1992; Tokino and Nakamura, 2000). Regardless of the type of stress, the final outcome of p53 activation is either cell cycle arrest and DNA repair or apoptosis, but the mechanism leading to the choice between these fates has not yet been elucidated (Vousden and Lu, 2002).

The p53 pathways can be divided into five parts (Levine et al., 2006):

1. The stress signals that activate the pathway
2. The upstream mediators that detect and interpret the upstream signals
3. The core regulation of p53 through its interaction with several proteins that modulate its stability
4. The downstream events, mainly transcriptional activation or protein-protein interactions
5. The final outcome, growth arrest, apoptosis, or DNA repair (Figure 5.3)

The role of p53 mutations in cellular transformation was postulated as early as 1990 (Eliyahu et al., 1989; Lane and Benchimol, 1990). Most of the tumor suppressor genes are inactivated by frameshift or nonsense mutations; however, almost 90% of p53 gene mutations are missense mutations leading to the synthesis of a stable protein, which lacks specific DNA binding function, and thus accumulates in the nucleus of tumor cells (Soussi and Béroud, 2001).

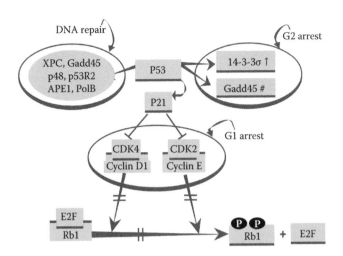

FIGURE 5.3   Cell cycle arrest and DNA repair by p53.

The role of p53 in the context of cancer has undergone a dramatic transition from oncogene to tumor suppressor gene, which is supported and refuted in various scientific studies.

5.3   P53 AS AN ONCOGENE

Early studies implicated p53 as a positive regulator of cell proliferation. The increase in p53 mRNA and the rate of p53 protein synthesis were observed to correlate with cell growth especially reaching its peak near the G1-S boundary prior to DNA replication (Reich and Levine, 1984). The role of p53 as a positive regulator for cell proliferation was strengthened by the work of Mercer et al. (1982, 1984). Microinjection of p53 antibody into the nucleus of quiescent Swiss 3T3 mouse cells inhibited the subsequent entry of the cell into the S phase after serum stimulation. This inhibition was effective only when microinjection was performed at or around the time of growth stimulation, suggesting that p53 is critical for G0-G1 transition. Similar results were demonstrated using the methylcholanthrene-transformed mouse cells, which express mutant (mt) p53 (Deppert et al., 1990; Steinmeyer et al., 1990).

5.4   P53 AS A TUMOR SUPPRESSOR

The role of p53 as a tumor suppressor gene was proposed as early as 1985 (Mowat et al., 1985), wherein it was observed that, in tumors induced by the Friend virus, the p53 gene was quite often rearranged, which led to an absence of expression or the synthesis of a truncated or mutant protein. Such mutation in p53 often affects one of the conserved blocks of the protein (Munroe et al., 1988). The second allele is either lost through loss of the chromosome or inactivated by deletion. Such functional inactivation of the p53 gene seems to confer a selective growth advantage to erythroid cells during the development of Friend leukemia in vivo. The tumor suppressor role of p53 was further corroborated by the experiment in which the reduction of transformation potential of a plasmid encoding p53 and an activated Ha-*ras* gene was observed when cotransfected with a plasmid encoding wild-type (wt) p53 (Eliyahu et al., 1989; Finlay et al., 1989; Khan and Bisen, 2013).

5.5 MUTATIONS IN P53

p53 expression is often high in different human cancers; however, a supporting explanation for this expression pattern does not exist. Genetic analysis of colorectal cancer reveals a very high rate of heterozygous loss of the short arm of chromosome 17, which carries the p53 gene (Vogelstein et al., 1988). PCR analysis and sequencing of the remaining p53 allele show that it often contains a point mutation (Baker et al., 1989). Similar observations have been made in the case of lung cancer (Takahashi et al., 1989).

Transgenic mice with a mutant p53 gene develop many types of cancer, with a high proportion of sarcomas (Lavigueur et al., 1989). This observation led various authors to study patients with Li-Fraumeni syndrome. This syndrome presents as a familial association of a broad spectrum of cancers, including osteosarcoma, breast cancer, soft tissue sarcoma, and leukemias, appearing at a very early age. Statistical analysis predicts that 50% of these individuals will have a tumor before the age of 30, and 90% before the age of 70. Germ-line mutations in the p53 gene have been found in several families with this syndrome (Malkin et al., 1990; Srivastava et al., 1990). A strong correlation between transmission of the mutant allele and development of cancer is evident from all these studies.

Such observations made it possible to define the p53 gene as a tumor suppressor gene and implicated mutant p53 as an oncogene. It has been shown that some p53 mutants (depending on the site of mutation) exhibit a transdominant phenotype, and are able to associate with wild-type p53 (expressed by the remaining wild-type allele) to induce the formation of an inactive heteroligomer (Milner and Medcalf, 1991). Moreover, cotransfection of mutant p53 with an activated *ras* gene shows that some p53 mutants have high, dominant oncogenic activity (Halevy et al., 1990). These observations led to the proposal that several classes of mutant p53 exist, according to the site of mutation and its phenotype (Michalovitz et al., 1991):

1. Null mutations with totally inactive p53 that do not directly intervene in transformation
2. Dominant negative mutations with a totally inactive p53 that is still able to interfere with wild-type p53 expressed from the wild-type allele
3. Positive dominant mutations where the normal function of p53 is altered, but in this case the mutant p53 acquires an oncogenic activity that is directly involved in transformation

5.6 APOPTOTIC REGULATION BY P53

Apoptosis is initiated by a number of factors that primarily act through two apoptotic pathways: extrinsic and intrinsic apoptotic pathways (Green and Kroemer, 2004; Green, 2005).

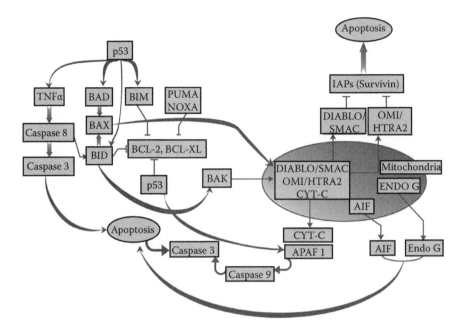

FIGURE 5.4   Apoptosis regulation by p53.

Apoptosis via extrinsic pathways is mediated through interaction of death ligands with extracellular death receptors. Activated death receptor, along with an adaptor protein such as FADD, and procaspase-8 form the death-inducible signaling complex (DISC), which in turn leads to the activation of apoptosis effector protein caspase-3 through caspase-8. Caspase-8 can also lead to the formation of caspase-3 via an indirect mechanism, through cleaving of Bid (tBid). The resulting tBid initiates apoptotic machinery involved in the intrinsic pathway to produce caspase-3 (Figure 5.4).

Apoptosis in response to the severe DNA damage, cytotoxic stress, is initiated by the intrinsic or mitochondrial pathway. Permeation of the outer layer of mitochondria plays a pivotal role in the intrinsic pathway. Intrinsic pathways are primarily regulated by the Bcl-2 protein family, which consists of both anti-apoptotic and pro-apoptotic factors. The anti-apoptotic factors (e.g., Bcl-xL, Bcl-2, and Mcl-1) are responsible for maintaining mitochondrial membrane integrity, whereas pro-apoptotic factors (e.g., Bax, Bak, Noxa, and PUMA) are responsible for increasing mitochondrial membrane permeability. The presence and activation of Bak or Bax is required for the induction of the intrinsic apoptotic pathway (Wei et al., 2001). The activation of Bak and Bax is achieved either by direct interaction with tBid or through neutralization of Bcl-2 family proteins, which keeps check on Bax and Bak. BH3-only proteins like PUMA, Noxa, Bid, and Bad are some of the important factors that are involved in neutralization of anti-apoptotic Bcl-2 proteins (Taylor et al., 2008; Khan and Bisen, 2013). Activated Bax and Bak promote the release of cytochrome c from mitochondria. Once released into the cytoplasm, cytochrome c binds to an adaptor protein Apaf1 to create apoptosome, a complex that activates procaspase-9 (Hockenbery et al., 1990;

Cory and Adams, 2002; Pelengaris and Khan, 2006). Caspase-9 is activated in the presence of nucleotide dATP/ATP, and activated caspase-9 in turn activates caspase-3 and other caspases, leading to caspase cascade, which eventually leads to apoptosis (Hockenbery et al., 1990).

Some other proteins, like inhibitors of IAPs such as DIABLO/Smac and Omi/HtrA2, AIF (apoptosis-inducing factor), and endonuclease G, are released by mitochondria. These proteins can contribute to apoptosis via caspase-independent pathways to cell death. The p53-induced apoptosis typically follows the intrinsic pathway; however, it also modulates apoptosis through the extrinsic pathway. p53 brings out apoptosis mainly through transcriptional regulation of apoptosis effector genes. p53 activates transcription of various pro-apoptotic genes like Noxa, PUMA, Bax, Bad, and Bim. p53 is also involved in transcriptional repression of anti-apoptotic factors like survivin, Bcl-2, and Bcl-xL. p53 also induces apoptosis by transcriptional activation of APAF1, which contributes toward the formation of apoptosome. p53 also activates death receptors (TNFα, Fas, and DR5), thereby promoting the extrinsic pathway. Transcription-dependent apoptosis regulation of p53 is well established; however, the transcription-independent route of p53-induced apoptosis is also gaining strength, which attributes regulatory function to other factors, like direct physical interaction of p53 with apoptotic proteins (Speidel, 2010).

5.7   INTERPLAY BETWEEN SURVIVIN AND P53

Survivin is a key member of the inhibitor of apoptosis (IAP) protein family. Survivin negatively regulates apoptosis by inhibiting caspase activation. Survivin is reported to be overexpressed in many human cancers, including oral cancer (De Maria et al., 2009; Khan et al., 2009; Su et al., 2010).

p53 is reported to repress survivin expression at the transcriptional level. Overexpression of survivin and complete absence of wild-type p53 are signature events of most of the tumors (Mirza et al., 2002). This could be one of the pathways through which p53 induces apoptosis by selectively inhibiting IAPs like survivin; however, the exact mechanism of regulation of survivin by p53 has not been worked out.

5.8   CLINICOPATHOLOGICAL SIGNIFICANCE OF P53

p53 is aberrantly expressed in various cancers, including oral cancer. Mutation in p53 is quite common in various cancers, and it has been reported to be associated with prognosis in oral cancer; however, p53 overexpression cannot be regarded as an independent prognostic factor (Mineta et al., 1998). It has been reported that the single-nucleotide polymorphism (SNP), arginine or proline at codon 72 of the p53 gene, is associated with the risk for development of various cancers (Nenutil et al., 2005). However, this association is still controversial (Zhuo et al., 2009). The SNP at codon 72 in wild-type p53 also influences response toward anticancer agents (Sullivan et al., 2004). The loss of function of p53 mutant proteins may predict a significant low pathological complete response rate and suboptimal response to cisplatin-based neoadjuvant chemotherapy in patients with oral cavity squamous cell carcinoma (SCC)

(Perrone et al., 2010). The credibility of p53 as an independent prognostic factor is still not fully established; however, it has been reported in several studies to be a significant clinical factor when used in combination with other markers, such as cyclin D1 and epidermal growth factor receptor (EGFR) (Shiraki et al., 2005).

5.9 THERAPEUTIC STRATEGIES TARGETING P53

Molecular understanding of p53's role in cancer development has enabled the formulation of strategies for better diagnosis, prognosis, and treatment of cancers. p53 plays a major role in the cellular response to stress, and consequently is a barrier to tumorigenesis. Under cancerous conditions p53 becomes dysfunctional, thereby leading to tumor development. Therefore, the restoration of wt p53's functionality is a potential strategy in anticancer therapy.

MDM2 negatively regulates p53, and its hyperactivity is implicated in inhibition of p53, and thereby promotion of tumor development. It is postulated that inhibiting the E3 activity of MDM2 and blocking the interaction of p53 with MDM2 are effective strategies for killing certain tumor cells selectively by restoring the function of wt p53 (Yang et al., 2004). Blocking of p53/MDM2 interaction by a variety of agents like small-molecule inhibitors, synthetic peptides, and benzodiazepinedione antagonists (Midgley et al., 2000; Chene et al., 2000a, 2000b; Grasberger et al., 2005) has led to p53-dependent apoptosis in cancer cells. The activation of mt p53 by small-molecule activator offers the opportunity to inhibit tumor growth through p53-mediated pathways. Some of the wt p53 activators are RITA (Issaeva et al., 2004), Nutlin-3 (Vassilev et al., 2004), MI-219 (Shangary et al., 2008), BDA (Grasberger et al., 2005), HLI98C (Yang et al., 2005), tenovin-1 (Lain et al., 2008), and JJ78:12 (Staples et al., 2008).

p53 is known to be inactivated in the presence of human papillomavirus (HPV). In cervical carcinomas, p53 is targeted by HPV-encoded E6 protein, which potentiates p53 degradation and inactivates its function in 90% of cervical cancer (Vousden, 1993). Inhibition of E6 offers selective therapeutic potential to reactivate p53, and thereby restore the tumor-silencing activity of p53. E6 expression has been reported to be inhibited by leptomycin B and actinomycin D, consequently stabilizing p53 and inducing apoptosis (Hietanen et al., 2000).

Apoptosis is one of the major routes through which p53 controls/inhibits tumor growth. p53-induced apoptosis is manifested by the host of downstream apoptotic genes; therapeutic strategies focused on the desired modulation of these genes offer attractive potential strategy for anticancer treatment. Moreover, these downstream apoptotic targets of p53 are rarely mutated in case of human cancers (Vogelstein et al., 2000). Some of the p53 apoptotic targets, such as *bax*, PUMA, p53AIP1, Noxa, and others, can potentially be used as targets for gene therapy (Yu and Zhang, 2003). Adenoviral gene transfer of *bax* can act synergistically with chemotherapy to induce apoptosis in tumors (Tai et al., 1999). One of the studies has demonstrated that siRNA targeting of survivin, a negative regulator of apoptosis that is downregulated by p53, could be potentially useful for increasing sensitivity to anticancer drugs, especially in drug-resistant cells with mutated p53 (Yonesaka et al., 2006).

Apart from reactivation of wt p53 in tumors, another strategy is the introduction of wt p53 into tumors (Liu et al., 1994). Delivery of the wt p53 gene by replication of defective adenoviruses in p53 null tumors can directly induce apoptosis and restore sensitivity to chemotherapeutic drugs. Irreversible cell cycle arrest is sufficient to elicit tumor regression after transfer of the p53 gene in p53-deficient tumor cells (Dubrez et al., 2001). Gene therapy based on the introduction of wt p53 has been undergoing clinical trials, and some of the results of the clinical trials have been promising, such as in non-small cell lung cancer and in ovarian cancer (Wen et al., 2003). New forms of p53 that have increased DNA binding to promoters of apoptosis-inducing genes, resistance to degradation, and enhanced thermodynamic stability will likely be more therapeutically active (Nikolova et al., 1998). A strategy that improves the antitumor efficacy using an adenovirus expressing p53 fusion to VP22 protein of herpes simplex virus type 1 has already been developed (Wills et al., 2001).

The restoration of normal functional activity of mt p53 offers yet another therapeutic strategy to control cancer cells. Mutant p53 is unable to perform its function due to the defect in its folding, which is produced by any one of many single amino acid substitutions. Several approaches aimed at reversing this defect and restoring the function of mt p53 haves been tried out during the past few years. One such potential approach is the use of several peptides and small-molecule compounds that can act to stabilize the structure of mt p53, and thus restore the specific DNA binding, transcription, and apoptosis functions to mt p53. They include synthetic peptides derived from the C terminus of p53 (Selivanova et al., 1997), as well as peptides such as CDB3, and compounds isolated from chemical library screening such as CP-31398 and PRIMA-1 (p53 reactivation and induction of massive apoptosis). CDB3 stabilizes the structure of mt p53 proteins (Friedler et al., 2002), and it binds mt p53 and efficiently induces the refolding of two hot spot p53 mutants, His273 and His175, in cancer cells. The transactivation activity of p53 can also be rescued by CDB3 (Issaeva et al., 2003). PRIMA-1 selectively inhibits the growth of tumor cells by provoking apoptosis in a transcription-dependent fashion through conformational manipulation of p53 mutants to restore sequence-specific DNA binding (Bykov et al., 2002). CP-31398 is a small synthetic molecule with the capacity to restore wt p53 function to mutants (Foster et al., 1999, Wang et al., 2003). It has been suggested that it triggers apoptosis of human cancer cells through the intrinsic Bax/mitochondrial/caspase-9 pathway (Luu et al., 2002) and can stabilize wt p53 protein (Takimoto et al., 2002). Some of the other small-molecule reactivators of mt p53 are MIRA-1 (Bykov et al., 2005), ellipiticine (Xu et al., 2008), P53R3 (Weinmann et al., 2008), and WR-1065 (North et al., 2002).

5.10 CONCLUSION

The protective role of p53 against cancer development is now well established. It is one of the most frequently dysregulated biomolecules in various cancers, including oral cancer. It regulates myriad functions critically linked to normal cellular development. p53 controls tumor development through multiple pathways involved in apoptosis and cell division. It controls a host of factors (genes/

proteins) through transactivation activity and most probably also through physical interaction. This versatility makes p53 a very important candidate for therapeutic intervention in various cancers, including oral cancer. We have discussed some of the exciting strategies in this chapter, where efforts focused around therapeutic and diagnostic applications of p53 are in various stages of clinical use. Looking at the importance of p53 in cancer biology, we can expect that many more exciting hypotheses around p53 will be successfully translated into its clinical use.

The rate of enrichment of knowledge is an ever-increasing trend, which can be attributed to the advancements in the field of technology, which allows the researcher to validate a putative hypothesis unambiguously with relatively less effort and time. The model of translational research has gained widespread acceptance and encouragement, which essentially will expedite the process of realization of inventions in bench studies to be put in actual clinical use. Such translational research focused around p53 can definitely help us in better management of oral cancer.

REFERENCES

Baker, S.J., E.R. Fearon, J. Nigro, et al. 1989. Chromosome 17 deletions and p53 gene mutations in colorectal carcinomas. *Science* 244: 217–221.

Bykov, V.J., N. Issaeva, A. Shilov, et al. 2002. Restoration of the tumor suppressor function to mutant p53 by a low-molecular weight compound. *Nat Med* 8: 282–288.

Bykov, V.J., N. Issaeva, N. Zache, et al. 2005. Reactivation of mutant p53 and induction of apoptosis in human tumor cells by maleimide analogs. *J Biol Chem* 280: 30384–30391.

Caron de Fromentel, C., F. May-Levin, H. Mouriesse, et al. 1987. Presence of circulating antibodies against cellular protein p53 in a notable proportion of children with B-cell lymphoma. *Int J Cancer* 39: 185–189.

Chang, C., D.T. Simmons, M.A. Martin, et al. 1979. Identification and partial characterization of new antigens from simian virus 40-transformed mouse cells. *J Virol* 31: 463–471.

Chen, X., L.J. Ko, L. Jayaraman, et al. 1996. p53 levels, functional domains, and DNA damage determine the extent of the apoptotic response of tumor cells. *Genes Dev* 10: 2438–2451.

Chene, C.A., J.M. Desterro, M.K. Saville, et al. 2000a. An N-terminal p14ARF peptide blocks Mdm2-dependent ubiquitination in vitro and can activate p53 in vivo. *Oncogene* 19: 2312–2323.

Chene, P., J. Fuchs, J. Bohn, et al. 2000b. A small synthetic peptide, which inhibits the p53-hdm2 interaction, stimulates the p53 pathway in tumour cell lines. *J Mol Biol* 299: 245–253.

Cho, Y., S. Gorina, P.D. Jeffrey, et al. 1994. Crystal structure of a p53 tumor suppressor-DNA complex: Understanding tumorigenic mutations. *Science* 265: 346–355.

Clayman, G.L., A.K. el-Naggar, S.M. Lippman, et al. 1998. Adenovirus-mediated p53 gene transfer in patients with advanced recurrent head and neck squamous cell carcinoma. *J Clin Oncol* 16: 2221–2232.

Cory, S., and J.M. Adams. 2002. The Bcl2 family: Regulators of the cellular life-or-death switch. *Nat Rev Cancer* 2: 647–656.

Crawford L.V., D.C. Pim, and R.D. Bulbrook. 1982. Detection of antibodies against the cellular protein p53 in sera from patients with breast cancer. *Int J Cancer* 30: 403–408.

De Leo, A.B., G. Jay, E. Appella, et al. 1979. Detection of a transformation-related antigen in chemically induced sarcomas and other transformed cells of the mouse. *Proc Natl Acad Sci USA* 76: 2420–2424.

De Maria, S., G. Pannone, P. Bufo, et al. 2009. Survivin gene-expression and splicing isoforms in oral squamous cell carcinoma. *J Cancer Res Clin Oncol* 135: 107–116.

Deppert, W., G. Buschhausendenker, T. Patschinsky, et al. 1990. Cell cycle control of p53 in normal (3T3) and chemically transformed (Meth-A) mouse cells. 2. Requirement for cell cycle progression. *Oncogene* 5: 1701–1706.

Dubrez, L., J.L. Coll, A. Hurbin, et al. 2001. Cell cycle arrest is sufficient for p53-mediated tumor regression. *Gene Ther* 8: 1705–1712.

El-Deiry, W.S., S.E. Kern, J.A. Pietenpol, et al. 1992. Definition of a consensus binding site for p53. *Nat Genet* 1: 45–49.

Eliyahu, D., D. Michalovitz, S. Eliyahu, et al. 1989. Wild-type p53 can inhibit oncogene-mediated focus formation. *Proc Natl Acad Sci USA* 86: 8763–8767.

Fields, S., and S.K. Jang. 1990. Presence of a potent transcription activating sequence in the p53 protein. *Science* 249: 1046–1049.

Finlay, C.A., P.W. Hinds, and A.J. Levine. 1989. The p53 proto-oncogene can act as a suppressor of transformation. *Cell* 57: 1083–1093.

Foster, B.A., H.A. Coffey, M.J. Morin, et al. 1999. Pharmacological rescue of mutant p53 conformation and function. *Science* 286: 2507–2510.

Friedler, A., L.O. Hansson, D.B. Veprintsev, et al. 2002. A peptide that binds and stabilizes p53 core domain: Chaperone strategy for rescue of oncogenic mutants. *Proc Natl Acad Sci USA* 99: 937–942.

Grasberger, B.L., T. Lu, C. Schubert, et al. 2005. Discovery and cocrystal structure of benzodiazepinedione HDM2 antagonists that activate p53 in cells. *J Med Chem* 48: 909–912.

Green, D.R. 2005. Apoptotic pathways: Ten minutes to dead. *Cell* 121: 671–674.

Green, D.R., and G. Kroemer. 2004. The pathophysiology of mitochondrial cell death. *Science* 305: 626–629.

Halevy, O., D. Machalovitz, and M. Oren. 1990. Different tumor-derived p53 mutants exhibit distinct biological activities. *Science* 250: 113–116.

Hietanen, S., S. Lain, E. Krausz, et al. 2000. Activation of p53 in cervical carcinoma cells by small molecules. *Proc Natl Acad Sci USA* 97: 8501–8506.

Hockenbery, D., G. Nuñez, C. Milliman, et al. 1990. Bcl-2 is an inner mitochondrial membrane protein that blocks programmed cell death. *Nature (London)* 348: 334–336.

Irwin, M.S., and W.G. Kaelin. 2001. p53 family update: p73 and p63 develop their own identities. *Cell Growth Differ* 12: 337–349.

Issaeva, N., P. Bozko, M. Enge, et al. 2004. Small molecule RITA binds to p53, blocks p53-HDM-2 interaction and activates p53 function in tumors. *Nat Med* 10: 1321–1328.

Issaeva, N., A. Friedler, P. Bozko, et al. 2003. Rescue of mutants of the tumor suppressor p53 in cancer cells by a designed peptide. *Proc Natl Acad Sci USA* 100: 13303–13307.

Kaghad, M., H. Bonnet, A. Yang, et al. 1997. Monoallelically expressed gene related to p53 at 1p36, a region frequently deleted in neuroblastoma and other human cancers. *Cell* 90: 809–819.

Khan, Z., R.P. Tiwari, R. Mulherkar, et al. 2009. Detection of survivin and p53 in human oral cancer: Correlation with clinicopathologic findings. *Head Neck* 31: 1039–1048.

Khan, Z., and Bisen, P.S. 2013. Oncoapoptotic signaling and deregulated target genes in cancers: Special reference to oral cancer. *Biochim Biophy Acta Reviews on Cancer*, dx.doi.org/10.1016/j.bbcan.2013.04.002.

Kress, M., E. May, R. Cassingena, et al. 1979. Simian virus 40-transformed cells express new species of proteins precipitable by anti-simian virus 40 serum. *J Virol* 31: 472–483.

Lain, S., J.J. Hollick, J. Campbell, et al. 2008. Discovery, in vivo activity, and mechanism of action of a small-molecule p53 activator. *Cancer Cell* 13: 454–463.

Lane, D.P. 1992. Cancer. P53, guardian of the genome. *Nature (London)* 358: 15–16.

Lane, D.P., and S. Benchimol. 1990. p53: Oncogenes and anti-oncogene. *Gene Dev* 4: 1–8.

Lane, D.P., and L.V. Crawford. 1979. T antigen is bound to a host protein in SV40-transformed cells. *Nature (London)* 278: 261–263.

Lavigueur, A., V. Maltby, D. Mock, et al. 1989. High incidence of lung, bone, and lymphoid tumors in transgenic mice overexpressing mutant alleles of the p53 oncogene. *Mol Cell Biol* 9: 3982–3991.

Levine, A.J., W. Hu, and Z. Feng. 2006. The P53 pathway: What questions remain to be explored? *Cell Death Differ* 13: 1027–1036.

Lin, J., J. Chen, B. Elenbaas, et al. 1994. Several hydrophobic amino acids in the p53 amino-terminal domain are required for transcriptional activation, binding to mdm-2 and the adenovirus 5 E1B 55-kD protein. *Genes Dev* 8: 1235–1246.

Linzer, D.I.H., and A.J. Levine. 1979. Characterization of a 54 K dalton cellular SV40 tumor antigen present in SV40-transformed cells and in infected embryonal carcinoma cells. *Cell* 1: 43–52.

Liu, T.J., W.W. Zhang, D.L. Taylor, et al. 1994. Growth suppression of human head and neck cancer cells by the introduction of a wild-type p53 gene via a recombinant adenovirus. *Cancer Res* 54: 3662–3667.

Luu, Y., J. Bush, K.J. Cheung Jr., et al. 2002. The p53 stabilizing compound CP-31398 induces apoptosis by activating the intrinsic Bax/mitochondrial/caspase-9 pathway. *Exp Cell Res* 276: 214–222.

Malkin D., F.P. Li, L.C. Strong, et al. 1990. Germ line p53 mutations in a familial syndrome of breast cancer, sarcomas, and other neoplasms. *Science* 250: 1233–1238.

Melero, J.A., D.T. Stitt, W.F. Mangel, et al. 1979. Identification of new polypeptide species (48–55K) immunoprecipitable by antiserum to purified large T antigen and present in simian virus 40-infected and transformed cells. *J Virol* 93: 466–480.

Melino, G., V. De Laurenzi, and K.H. Vousden. 2002. p73: Friend or foe in tumerogenesis. *Nat Rev Cancer* 2: 605–615.

Mercer, W.E., C. Avignolo, and R. Baserga. 1984. Role of the p53 protein in cell proliferation as studied by microinjection of monoclonal antibodies. *Mol Cell Biol* 4: 276–281.

Mercer, W.E., D. Nelson, A.B. DeLeo, et al. 1982. Microinjection of monoclonal antibody to protein p53 inhibits serum-induced DNA synthesis in 3T3 cells. *Proc Natl Acad Sci USA* 79: 6309–6312.

Michalovitz, D., O. Halevy, and M. Oren. 1991. p53 mutations—Gains or losses. *J Cell Biochem* 45: 22–29.

Midgley, C.A., J.M. Desterro, and M. Saville. 2000. An N-terminal p14ARF peptide blocks Mdm2-dependent ubiquitination in vitro and can activate p53 in vivo. *Oncogene* 19: 2312–2313.

Milner, J., and E.A. Medcalf. 1991. Cotranslation of activated mutant p53 with wild type drives the wild-type p53 protein into the mutant conformation. *Cell* 65: 765–774.

Mineta, H, A. Borg, M. Dictor, et al. 1998. p53 mutation, but not p53 overexpression, correlates with survival in head and neck squamous cell carcinoma. *Br J Cancer* 78: 1084–1090.

Mirza, A., M. McGuirk, T.N. Hockenberry, et al. 2002. Human survivin is negatively regulated by wild-type p53 and participates in p53-dependent apoptotic pathway. *Oncogene* 21: 2613–2622.

Moll, U.M., and L.M. Schramm. 1998. p53—An acrobat in tumorigenesis. *Crit Rev Oral Biol Med* 9: 23–37.

Mowat, M., A. Cheng, N. Kimura, et al. 1985. Rearrangements of the cellular p53 gene in erythroleukaemic cells transformed by Friend virus. *Nature (London)* 314: 633–636.

Munroe, D.G., B. Rovinski, A. Bernstein, et al. 1988. Loss of highly conserved domain on p53 as a result of gene deletion during friend virus-induced erythroleukemia. *Oncogene* 2: 621–624.

Nenutil, R., J. Smardova, S. Pavlova, et al. 2005. Discriminating functional and non-functional p53 in human tumours by p53 and MDM2 immunohistochemistry. *J Pathol* 207: 251–259.

Nikolova, P.V., J. Henckel, D.P. Lane, et al. 1998. Semirational design of active tumor suppressor p53 DNA binding domain with enhanced stability. *Proc Natl Acad Sci USA* 95: 14675–14680.

North, S., O. Pluquet, D. Maurici, et al. 2002. Restoration of wild-type conformation and activity of a temperature-sensitive mutant of p53 (p53(V272M)) by the cytoprotective aminothiol WR1065 in the esophageal cancer cell line TE-1. *Mol Carcinog* 33: 181–188.

Pelengaris, S., and M. Khan. 2006. In *The molecular biology of cancer*, ed. S. Pelengaris and M. Khan, 251–278. Blackwell, Boston.

Perrone, F., P. Bossi, B. Cortelazzi, et al. 2010. TP53 mutations and pathologic complete response to neoadjuvant cisplatin and fluorouracil chemotherapy in resected oral cavity squamous cell carcinoma. *J Clin Oncol* 28: 761–766.

Reich, N.C., and A.J. Levine. 1984. Growth regulation of a cellular tumour antigen, p53, in non transformed cells. *Nature (London)* 308: 199–201.

Sakamuro, D., P. Sabbatini, E. White, et al. 1997. The polyproline region of p53 is required to activate apoptosis but not growth arrest. *Oncogene* 15: 887–898.

Schmale, H., and C. Bamberger. 1997. A novel protein with strong homology to the tumor suppressor p53. *Oncogene* 15: 1363–1367.

Selivanova G., V. Iotsova, I. Okan, et al. 1997. Restoration of the growth suppression function of mutant p53 by a synthetic peptide derived from the p53 C-terminal domain. *Nat Med* 3: 632–638.

Shangary, S., D. Qin, D. McEachern, et al. 2008. Temporal activation of p53 by a specific MDM2 inhibitor is selectively toxic to tumors and leads to complete tumor growth inhibition. *Proc Natl Acad Sci USA* 105: 3933–3938.

Sharpless, N.E., and R.A. DePinho. 2002. p53: Good cop/bad cop. *Cell* 110: 9–12.

Shiraki, M., T. Odajima, T. Ikeda, et al. 2005. Combined expression of p53, cyclin D1 and epidermal growth factor receptor improves estimation of prognosis in curatively resected oral cancer. *Mod Pathol* 18: 1482–1489.

Soussi, T., and C. Béroud. 2001. Assessing TP53 status in human tumours to evaluate clinical outcome. *Nat Rev Cancer* 1: 233–240.

Speidel, D. 2010. Transcription-independent p53 apoptosis: An alternative route to death. *Trends Cell Biol* 20: 14–24.

Srivastava, S., Z.Q. Zou, K. Pirollo, et al. 1990. Germ-line transmission of a mutated p53 gene in a cancer-prone family with Li-Fraumeni syndrome. *Nature (London)* 348: 747–749.

Staples, O.D., J.J. Hollick, J. Campbell, et al. 2008. Characterization, chemical optimization and anti-tumour activity of a tubulin poison identified by a p53-based phenotypic screen. *Cell Cycle* 7: 3417–3427.

Steinmeyer, K., H. Maacke, and W. Deppert. 1990. Cell cycle control by p53 in normal (3T3) and chemically transformed (Meth A) mouse cells. I. Regulation of p53 expression. *Oncogene* 5: 1691–1699.

Su, L., Y. Wang, M. Xiao, et al. 2010. Up-regulation of survivin in oral squamous cell carcinoma correlates with poor prognosis and chemoresistance. *Oral Surg Oral Med Oral Pathol Oral Radiol Endod* 110: 484–491.

Sullivan, A., N. Syed, M. Gasco, et al. 2004. Polymorphism in wild-type p53 modulates response to chemotherapy in vitro and in vivo. *Oncogene* 23: 3328–3337.

Tai, Y.T., T. Strobel, D. Kufe, et al. 1999. In vivo cytotoxicity of ovarian cancer cells through tumor-selective expression of the BAX gene. *Cancer Res* 59: 2121–2126.

Takahashi, T., M.M. Nau, I. Chiba, et al. 1989. p53—A frequent target for genetic abnormalities in lung cancer. *Science* 246: 491–494.

Takimoto, R., W. Wang, D.T. Dicker, et al. 2002. The mutant p53-conformation modifying drug, CP- 31398, can induce apoptosis of human cancer cells and can stabilize wild-type p53 protein. *Cancer Biol Ther* 1: 47–55.

Taylor, R.C., B.E. Klein, J. Stein, et al. 2008. Apoptosis: Controlled demolition at the cellular level. *Nat Rev Mol Cell Biol* 9: 231–241.

Tokino, T., and Y. Nakamura. 2000. The role of p53-target genes in human cancer. *Crit Rev Oncol Hematol* 33: 1–6.

Vassilev, L.T., B.T. Vu, B. Graves, et al. 2004. In vivo activation of the p53 pathway by small-molecule antagonists of MDM2. *Science* 303: 844–848.

Vogelstein, B., E.R. Fearon, S.R. Hamilton, et al. 1988. Genetic alterations during colorectal-tumor development. *N Engl J Med* 319: 525–532.

Vogelstein, B., D. Lane, and A.J. Levine. 2000. Surfing the p53 network. *Nature (London)* 408: 307–310.

Vousden, K. 1993. Interactions of human papillomavirus transforming proteins with the products of tumor suppressor genes. *FASEB J* 7: 872–879.

Vousden, K.H. 2000. p53: Death star. *Cell* 103: 691–694.

Vousden, K.H., and X. Lu. 2002. Live or let die: The cell's response to p53. *Nat Rev Cancer* 2: 594–604.

Wang, W., R. Takimoto, F. Rastinejad, et al. 2003. Stabilization of p53 by CP-31398 inhibits ubiquitination without altering phosphorylation at serine 15 or 20 or MDM2 binding. *Mol Cell Biol* 23: 2171–2181.

Wei, M.C., W.X. Zong, E.H. Cheng, et al. 2001. Proapoptotic BAX and BAK: A requisite gateway to mitochondrial dysfunction and death. *Science* 292: 727–730.

Weinmann, L., J. Wischhusen, M.J. Demma, et al. 2008. A novel p53 rescue compound induces p53-dependent growth arrest and sensitises glioma cells to Apo2L/TRAIL-induced apoptosis. *Cell Death Differ* 15: 718–729.

Wen, S.F., V. Mahavni, E. Quijano, et al. 2003. Assessment of p53 gene transfer and biological activities in a clinical study of adenovirus-p53 gene therapy for recurrent ovarian cancer. *Cancer Gene Ther* 10: 224–238.

Wills, K.N., I.A. Atencio, J.B. Avanzini, et al. 2001. Intratumoral spread and increased efficacy of a p53-VP22 fusion protein expressed by a recombinant adenovirus. *J Virol* 75: 8733–8741.

Xu, G.W., I.A. Mawji, C.J. Macrae, et al. 2008. A high-content chemical screen identifies ellipticine as a modulator of p53 nuclear localization. *Apoptosis* 13: 413–422.

Yang, Y., C.C. Li, and A.M. Weissman. 2004. Regulating the p53 system through ubiquitination. *Oncogene* 23: 2096–2106.

Yang, Y., R.L. Ludwig, J.P. Jensen, et al. 2005. Small molecule inhibitors of HDM2 ubiquitin ligase activity stabilize and activate p53 in cells. *Cancer Cell* 7: 547–559.

Yonesaka, K., K. Tamura, T. Kurata, et al. 2006. Small interfering RNA targeting survivin sensitizes lung cancer cell with mutant p53 to adriamycin. *Int J Cancer* 118: 812–820.

Yu, J., and L. Zhang. 2003. No PUMA, no death: Implications for p53 dependent apoptosis. *Cancer Cell* 4: 248–249.

Zeimer, M.A., A. Mason, and D.M. Carlson. 1982. Cell-free translations of proline-rich protein mRNAs. *J Biol Chem* 257: 11176–11780.

Zhuo, X.L., Q. Li, Y. Zhou, et al. 2009. Study on TP53 codon 72 polymorphisms with oral carcinoma susceptibility. *Arch Med Res* 40: 625–634.

6 Diagnostic and Therapeutic Potential of Apoptotic Marker

KEY WORDS

Apoptosis
Gene silencing
Gene therapy
Inhibitor of apoptosis protein (IAP)
Oncolytic virus

6.1 INTRODUCTION

Apoptosis is an evolutionary conserved process that maintains a subtle balance between progression and cell death. Apoptosis is tightly regulated by various molecular factors in normal cells. Dysregulation of apoptosis is associated with diseases like liver failure, stroke, heart attack, cancer, etc. Cancer cells survive and thrive in the body by adapting various survival mechanisms, by which they make the cellular environment conducive for their growth. Evasion of apoptosis is one of the common hallmarks of human cancers (Hanahan and Weinberg, 2000). Killing of cancerous cells is the endpoint of various therapeutic interventions for cancer, and such cell killing invariably converges into the apoptotic pathway. Therefore, dysregulation of apoptosis is also responsible for chemoresistance.

Apoptosis is initiated by a number of factors that primarily act through two apoptotic pathways: (1) extrinsic and (2) intrinsic apoptotic pathways (Green and Kroemer, 2004; Green 2005). Apoptosis via extrinsic pathways is mediated through interaction of death ligands with extracellular death receptors. Activated death receptor, along with an adaptor protein such as FADD, and procaspase-8 form the death-inducible signaling complex (DISC), which in turn leads to the activation of apoptosis effector protein caspase-3 through caspase-8. Caspase-8 can also lead to the formation of caspase-3 via an indirect mechanism, through the cleaving of Bid (tBid). The resulting tBid initiates or activates apoptotic machinery involved in the intrinsic pathway, to produce caspase-3. Apoptosis in response to severe DNA damage, cytotoxic stress, is initiated by the intrinsic or mitochondrial pathway. Permeation of the outer layer of mitochondria plays a pivotal role in the intrinsic pathway. Intrinsic pathways are primarily regulated by the Bcl-2 protein family, which consists of both anti-apoptotic and pro-apoptotic factors. The anti-apoptotic factors (e.g., Bcl-xL, Bcl-2, and Mcl-1) are responsible for maintaining mitochondrial membrane integrity,

163

whereas pro-apoptotic factors (e.g., Bax, Bak, Noxa, and PUMA) are responsible for increasing mitochondrial membrane permeability. The presence and activation of Bak or Bax are required for the induction of the intrinsic apoptotic pathway (Wei et al., 2001). Activation of Bak and Bax is achieved either by direct interaction with tBid or through neutralization of Bcl-2 family proteins, which keep check on Bax and Bak. BH3-only proteins like PUMA, Noxa, Bid, and Bad are some of the important factors that are involved in the neutralization of anti-apoptotic Bcl-2 proteins (Taylor et al., 2008). Activated Bax and Bak promote the release of cytochrome c from the mitochondria. Once released into the cytoplasm cytochrome c binds to an adaptor protein Apaf1 to create apoptosome, a complex that activates procaspase-9 (Hockenbery et al., 1990; Cory and Adams, 2002; Pelengaris and Khan, 2006). Caspase-9 is activated in the presence of nucleotide dATP/ATP, and activated caspase-9 in turn activates caspase-3 and other caspases, leading to a caspase cascade, which eventually leads to apoptosis (Hockenbery et al., 1990). Some other proteins, like inhibitors of IAPs, such as DIABLO/Smac and Omi/HtrA2, AIF (apoptosis-inducing factor), and endonuclease G are released by mitochondria. These proteins can contribute to apoptosis via caspase-independent pathways to cell death. p53 controls the apoptosis pathway through transcriptional regulation of apoptosis effector genes (Khan and Bisen, 2013).

The above-mentioned key players of apoptosis can be exploited for therapeutic or diagnostic application in oral cancer. Therapeutic strategies by leveraging apoptotic machinery are focused around promoting apoptosis through activation of the death receptors, pro-apoptotic factors (caspases and p53), and inhibition of anti-apoptotic factors like IAPs and Bcl-2. Dysregulation of factors of apoptosis pathways can be used as diagnostic or prognostic markers for oral cancer.

6.2 APOPTOTIC MARKERS IN ORAL CANCER THERAPEUTICS

6.2.1 GENE THERAPY

Gene therapy essentially involves the transfer of genetic material into targeted cells, and manipulation of the existing genetic material. Owing to the advancements in science, we have a better mechanistic understanding of the events associated with the development of oral cancer. This understanding of disease etiology sets the premise for genetic manipulation of cancer cells, with an assumption that "the normal physiological state can be restored when the gene exerts its normal function."

The success of gene therapy depends on the following major factors:

- Clear understanding of the genetic basis of genesis and progression of disease
- Manageable side effects
- Accessibility of target organ/tissue for delivering therapeutic material
- In vivo stability of therapeutic material to produce the desired effect

Oral squamous cell carcinoma (OSCC) is a good candidate for gene therapy, because of easy accessibility of primary and secondary lesions, for delivery of therapeutic material through injection or topical administration (Xi and Grandis, 2003).

The gene therapy procedure typically involves the following steps:

1. Identification of the target gene, which is involved in disease and qualifies to be the candidate for gene therapy
2. Isolation and amplification of the candidate gene
3. Cloning of the candidate gene on a specific vector
4. Delivery of candidate genes through one of the following processes:
    a. Ex vivo:
        i. Extraction and in vitro culture of tissue from the patient
        ii. Introduction/transfer of amplified candidate genes into cultured cells via a vector
        iii. Screening of cultured cells to check successful integration of candidate genes and check for any possible contamination
        iv. Selected genes containing cells are transferred back into the patient
    b. In vivo: The candidate gene cloned with the vector is directly injected into the target tissue of the patient.

6.2.2 Gene Therapy Strategies to Induce Apoptosis in Oral Cancer Cells

6.2.2.1 Gene Addition Therapy

Gene addition therapy is carried on with an objective of controlling tumorigenesis through the introduction of tumor suppressor genes like p53. p53 is known to be aberrantly expressed in various cancers, including oral cancer. Alteration of p53 is known to be one of the earliest events in the progressive development of oral cancer from precancerous lesions (Lopez-Martinez et al., 2002). The feasibility and efficacy of adenoviral-mediated p53 (Ad-p53) gene transfer was found to be promising in the treatment of advanced head and neck cancer (Liu et al., 1994; Clayman et al., 1998). These results formed the basis of human trials for designing a clinical therapeutic regime of adenoviral-mediated p53 gene therapy (ClinicalTrials.gov number NCT00064103).

6.2.2.2 Oncolytic Viruses

It has been observed that cancer patients undergo regression when infected with viruses; such observations have led to the foundation of one of the most promising applications of a virus in cancer therapy. The advancements in genetic manipulation techniques have made it possible to design viruses that can selectively target and kill tumor cells.

The replication of a virus is critically dependent on E1B viral protein, and the absence of the functional p53 protein. The viruses without E1B protein are unable to replicate in the presence of the functional p53 pathway in the host cell. The p53 pathway becomes dysfunctional in oral cancer cells; this fact is leveraged in the treatment strategy of utilizing oncolytic virus ONYX-015, which lacks the viral E1B protein (Rudin et al., 2003). In the absence of the functional p53 pathway in tumor cells, ONYX-015 replicates in tumor cells and thereby selectively lyses tumor cells. The efficacy of ONYX-015 in cancer treatment can be further enhanced in combination with chemotherapy (Heise and Kirn, 2000; Nemunaitis et al., 2003).

6.2.2.3 Suicide Gene Therapy

The objective of suicide gene therapy is to make a modulation of the cellular environment of cancer cells, by transforming nontoxic pro-drugs into cytotoxic compounds (Neves et al., 2006). The thymidine kinase gene of herpes simplex virus (HSVtk) converts nontoxic ganciclovir into toxic ganciclovir phosphate, which increases apoptosis of cancer cells. Ganciclovir is phosphorylated by HSVtk, which terminates DNA synthesis and thereby induces apoptosis.

6.2.2.4 Antisense RNA

Apoptosis in cancer cells can be promoted by selectively inhibiting survival factors like Bcl-2, IAPs like XIAP, and survivin. Antisense technology offers a potential approach for such sequence-specific selective silencing of the genes involved in the survival mechanism. Survivin is a well-established endogenous apoptosis inhibitor that is overexpressed in oral cancer (Su et al., 2010). siRNA targeting survivin in the oral cancer cell line was observed to be effective in inhibiting growth of oral cancer cells, through induction of apoptosis (Li and Xiao-Hong, 2011).

6.2.3 TARGETED THERAPY

Chemotherapy indiscriminately kills normal proliferating cells along with its intended cancer cells. Therefore, such chemotherapeutic treatments are associated with unavoidable toxicities. Owing to scientific advancements made in the last couple of decades, now we are in a better position as far as molecular understanding of oral carcinogenesis is concerned. The detailed study and understanding of molecular events of oral carcinogenesis has gifted us with molecular targets, which could form the base for therapeutic intervention.

In comparison to other cancer-related malignancies, like breast and lung cancer, oral cancer has very few targeted therapies to date. Targeted therapies for oral cancer are available for a couple of molecular targets like EGFR (Cetuximab) (Bonner et al., 2010), CD44 (GSK1120212) (Judd et al., 2012), PPARgamma (Pioglitazone) (ClinicalTrials.gov number NCT00951379), and EpCAM (Proxinium) (ClinicalTrials.gov number NCT00272181). There is a host of unexplored potential therapeutic targets involved in various other molecular events in oral carcinogenesis, like cell growth, angiogenesis, apoptosis, metastasis, and inflammation. We have attempted to briefly discuss potential molecular targets involved in apoptosis, which could be exploited for therapeutic intervention in the future.

6.2.3.1 Targets of Extrinsic Pathway

Apoptosis through the extrinsic pathway is mediated through activation of the death receptor by death ligands like tumor necrosis factor-α (TNFα), CD95L, and TNF-related apoptosis-inducing ligand (TRAIL). Esculetin was reported to enhance TRAIL mediated apoptosis in the oral cancer cell line (Kok et al., 2009). Monoclonal antibodies (HGS-ETR1, HGS-ETR2, HGS-TR2J, and TRA-8) with agonist function at death receptors DR4 (TRAIL-R1) and DR5 (TRAIL-R2) have also been used to induce apoptosis; however, they have not been tested for oral cancer treatment.

6.2.3.2 Targets of Intrinsic Pathway

The proteins of the Bcl-2 family form the core component of the intrinsic apoptotic pathway. The anti-apoptotic factors (e.g., Bcl-xL, Bcl-2, and Mcl-1) are responsible for maintaining mitochondrial membrane integrity, whereas pro-apoptotic factors (e.g., Bax, Bak, Noxa, and PUMA) are responsible for increasing mitochondrial membrane permeability. BH3-only proteins like PUMA, Noxa, Bid, and Bad are some of the important factors that are involved in neutralization of anti-apoptotic Bcl-2 proteins (Taylor et al., 2008). Some of the strategies used in inducing apoptosis through Bcl-2 family proteins are: (1) silencing or downregulation of pro-apoptotic proteins by antisense techniques; (2) release of pro-apoptotic protein from the Bcl-2 complex, through application of the BH3 domain peptide; and (3) release of pro-apoptotic protein from the Bcl-2 complex, through application of synthetic small-molecule drugs like tetrocarcin A (TC-A) (Nakashima et al., 2000; Tinhofer et al., 2002), antimycin A_3 (Tzung et al., 2001), and chelerythrine (Chan et al., 2003).

Caspases are key initiators (caspase-8, -9, and -10) and executioners (caspases-3, 4, and -7) of apoptosis. Many groups have attempted to find the caspase activator based on high-throughput screening (HTS) studies. In one of such screening study, dichlorobenzyl carbamates and indolones were detected as strong caspase activators (Nguyen and Wells, 2003). Maxim Pharmaceuticals' MX-2060 series of caspase-activating compounds is reported to have caspase-activating attribute (Zhang et al., 2004).

The inhibitors of apoptosis proteins (IAPs) are a class of regulatory proteins involved in caspase-inhibiting activity. IAPs are characterized by a conserved sequence domain, known as the baculoviral IAP repeat (BIR) domain. Survivin and X chromosome-linked inhibitor of apoptosis protein (XIAP) are prominent members of the IAP family of proteins. Survivin is single BIR domain containing IAP, which negatively regulates apoptosis by inhibiting caspase activation. Survivin is reported to be overexpressed in many human cancers, including oral cancer (De Maria et al., 2009; Khan et al., 2009; Su et al., 2010).

The IAPs inhibit caspases by masking substrate binding sites of caspase through interaction with their BIR domain. XIAP contains three BIR domains, which are involved in inhibiting caspases-3, -7, and -9. Cisplatin induces apoptosis in oral cancer cells through inhibition of XIAP (Matsumiya et al., 2001). Caspases-3 and -7 are inhibited through binding with BIR-1, and the BIR-2 domain of XIAP, whereas domain BIR-3 is exclusively involved in binding with caspase-9 (Huang et al., 2001; Riedl et al., 2001; Deveraux et al., 1999; Sun et al., 2000). Caspase-9 contains the IAP binding motif (IBM) through which it binds with the BIR-3 domain of XIAP. The domain homologous to IBM is found in mitochondrial proteins Smac and Omi/HtrA2. Smac interacts with XIAP and releases caspase-9 for inducing apoptosis. Smac mimetic drugs hold potential in cancer treatment.

6.2.3.3 Targeting p53

Under cancerous conditions p53 becomes dysfunctional, which promotes tumor development; restoration of wt-p53 is a potential strategy in anticancer therapy.

MDM2 is negatively regulated by p53; its hyperactivity is implicated in inhibition of p53, and thereby promotion of tumor development. Blocking of p53/MDM2 interaction by a variety of agents like small-molecule inhibitors, synthetic peptides, and benzodiazepinedione antagonists (Midgley et al., 2000; Chene et a l., 2000; Grasberger et al., 2005) has led to p53-dependent apoptosis in cancer cells. Activation of mt p53 by a small-molecule activator offers an opportunity to inhibit tumor growth through p53-mediated pathways. Some of the wt p53 activators are RITA (Issaeva et al., 2004), Nutlin-3 (Vassilev et al., 2004), MI-219 (Shangary et al., 2008), BDA (Grasberger et al., 2005), HLI98C (Yang et al., 2005), tenovin-1 (Lain et al., 2008), and JJ78:12 (Staples et al., 2008).

Restoration of normal functional activity of mt p53 offers yet another therapeutic strategy to control cancer cells. Mutant p53 is unable to perform its function due to the defect in its folding, which is produced by any one of many single amino acid substitutions. Several approaches aimed at reversing this defect and restoring the function of mt p53 have been tried during the past few years. One such potential approach is the use of several peptides and small-molecule compounds that can act to stabilize the structure of mt p53, and thus restore the specific DNA binding, transcription, and apoptosis functions of the mt p53. They include synthetic peptides derived from the C-terminus of p53 (Selivanova et al., 1997), as well as peptides such as CDB3, and compounds isolated from chemical library screening such as CP-31398 and PRIMA-1 (p53 reactivation and induction of massive apoptosis). CDB3 stabilizes the structure of mt p53 proteins (Friedler et al., 2002), and it binds mt p53 and efficiently induces the refolding of two hot spot p53 mutants, His273 and His175, in cancer cells. The transactivation activity of p53 can also be rescued by CDB3 (Issaeva et al., 2003). PRIMA-1 selectively inhibits the growth of tumor cells by provoking apoptosis in a transcription-dependent fashion through conformational manipulation of p53 mutants to restore sequence-specific DNA binding (Bykov et al., 2002). CP-31398 is a small synthetic molecule (Foster et al., 1999) with the capacity to restore wt p53 function to mutants (Wang et al., 2003). It has been suggested that it triggers apoptosis of human cancer cells through the intrinsic Bax/mitochondrial/caspase-9 pathway (Luu et al., 2002) and can stabilize wt p53 protein (Takimoto et al., 2002). Some of the other small-molecule reactivators of mt p53 are MIRA-1 (Bykov et al., 2005), ellipiticine (Xu et al., 2008), P53R3 (Weinmann et al., 2008), and WR-1065 (North et al., 2002).

6.3   APOPTOTIC MARKERS IN ORAL CANCER DIAGNOSTICS

The members of Bcl-2 family proteins carry high prognostic significance in oral cancer. Bcl-2, Bax, and the Bcl-2/Bax ratio can be used as effective biomarkers to predict prognosis of oral cancer (Xie et al., 1999; Zhang et al., 2009). Caspases-8 and -9 were prominently expressed in OSCC, and the expression of caspase-7 was reported to be a predictor of locoregional recurrence of OSCC (Coutinho-Camillo et al., 2011). Fas ligand is overexpressed and Fas receptor is underexpressed in OSCC (Chen et al., 1999; Das et al., 2011).

Survivin is a key member of the IAP protein family. Survivin negatively regulates apoptosis by inhibiting caspase activation. Survivin is reported to be overexpressed in many human cancers, including oral cancer (De Maria et al., 2009; Khan et al., 2009). Upregulation of survivin is associated with poor prognosis and chemoresistance (Su et al., 2010). XIAP has been reported recently as a prognostic and treatment response marker (Yang et al., 2012).

Mutation in p53 is quite common in cancer, and it has been reported to be associated with prognosis in oral cancer; however, p53 overexpression cannot be regarded as an independent prognostic factor (Mineta et al., 1998). The loss of function of p53 mutant proteins may predict a significant low pathological complete response rate and suboptimal response to cisplatin-based neoadjuvant chemotherapy in patients with oral cavity SCC (Perrone et al., 2010).

6.4 CONCLUSION

Mechanistic details of the apoptotic pathway have gifted us with knowledge about key regulators, which can be used for designing diagnostic and therapeutic applications. Entities of apoptotic pathways like p53, survivin, Bcl-2/Bax, and XIAP hold potential to qualify as successful diagnostic/prognostic biomarkers. However, the road for early detection of oral cancer by these biomarker is full of challenges, like: (1) the biomarker should show its activity or expression in the case of oral cancer, and its expression should be absent in the absence of oral cancer; (2) biomarkers should be expressed much before clinical manifestation of oral cancer can be detected by physical examination; i.e., they should be able to detect early molecular changes leading to oral cancer; (3) biomarkers should be available in body fluids like blood and saliva for easy detection through noninvasive tests; and (4) biomarkers should be validated thoroughly with a large and diverse sample pool, possibly by an independent authority. It is very challenging to discovery biomarker(s) that qualify all these essential features. These biomarkers should also be studied in various combinations to explore any hidden pattern that can form the basis of designing diagnostic applications.

Current treatment options for oral cancer heavily rely on traditional methods, like surgery, radiation therapy, and chemotherapy. However, use of gene therapies like ONYX-015, Ad-p53, and antisense techniques should be treated as the turning point for oral cancer treatment. Targeted therapies require a detailed understanding of the underlying disease process. With the accumulation of knowledge about the role of apoptosis in oral carcinogenesis, we should expect the availability of better-targeted drugs for treatment of oral cancer.

REFERENCES

Bonner, J.A., P.M. Harari, J. Giralt, et al. 2010. Radiotherapy plus cetuximab for locoregionally advanced head and neck cancer: 5-year survival data from a phase 3 randomised trial, and relation between cetuximab-induced rash and survival. *Lancet Oncol* 11: 21–28.

Bykov, V.J., N. Issaeva, A. Shilov, et al. 2002. Restoration of the tumor suppressor function to mutant p53 by a low-molecular weight compound. *Nat Med* 8: 282–288.

Bykov, V.J., N. Issaeva, N. Zache, et al. 2005. Reactivation of mutant p53 and induction of apoptosis in human tumor cells by maleimide analogs. *J Biol Chem* 280: 30384–30391.

Chan, S.L., M.C. Lee, K.O. Tan, et al. 2003. Identification of chelerythrine as an inhibitor of BclXL function. *J Biol Chem* 278: 20453–20456.

Chen, Q., L.P. Samaranayake, X. Zhen, et al. 1999. Up-regulation of Fas ligand and down-regulation of Fas expression in oral carcinogenesis. *Oral Oncol* 35: 548–553.

Chene, P., J. Fuchs, J. Bohn, et al. 2000. A small synthetic peptide, which inhibits the p53-hdm2 interaction, stimulates the p53 pathway in tumour cell lines. *J Mol Biol* 299: 245–253.

Clayman, G.L., A.K. el-Naggar, S.M. Lippman, et al. 1998. Adenovirus-mediated p53 gene transfer in patients with advanced recurrent head and neck squamous cell carcinoma. *J Clin Oncol* 16: 2221–2232.

Cory, S., and J.M. Adams. 2002. The Bcl2 family: Regulators of the cellular life-or-death switch. *Nat Rev Cancer* 2: 647–656.

Coutinho-Camillo, C.M., S.V. Lourenço, I.N. Nishimoto, et al. 2011. Caspase expression in oral squamous cell carcinoma. *Head Neck* 33: 1191–1198.

Das, S.N., P. Khare, M.K. Singh, et al. 2011. Fas receptor (CD95) and Fas ligand (CD178) expression in patients with tobacco-related intraoral squamous cell carcinoma. *Indian J Med Res* 134: 54–60.

De Maria, S., G. Pannone, P. Bufo, et al. 2009. Survivin gene-expression and splicing isoforms in oral squamous cell carcinoma. *J Cancer Res Clin Oncol* 135: 107–116.

Deveraux, Q.L., E. Leo, H.R. Stennicke, et al. 1999. Cleavage of human inhibitor of apoptosis protein XIAP results in fragments with distinct specificities for caspases. *EMBO J* 18: 5242–5251.

Foster, B.A., H.A. Coffey, M.J. Morin, et al. 1999. Pharmacological rescue of mutant p53 conformation and function. *Science* 286: 2507–2510.

Friedler, A., L.O. Hansson, D.B. Veprintsev, et al. 2002. A peptide that binds and stabilizes p53 core domain: Chaperone strategy for rescue of oncogenic mutants. *Proc Natl Acad Sci USA* 99: 937–942.

Grasberger, B.L., T. Lu, C. Schubert, et al. 2005. Discovery and cocrystal structure of benzo-diazepinedione HDM2 antagonists that activate p53 in cells. *J Med Chem* 48: 909–912.

Green, D.R. 2005. Apoptotic pathways: Ten minutes to dead. *Cell* 121: 671–674.

Green D.R., and G. Kroemer. 2004. The pathophysiology of mitochondrial cell death. *Science* 305: 626–629.

Hanahan, D., and R.A. Weinberg. 2000. The hallmarks of cancer. *Cell* 100: 57–70.

Heise, C., and D.H. Kirn. 2000. Replication-selective adenoviruses as oncolytic agents. *J Clin Invest* 105: 847–851.

Hockenbery, D., G. Nuñez, C. Milliman, et al. 1990. Bcl-2 is an inner mitochondrial membrane protein that blocks programmed cell death. *Nature (London)* 348: 334–336.

Huang, Y., Y.C. Park, R.L. Rich, et al. 2001. Structural basis of caspase inhibition by XIAP: Differential roles of the linker versus the BIR domain. *Cell* 104: 781–790.

Issaeva, N., P. Bozko, M. Enge, et al. 2004. Small molecule RITA binds to p53, blocks p53-HDM-2 interaction and activates p53 function in tumors. *Nat Med* 10: 1321–1328.

Issaeva, N., A. Friedler, P. Bozko, et al. 2003. Rescue of mutants of the tumor suppressor p53 in cancer cells by a designed peptide. *Proc Natl Acad Sci USA* 100: 13303–13307.

Judd, N.P., A.E. Winkler, O. Murillo-Sauca, et al. 2012. ERK1/2 regulation of CD44 modulates oral cancer aggressiveness. *Cancer Res* 72: 365–374.

Khan, Z., R.P. Tiwari, R. Mulherkar, et al. 2009. Detection of survivin and p53 in human oral cancer: Correlation with clinicopathologic findings. *Head Neck* 31: 1039–1048.

Khan, Z., and Bisen, P.S. 2013. Oncoapoptotic signaling and deregulated target genes in cancers: Special reference to oral cancer. *Biochim Biophy Acta Reviews on Cancer*, dx.doi.org/10.1016/j.bbcan.2013.04.002.

Kok, S.H., C.C. Yeh, M.L. Chen, et al. 2009. Esculetin enhances TRAIL-induced apoptosis through DR5 upregulation in human oral cancer SAS cells. *Oral Oncol* 45: 1067–1072.

Lain, S., J.J. Hollick, J. Campbell, et al. 2008. Discovery, in vivo activity, and mechanism of action of a small-molecule p53 activator. *Cancer Cell* 13: 454–463.

Li, G., and T. Xiao-Hong. 2011. Inhibitory effect on growth of oral cancer cell lines KB induced by survivin RNAi. *China Res Prev Treat* 38: 257–260.

Liu, T.J., W.W. Zhang, D.L. Taylor, et al. 1994. Growth suppression of human head and neck cancer cells by the introduction of a wild-type p53 gene via a recombinant adenovirus. *Cancer Res* 54: 3662–3667.

Lopez-Martinez, M., M. Anzola, N. Cuevas, et al. 2002. Clinical applications of the diagnosis of p53 alterations in squamous cell carcinoma of the head and neck. *Med Oral* 7: 108–120.

Luu, Y., J. Bush, K.J. Cheung Jr., et al. 2002. The p53 stabilizing compound CP-31398 induces apoptosis by activating the intrinsic Bax/mitochondrial/caspase-9 pathway. *Exp Cell Res* 276: 214–222.

Matsumiya, T., T. Imaizumi, H. Yoshida, et al. 2001. Cisplatin inhibits the expression of X-chromosome-linked inhibitor of apoptosis protein in an oral carcinoma cell line. *Oral Oncol* 37: 296–300.

Midgley, C.A., J.M. Desterro, M.K. Saville, et al. 2000. An N-terminal p14ARF peptide blocks Mdm2-dependent ubiquitination in vitro and can activate p53 in vivo. *Oncogene* 19: 2312–2323.

Mineta, H., A. Borg, M. Dictor, et al. 1998. p53 mutation, but not p53 overexpression, correlates with survival in head and neck squamous cell carcinoma. *Br J Cancer* 78: 1084–1090.

Nakashima, T., M. Miura, and M. Hara. 2000. Tetrocarcin A inhibits mitochondrial functions of Bcl-2 and suppresses its anti-apoptotic activity. *Cancer Res* 60: 1229–1235.

Nemunaitis, J., C. Cunningham, and A.W. Tong. 2003. Pilot trial of intravenous infusion of a replication-selective adenovirus (ONYX-015) in combination with chemotherapy or IL-2 treatment in refractory cancer patients. *Cancer Gene Ther* 10: 341–352.

Neves, S.S., A.B. Sarmento-Ribeiro, S.P. Simões, et al. 2006. Transfection of oral cancer cells mediated by transferrin-associated lipoplexes: Mechanisms of cell death induced by herpes simplex virus thymidine kinase/ganciclovir therapy. *Biochim Biophys Acta* 1758: 1703–1712.

Nguyen, J.T., and J.A. Wells. 2003. Direct activation of the apoptosis machinery as a mechanism to target cancer cells. *Proc Natl Acad Sci USA* 100: 7533–7538.

North S., O. Pluquet, D. Maurici, et al. 2002. Restoration of wild-type conformation and activity of a temperature-sensitive mutant of p53 (p53(V272M)) by the cytoprotective aminothiol WR1065 in the esophageal cancer cell line TE-1. *Mol Carcinog* 33: 181–188.

Pelengaris, S., and M. Khan. 2006. In *The molecular biology of cancer*, ed. S. Pelengaris and M. Khan, 251–278. Blackwell, Boston.

Perrone, F., P. Bossi, B. Cortelazzi, et al. 2010. TP53 mutations and pathologic complete response to neoadjuvant cisplatin and fluorouracil chemotherapy in resected oral cavity squamous cell carcinoma. *J Clinl Oncol* 28: 761–766.

Riedl, S.J., M. Renatus, R. Schwarzenbacher, et al. 2001. Structural basis for the inhibition of caspase-3 by XIAP. *Cell* 104: 791–800.

Rudin, C.M., E.E. Cohen, V.A. Papadimitrakopoulou, et al. 2003. EE: An attenuated adenovirus, ONYX-015, as mouthwash therapy for premalignant oral dysplasia. *J Clin Oncol* 21: 4546–4552.

Selivanova, G., V. Iotsova, I. Okan, et al. 1997. Restoration of the growth suppression function of mutant p53 by a synthetic peptide derived from the p53 C-terminal domain. *Nat Med* 3: 632–638.

Shangary, S., D. Qin, D. McEachern, et al. 2008. Temporal activation of p53 by a specific MDM2 inhibitor is selectively toxic to tumors and leads to complete tumor growth inhibition. *Proc Natl Acad Sci USA* 105: 3933–3938.

Staples, O.D., J.J. Hollick, J. Campbell, et al. 2008. Characterization, chemical optimization and anti-tumour activity of a tubulin poison identified by a p53-based phenotypic screen. *Cell Cycle* 7: 3417–3427.

Su, L., Y. Wang, M. Xiao, et al. 2010. Up-regulation of survivin in oral squamous cell carcinoma correlates with poor prognosis and chemoresistance. *Oral Surg Oral Med Oral Pathol Oral Radiol Endod* 110: 484–491.

Sun, C., M. Cai, R.P. Meadows, et al. 2000. NMR structure and mutagenesis of the third Bir domain of the inhibitor of apoptosis protein XIAP. *J Biol Chem* 275: 33777–33781.

Takimoto, R., W. Wang, D.T. Dicker, et al. 2002. The mutant p53-conformation modifying drug, CP-31398, can induce apoptosis of human cancer cells and can stabilize wild-type p53 protein. *Cancer Biol Ther* 1: 47–55.

Taylor, R.C., B.E. Klein, J. Stein, et al. 2008. Apoptosis: Controlled demolition at the cellular level. *Nat Rev Mol Cell Biol* 9: 231–241.

Tinhofer, I., G. Anether, M. Senfter, et al. 2002. Stressful death of T-ALL tumor cells after treatment with the anti-tumor agent Tetrocarcin-A. *FASEB J* 16: 1295–1297.

Tzung, S.P., K.M. Kim, G. Basanez, et al. 2001. Antimycin A mimics a cell-death-inducing Bcl-2 homology domain 3. *Nat Cell Biol* 3: 183–191.

Vassilev, L.T., B.T. Vu, B. Graves, et al. 2004. In vivo activation of the p53 pathway by small-molecule antagonists of MDM2. *Science* 303: 844–848.

Wang, W., R. Takimoto, F. Rastinejad, et al. 2003. Stabilization of p53 by CP-31398 inhibits ubiquitination without altering phosphorylation at serine 15 or 20 or MDM2 binding. *Mol Cell Biol* 23: 2171–2181.

Wei, M.C., W.X. Zong, E.H. Cheng, et al. 2001. Proapoptotic BAX and BAK: A requisite gateway to mitochondrial dysfunction and death. *Science* 292: 727–730.

Weinmann, L., J. Wischhusen, M.J. Demma, et al. 2008. A novel p53 rescue compound induces p53-dependent growth arrest and sensitises glioma cells to Apo2L/TRAIL-induced apoptosis. *Cell Death Differ* 15: 718–729.

Xi, S., and J.R. Grandis. 2003. Gene therapy for the treatment of oral squamous cell carcinoma. *J Dent Res* 82: 11–16.

Xie, X., O.P. Clausen, J. Sudbo, et al. 1999. Diagnostic and prognostic value of nucleolar organizer regions in normal epithelium, dyplasia and squamous cell carcinoma of the oral cavity. *Cancer* 79: 2200–2208.

Xu, G.W., I.A. Mawji, C.J. Macrae, et al. 2008. A high-content chemical screen identifies ellipticine as a modulator of p53 nuclear localization. *Apoptosis* 13: 413–422.

Yang, X.H., Z.E. Feng, M. Yan, et al. 2012. XIAP is a predictor of cisplatin-based chemotherapy response and prognosis for patients with advanced head and neck cancer. *PLoS One* 7: e31601.

Yang, Y., R.L. Ludwig, J.P. Jensen, et al. 2005. Small molecule inhibitors of HDM2 ubiquitin ligase activity stabilize and activate p53 in cells. *Cancer Cell* 7: 547–559.

Zhang, H.Z., S. Kasibhatla, Y. Wang, et al. 2004. Discovery, characterization and SAR of gambogic acid as a potent apoptosis inducer by a HTS assay. *Bioorg Med Chem* 12: 309–317.

Zhang, M., P. Zhang, C. Zhang, et al. 2009. Prognostic significance of Bcl-2 and Bax protein expression in the patients with oral squamous cell carcinoma. *J Oral Pathol Med* 38: 307–313.

7 Expression and Regulation of Survivin

KEY WORDS

Apoptosis
Cancer
Expression and regulation of survivin
p53
Structure of survivin
Survivin

7.1 INTRODUCTION

Survivin is a member of the inhibitor of apoptosis (IAP) family (Salvesen and Duckett, 2002). Survivin contains a single baculovirus IAP repeat and lacks a carboxy-terminal RINGfinger. The protein has been renamed BIRC5 (baculoviral IAP repeat-containing protein-5) (Ambrosini et al., 1997). Hence, survivin is encoded by the *BIRC5* gene. The survivin "apoptosis inhibitor 4" has its structural homology with the murine TIAP (Thiol inhibitor of apoptosis) and deterin of *Drosophila melanogaster*; however, the functional aspects are only partially unique. Similarly, the genomes of *Xenopus laevis*, *Xenopus tropicalis*, zebra fish, fugu pufferfish, and rainbow trout encode two different survivin genes (Su1 and Su2), which is contrary to mammalian genomes, which encode a single survivin gene (Sah et al. 2006; Du Pasquier et al., 2006).

7.2 MOLECULAR ORGANIZATION AND STRUCTURE OF SURVIVIN

Survivin is composed of a single BIR domain with an extended carboxy-terminal coil-coiled domain (LaCasse et al., 1998). The BIR domain consists of 70 amino acids, which are evolutionarily conserved. The human survivin gene, spanning 14.7 kb on the telomeric position of chromosome 17, contains four exons and three introns and produces a 16.5 kDa protein. Unlike other IAPs, survivin is small and has only a single N-terminal BIR domain, a long C-terminal α-helix coiled region, and forms a stable dimer in solution (Khan et al.). The BIR domain is thought to be critical for anti-apoptotic function, whereas the coiled domain probably interacts with tubulin structures (Wheatley and McNeish, 2005).

7.2.1 Isoforms and Functions of Survivin

The survivin gene locus encodes multiple genetic splice variants with unique properties and functions. Transcription from the *survivin* locus gives rise to spliced isoforms, identified in both humans and mice (Badran et al., 2004; Mahotka et al., 1999; Conway et al., 2000). Survivin-2B is generated by the insertion of an alternative exon 2B, while survivin-3B results from insertion of novel exon 3B, leading to frameshift and premature termination of the protein. Survivin is a bifunctional protein inhibiting apoptosis and regulating mitosis. The process of regulation of apoptosis by the survivin protein has been discussed in detail in Section 7.3.2.

The intracellular localization of these splice variants of survivin is through a Crm1-dependent nuclear export signal (NES) present in survivin(-2B) and survivin(-3B), but absent in survivin(-ΔEx-3) and survivin(-2α). However, survivin isoforms lack an active nuclear import signal, and hence enter the nucleus by passive diffusion. Only survivin(-3B) splice variants are cytoprotective, and hence efficiently interact with chromosomal passenger complex (CPC) proteins. The NES, together with efficient CPC formation of survivin, is required for the cytoprotective activity, correct localization, and function during cell division. Survivin and survivin(-2B) were overexpressed in breast, colorectal, head and neck cancer, lymphoma, and leukemia patients. However, survivin was the predominant form detected, and the other survivin isoforms were only expressed at low levels in tumors.

The BIR domain is assumed to have anti-apoptotic function, where the coil-coiled domain probably interacts with tubulin structures. The survivin molecule contains three separate and chemically distinct surfaces, including acidic and basic regions on the BIR domain and a hydrophobic helical surface on alpha 6. This arrangement functionally favors the relevant protein-protein interaction surfaces. The murine survivin gene, located on chromosome 11E2, contains four exons, which gives rise to three separate gene products through alternative splicing (Conway et al., 2000). The longest open reading frame, containing all four exons, produces a 140-amino acid protein that is similar to human full-length survivin. A second isoform, lacking a sequence derived from exon 2, gives rise to a 40-amino acid protein that lacks the BIR domain and the C-terminal coil-coiled domain. The third gene product results from the retention of intron 3, subsequently acquiring a new in-frame stop codon; however, the gene product is a 121-amino acid protein produced that contains the BIR domain but lacks the coil-coiled domain.

Ambrosini et al. (1998) and Mahotka et al. (1999, 2002) showed a survivin gene to span 14.7 kb on telomeric position of chromosome 17 and produce a 16.5 kDa protein. The isoforms of the survivin protein have been reported as spliced transcripts. The isoforms form almost the full length of survivin, survivin-2B, and survivin-DEx-3 with insertion and deletion of some of the coding and noncoding sequences. The full-length survivin gene consists of a three-intron, four-exon structure. The survivin-2B transcript resulted in the retention of a portion of intron 2, whereas removal of exon 3 was seen in the survivin-DEx-3 transcript. The sequence alterations produced from the splice variants markedly change the corresponding protein conformations and subsequently create differences in their ability to inhibit apoptosis. Insertion of exon 2B in survivin-2B, in the essential BIR domain, may lead to

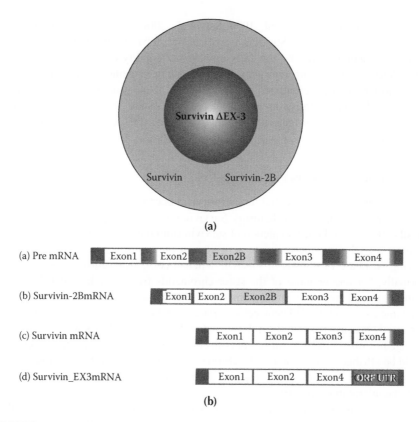

FIGURE 7.1   (a) Survivin and its splice variants, survivin-2B and survivin ΔEx. (b) Survivin isoforms: Survivin exists in three isoforms as a result of alternatively spliced transcripts. Survivin-DEx-3 is primarily nuclear in location, whereas full-length survivin and survivin-2B are predominantly within the cytoplasm.

a decrease in apoptotic activity. The removal of exon 3 in survivin-DEx-3 likewise interrupts the BIR domain and retains its ability to suppress apoptosis. In addition, loss of this exon results in a frameshift in exon 4, generating a novel COOH-terminal. There are also different subcellular localizations of survivin and its splice variants. Survivin and survivin-2B are predominantly cytoplasmic, whereas survivin-DEx-3 is primarily nuclear (Mahotka et al., 1999, 2002) (Figures 7.4 and 7.5). The different isoforms of survivin and their localizations in the cell make a regulatory balance between apoptosis and inhibition of apoptosis (Mahotka et al., 2002) (Figure 7.1a and b).

7.3   FUNCTIONS OF SURVIVIN

Survivin has two main functions: one as a chromosomal passenger protein and the other as an inhibitor of apoptosis. Survivin-2B has been shown to be a pro-apoptotic protein that sensitizes resistant leukemia cells to chemotherapy in a p53-dependent fashion. Survivin-Ex-3 functions as an anti-apoptotic protein and is upregulated in malignancies (Mahotka et al., 1999; Sah et al., 2006). No function has yet been described for survivin-3B. Survivin is critical for global normal embryonic development as demonstrated by the early embryonic lethality of mice with homozygous deletions in the survivin gene locus. Survivin proteins are virtually absent from most normal differentiated tissues; however, these proteins are expressed in certain highly proliferative areas within normal tissues. In contrast, survivin is highly expressed in the majority of human malignancies, derived from different cell origins.

7.3.1   ROLE OF SURVIVIN IN CELL DIVISION

The survivin protein is essential for cell survival because it counteracts in constitutive propensity apoptotic mechanisms. Survivin protein participates in multiple facets of cell division; hence, depletion of survivin caused defects in cell division, such as arrest of DNA synthesis due to activation of the tumor suppressor protein p53. During anaphase of mitosis in survivin-deficient cells, sister chromatids disjoined normally, but one or more of the sister chromatids frequently lagged behind the main mass of segregating chromosomes, probably because of merotelic kinetochore attachments. Survivin-deficient cells initiated but failed to complete cytokinesis, apparently because the spindle midzone and midbody microtubules were absent during late mitosis. The abnormalities of both chromosome segregation and cytokinesis could be attributed to a defect in the chromosomal passenger protein complex, with a consequent mislocalization of the kinesin-like motor protein MKLP-1 associated with the microtubule abnormalities.

The RNA interference studies on the depletion of aurora B recapitulated the importance of survivin in the proliferation of normal human cells by virtue of its contributions to accurate sister chromatid segregation and assembly/stabilization of microtubules in late mitosis. Survivin is also expected to function like other chromosomal passenger proteins (CPPs), such as the aurora B kinase, inner centromere protein, and telophase disk antigen (TD-60). These proteins function as part of a multiprotein complex and have multiple roles during cell division. Depletion of CPPs in human cells may cause apoptosis and pleiotropic defects in cell division.

The validation studies on the knockout embryos in *Caenorhabditis elegans* were superimposable with the earlier ones, and the studies confirmed that survivin molecules have a conserved role in cell division (Li et al., 1998, 1999). Recent reports reveal that survivin depletion results in the formation of abnormalities in chromosome segregation with lagging chromosomes and DNA bridges during anaphase and telophase, which results in the formation of a bilobed nucleus (Figure 7.6). Twenty-five percent of survivin-depleted cells during late telophase likely originating from lagging chromosomes exhibited a mini nuclei condition (Chen et al., 2000; Li et al., 1999; Li and Brattain, 2006).

Additional evidence for the crucial role of survivin in mitosis has been demonstrated in knockout mice. Homozygous disruption of the survivin gene resulted in embryonic death at 4 to 5 days, with the null embryos revealing failed cytokinesis and disrupted microtubule formation (Uren et al., 2000; Skoufias et al., 2000). However, it is clear that abnormal ploidy and multiple genetic defects could result after survivin depletion. Since the defects induced by survivin targeting are complex and pleiotropic, the exact pathway affected by the depletion of survivin remained obscure. Earlier studies hypothesized survivin to function like other (chromosomal passenger proteins) aurora proteins, and the studies were done to establish the connection of survin proteins with other CPPs involved in the regulation of mitosis.

7.3.1.1 Association of Survivin with Other Chromosomal Passenger Proteins

The connection between survivin and aurora has been extensively studied through initial colocalization experiments. The partial rescue of the *C. elegans* defect by survivin established the role of survivin in the organization of microtubule functions like nucleation, assembly, and dynamics. Survivin is clearly a microtubule-associated protein and a component of the centrosome in concert with the aurora B kinase and inner centromere protein (Bolton et al., 2002) (Figure 7.6). Survivin depletion in retinal pigment epithelia (RPE) cells disrupted association of the chromosomal passenger proteins, inner centromere protein, and aurora B, and the spindle midzone motor protein MKLP-1 with the spindle midzone, the equatorial cortex, and the midbody.

These findings suggest that the CPP complex is required for the recruitment of MKLP-1 to organize microtubules of the spindle midzone and midbody in mammalian cells, as in nematodes, and the failure of this recruitment is at least partially responsible for defects in cytokinesis displayed by survivin-depleted RPE cells. Survivin binding also enhanced the aurora B kinase activity (Bolton et al., 2002). The survivin and its alternatively spliced transcripts localized in the cytoplasmic and nucleoplasmic regions, in association with the CPP complex, regulate mitosis (Fortugno et al., 2002).

The chromosomal passenger complex (CPC) regulates numerous mitotic events, such as chromatin modification, correction of kinetochore-microtubule misattachments, regulation of sister chromatid cohesion, and completion of cytokinesis (Vagnarelli and Earnshaw, 2004). The active subunit of this complex is the aurora B kinase (Carmena et al., 2009), with the inner centromere protein (INCENP) exerting a regulatory/scaffolding role, while survivin and borealin are responsible for targeting the complex to its various sites of action during mitosis.

INCENP acts as a scaffold for CPC assembly through its N terminus, which forms a three-helix bundle with the N terminus of borealin and C terminus of survivin. This interaction is sufficient for centromere targeting of the three proteins and does not require complex formation with aurora B (Xu et al., 2009). The INCENP C terminus binds aurora B through its highly conserved inbox. This activates the aurora B kinase through a two-step feedback mechanism. When aurora B binds INCENP, this partially activates the kinase, and it phosphorylates a conserved Thr-Ser-Ser (TSS) motif near the INCENP C terminus (residues 813 to 815 in human INCENP). This is followed by the autophosphorylation of aurora B on its T-loop, which elicits full kinase activity.

The chromosomal passenger proteins, INCENP and aurora B, which can interact directly with survivin, were absent from the centromeres of survivin-depleted cells. These data contribute to the emerging picture that survivin operates together with

INCENP and aurora B to perform its mitotic duties. Some survivin-depleted cells eventually exhibited mitosis without completing cytokinesis. This resulted in a gradual increase in the percentage of multinucleated cells in culture. Immunofluorescence studies revealed that survivin-depleted cells were unable to stably maintain BubR1 (spindle assembly checkpoints) at the kinetochores in the presence of either taxol or nocodazole (Carvalho et al., 2003).

During mitosis, survivin is regulated by the kinases aurora B and Cdk1. Mutation of the aurora B phosphorylation site at Thr[117] or treatment with an aurora B inhibitor alters the affinity of survivin for centromeres and interferes with the error correction process facilitated by CPPs that ensure proper alignment of chromosomes at the metaphase plate. The phosphorylation of survivin by aurora B prevents the completion of cytokinesis, implying a critical requirement for dephosphorylation of this site, while Cdk1 phosphorylates survivin at Thr[34] in its BIR domain. Mutational analysis has shown that expression of a Thr[34] phosphomimic, T34E, greatly reduces the rate of cell proliferation and cannot support cell division in the absence of the endogenous protein, whereas expression of the nonphosphorylatable counterpart, T34A, supports cell growth (Barrett et al., 2009). Intriguingly, T34A sensitizes cells to apoptotic stimuli and is being explored as a potential therapeutic tool, whereas T34E potently inhibits cell death. Thus, phosphorylation by Cdk1 is one means of separating the mitotic and anti-apoptotic roles of survivin.

Survivin and Plk1 kinase (polo-like kinase) are important mediators of cell survival that are required for chromosome alignment, cytokinesis, and protection from apoptosis. Interference with either survivin or Plk1 activity manifests many similar outcomes: prometaphase delay/arrest, multinucleation, and increased apoptosis. Moreover, the expression of both survivin and Plk1 is deregulated in cancer.

Plk1 (polo-like kinase-1) is an essential multitasking protein deregulated in cancer. First identified in *Drosophila*, Plk1 regulates mitotic entry, centrosome separation, spindle assembly, chromosome alignment, APC/C activation, and cytokinesis, and has been implicated as a mediator of apoptosis. As expected for a protein with many roles, its loss has pleiotropic effects, including the generation of monopolar spindles, polyploidy, and increased apoptosis. Although the majority of Plk1 is centrosomal in early mitosis, a subpopulation associates with the kinetochores and has been implicated in mediating the spindle checkpoint. Mad2 and BubR1 are checkpoint proteins that are recruited to the kinetochores of chromosomes that are not properly attached to the spindle. Mad2 is recruited due to the absence of microtubule attachments, whereas BubR1 is recruited when paired kinetochores are not under tension.

Treatment of Plk1 or survivin-depleted cells with microtubule poisons has suggested that Plk1 stabilizes Mad2 recruitment at kinetochores (van Vugt et al., 2004), whereas survivin stabilizes BubR1 at these sites. Simultaneous depletion of survivin and Plk1 eliminates both spindle checkpoint signals, and consequently, cells exit mitosis inappropriately and undergo mitotic catastrophe (van Vugt et al., 2004). However, Matsumura et al. (2007) reported that Plk1 interacts directly with BubR1, and that phosphorylation of BubR1 by Plk1 is required for correct chromosome orientation during prometaphase, but not for its recruitment to kinetochores or for spindle checkpoint activation. Thus, although Plk1 and survivin may have complementary roles in the maintenance of the spindle checkpoint, direct links between Plk1 and BubR1 also exist that facilitate chromosome biorientation. In cells that enter anaphase normally, Plk1 is found at the central spindle and midbody, where it colocalizes with the CPPs and is required to facilitate cytokinesis through communication with the microtubule organizers, MKLP1, MKLP2, and PRC1, and the RhoA signaling cascade.

7.3.1.2   Cell Division Network of Survivin

7.3.1.2.1   Chromosomal Complex

Survivin, along with other chromosomal passenger proteins, such as aurora kinase B (AURKB), inner centromere protein (INCENP) antigens, and borealin (also known as CDCA8), has affinity to kinetochores. The affinity of each of the proteins of the complex is independent; however, all the pathways are also centered on survivin (Jeyaprakash et al., 2007). Chromosome segregation is the major process affected by the deletion of any of the chromosomal passenger proteins.

In yeast cells, the chromosome segregation mechanism includes the binding of survivin to a regulator called shugoshin 2 and recruitment of the mitotic exit network, leading to inactivation of cyclin-dependent kinases (CDKs), completion of cytokinesis, and initiation of G1 gene expression (Vanoosthuyse et al., 2007). In human cells, regulatory phosphorylation of survivin by aurora kinase B and sequential cycles of survivin ubiquitylation and deubiquitylation by the enzyme hFAM are seen. Survivin orthologs in different model organisms have evolved considerable diversity in molecular interactions and cellular functions. The next process of the survivin is its binding with exportin 1 (XPO1) (the human homolog of yeast Crm1 (chromosomal region maintenance)) through the RAN-GTP pathway (mitotic spindle formation pathway). Survivin affects the RAN (Ras-related nuclear protein), an effector molecule that regulates kinetochore fiber assembly, and recruitment of the RCC1/RCC2 family protein TD-60 (molecule located inside the nucleus) (Mollinari et al., 2003).

7.3.1.2.2   Spindle Formation and Checkpoint Control

Once localized in the chromosomal passenger complex, survivin contributes to chromatin-associated spindle formation (Sampath et al., 2004). This process involves the stimulation of aurora kinase B activity, which in turn phosphorylates the mitotic centromere-associated kinesin (MCAK) (also known as KIF2C) and removes its microtubule depolymerizing activity. Survivin has been characterized as a sensor of kinetochore-microtubule attachment (Sandall et al., 2006), a component of the spindle assembly checkpoint that is activated by a lack of microtubule tension.

The survivin network further intersects with mechanisms of apoptosis regulation. Activation of the checkpoint kinase CHK2 (encoded by CHEK2) by DNA damage stimulates a rapid discharge of the mitochondrial pool of survivin in the cytosol. This pathway does not seem to participate in cell division, but preserves the viability of tumor cells during a protracted G2 block by antagonizing DNA damage-induced apoptosis. A parallel survivin-p53 subsystem has evolved to oppose this effect. DNA damage also stabilizes p53, which functions as an efficient repressor of

BIRC5 transcription, through occupancy of a binding site in the *BIRC5* promoter, changes in chromatin structure affecting promoter accessibility, or epigenetic modifications such as DNA cytosine methyltransferase-1. The net effect abruptly lowers survivin levels, which itself is a stimulus to stabilize p53 and further reduce survivin expression. Therefore, balancing a survival effect of CHK2, the p53 subsystem aims to eliminate survivin expression during DNA damage, thus tilting the balance toward stable cell cycle arrest and apoptosis. Loss of p53 occurs frequently in human cancer and might cause unrestrained survivin expression resulting in enhanced cell viability, impaired checkpoint function, and increased propensity to aneuploidy.

7.3.1.2.3 Microtubule-Associated Survivin

After localization in the chromosomal passenger complex, a fraction of mitotic survivin directly assembles on polymerized microtubules. Similarly to kinetochore survivin, microtubule-associated survivin contributes to spindle formation, but this involves a different pathway of enhanced microtubule stability through suppression of microtubule dynamics, reduction of microtubule nucleation from centrosomes, and increased acetylated tubulin content that is independent of aurora kinase B. This survivin network also branches out to parallel pathways of genomic integrity and intersects mechanisms of apoptosis. Through its association with CDK1 (also known as CDC2), microtubule-bound survivin becomes phosphorylated at mitosis on Thr34.

This step is crucial to stabilize survivin at mitosis and efficiently counteract apoptosis of dividing cells, or in response to spindle poisons. The anti-apoptotic environment created by CDK1 is not limited to its effects on survivin. CDK1 phosphorylation of caspase, an upstream initiator of mitochondrial cell death, abolishes its anti-apoptotic activity and antagonizes cell death induced by anti-mitotic agents (Figure 7.2).

7.3.2 ROLE OF SURVIVIN IN APOPTOSIS

Since its discovery in 1997, survivin has attracted attention as a unique member of the IAP gene family with a potential dual role in apoptosis inhibition and regulation of mitosis. The role of survivin in regulation of mitosis has recently become better understood, and linked to multiple spindle microtubule functions and mitotic checkpoints. In contrast, despite extensive experimental evidence in vitro, and in transgenic animals in vivo, the precise mechanism(s) by which survivin interferes with apoptosis has not been elucidated. The evidence presented in this chapter defines a novel anti-apoptotic pathway mediated by survivin, functionally separable from its role in mitosis. Cytoprotection by survivin is more selective than that by other IAPs and is specifically targeted at the initiation of mitochondrial apoptosis to prevent caspase-9 activation (Figure 7.3).

Apoptosis occurs in human cancer as a result of loss of tumor suppressor genes, angiogenic changes, and immortalization (Evan and Vousden, 2001). The mechanism of apoptosis is counteracted by several regulators implicated in cancer cell survival. Bcl-2 family proteins are one such inhibitor of apoptosis (Cory and Adams, 2002) and act at the mitochondrion (Zamzami and Kroemer, 2001) to control the

Kitelectrons

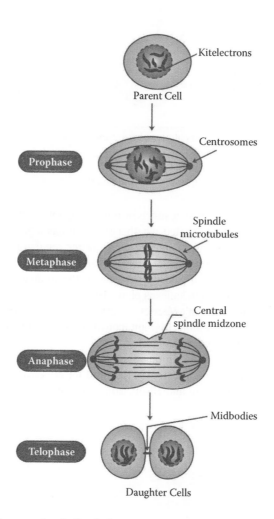

FIGURE 7.2   Immunochemically distinct survivin pools localize to multiple components of the mitotic apparatus, including centrosomes, metaphase and anaphase spindle microtubules, and midbodies.

release of apoptogenic proteins, notably cytochrome c and Smac (Martinou and Green, 2001). The inhibitor of apoptosis (IAP) gene family functions as endogenous inhibitors of caspases (Salvesen and Duckett, 2002), the enzymatic effectors of apoptosis (Hengartner, 2000; Khan and Bisen, 2013) (Figure 7.4).

Survivin is a bifunctional IAP involved in protection from apoptosis and regulation of mitosis (Altieri, 2003). The essential role of survivin at cell division has been linked to centrosomal function (Li et al., 1999), metaphase and anaphase microtubule assembly (Giodini et al., 2002; Wheatley et al., 2001), and spindle checkpoint

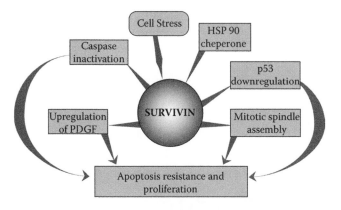

FIGURE 7.3   General scheme of survivin apoptotic function.

regulation; the mechanism by which survivin inhibits apoptosis has remained elusive. In particular, it has been debated whether survivin has a genuine function in apoptosis inhibition independent of its role as a master regulator of mitosis, and whether this is important in disease pathogenesis. Validating a cytoprotective mechanism of survivin has become a priority because of the dramatic exploitation of this pathway in human tumors (Velculescu et al., 1999) its frequent association with unfavorable disease outcome (van't Veer et al., 2002; van de Wetering et al., 2002) and the recent identification of molecular antagonists of survivin that are approaching clinical testing in cancer patients.

Survivin opposes death receptor-initiated apoptosis, in a cell type-specific fashion, in which mitochondrial amplification is required for effective execution of cell death (Scaffidi et al., 1998; Khan and Bisen, 2013). Survivin occurs abundantly in

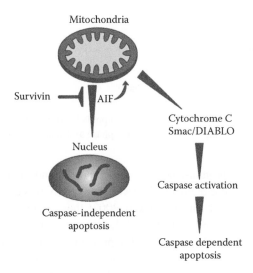

FIGURE 7.4   Role of caspase in mitochondrial apoptosis mediated by survivin.

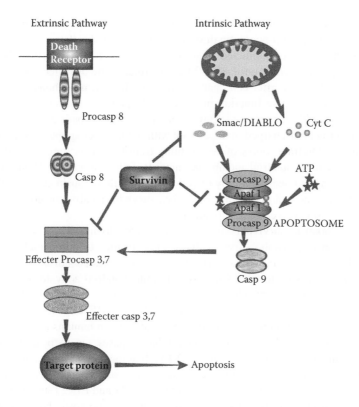

FIGURE 7.5   Apoptotic pathways and the site of survivin anti-apoptotic action.

the mitochondrial pool, especially in intermitochondrial membrane space. In mammalian cells, the mitochondrion has emerged as a central region of apoptosis, eliciting arrays of cell death regulators. Several factors activate the cell death receptors in mitochondria; survivin IAPs (cIAP2 and XIAP) were identified in mitochondrial extracts of lung cancer cell lines in response to ionizing radiation (Kroemer and Reed, 2000). The mechanism(s) by which survivin, and possibly other IAPs, localizes to mitochondria is currently not known. Conversely, survivin was not present in mitochondrial fractions of normal tissues, which suggests that its localization to mitochondria may preferentially, or exclusively, be associated with oncogenic transformation. An attractive candidate for this pathway is the molecular chaperone Hsp90 (Young et al., 2003; Khan and Bisen, 2013), which participates in mitochondrial import of client proteins and associates with survivin and other IAPs in vivo, and this study is in agreement with the studies of Kamal et al. (2003) (Figure 7.5).

An intriguing feature of mitochondrial survivin is its dynamic expansion in response to cellular stress, which is in keeping with an emerging role of IAPs in cellular adaptation to stress. Accordingly, exposure to chemotherapeutic drugs, exposure to heat shock, and exposure to hypoxia have all been associated with increased IAP expression. For survivin, this appears to involve increased protein stability, and to result largely, if not exclusively, in an expansion of the mitochondrial pool. This may be envisioned as an adaptive response to environmental stresses, ideally suited to further elevate an anti-apoptotic threshold in transformed cells.

As might have been predicted from its localization to the intermitochondrial membrane space, survivin, unlike Bcl-2 family proteins, did not affect the release of Smac or cytochrome c after cell death stimulation. Rather, mitochondrial survivin exerted its cytoprotective function in the cytosol upon release from mitochondria during apoptosis. The kinetics of survivin release from mitochondria, with more than 70% of the protein discharged in the cytosol within 4 hours of apoptotic stimulation, closely mirror the time course of permeability transition. This suggests a model in which loss of mitochondrial membrane integrity following cell death stimulation would result in the release not only of cell death amplifiers (Susin et al., 1999), but also of at least one cell death antagonist, i.e., survivin, ideally positioned to limit apoptosome-associated caspase activation.

The mitochondrial survivin is specifically competent to inhibit apoptosis, whereas overexpression of cytosolic survivin that cannot localize to mitochondria, i.e., in INS-1 cells, is unable to mediate cytoprotection. The mitochondrial survivin was hypothesized to associate with Smac (second mitochondrial activator of caspases) and sequester it away from other IAP proteins, such as XIAP. The IAP family of proteins in humans has eight IAP proteins, such as NAIP, cIAP1, cIAP2, XIAP, Ts-XIAP, ML-IAP, apollon, and survivin. Several IAPs in humans and *Drosophila* have been shown to directly bind and inhibit caspases via their BIR domains (Deveraux and Reed, 1999). For example, XIAP binding and inhibition of caspase-9 is through its third BIR domain (BIR3), whereas caspase-3 and caspase-7 are suppressed via a region within the molecule between BIR1 and BIR2 (Sun et al., 1999). The precise mechanism by which survivin suppresses apoptosis, however, is still incompletely understood.

Direct suppression of caspase-3 by survivin has been speculated by some investigators; yet, survivin lacks structural components present in other IAPs that allow their direct binding to caspase-3. There is also speculation that survivin binds to caspase-9, but there are likewise problems with this theory. Phosphorylation of survivin on the threonine at position 34 (Thr34) is critical for a functional survivin molecule. However, phosphorylation of survivin fails to explain why this would promote interaction with caspase-9. Finally, survivin may indirectly inhibit caspases via intermediate proteins. Survivin binds to Smac/DIABLO, which is a pro-apoptotic protein that binds IAPs, and thus prevents them from inhibiting caspases. Another example is the taxol-induced apoptosis in HeLa cells (Song et al., 2003), in which survivin colocalizes with Smac/DIABLO in the cytosol and thereby protects against apoptosis. A point mutation at Asp-71 in survivin results in the failure of survivin to complex with Smac/DIABLO. Consequently, this was shown to abolish survivin's ability to protect against cancers (Figure 7.5). Survivin can directly interact with caspases and inhibit them to suppress apoptosis. XIAP is a strong inhibitor of apoptosis, interacts directly with caspases, and inhibits them. Smac/DIABLO is the negative regulator of XIAP. Survivin may interact with Smac/DIABLO, which ultimately leads to inhibition of apoptosis (Khan and Bisen, 2013).

The role of survivin and its function in cell division has been explored in vivo through the use of transgenic (K14-survivin) mice. However, K14-survivin mice had a 60% reduction in apoptosis compared with the control group after UV-B exposure. These results demonstrate that constitutive expression of survivin inhibits keratinocyte apoptosis; yet, cell proliferation remains unaffected. In vitro, K14-survivin keratinocytes also have increased resistance to UV-B-induced apoptosis. The transgenic cells, however, are more susceptible to Fas-mediated (extrinsic) apoptosis, indicating that the cytoprotection offered by survivin is ineffective against the Fas-induced apoptotic pathway. K14-survivin mice crossed with p53-deficient mice, creating K14-survivin mice lacking one p53 allele (p53/2), are equally resistant to UV-B-induced apoptosis as p532/2 keratinocytes. Thus, it appears that survivin expression can substitute for the loss of one p53 allele in suppressing apoptosis.

7.4   REGULATION OF SURVIVIN

7.4.1   REGULATION BY P53

The balance between cell cycle procession and apoptosis is crucial for normal tissue homeostasis. Any breaks of this balance, which leads to inappropriate cell death or abnormal proliferation, can result in diseases, even tumorigenesis (Strasser et al., 1990). Numerous works have been done to discover the mechanism of balancing apoptosis and cell cycle progression, and a number of observations have indicated that regulation of the apoptotic process is tightly coupled with surveillance mechanisms of the cell cycle checkpoints. Dual roles of these molecules, such as c-Myc, p53, survivin, and Bcl-2, provide a rational linkage between cell cycle and apoptosis (Vermeulen et al., 2003). Among these factors, tumor suppressor p53 performs a pivotal role switching between cell cycle regulation and apoptosis induction (Vogelstein et al., 2000). Functional p53 mainly contributes as a transcriptional factor with both capabilities of transactivation and transrepression. Hundreds of p53-responsive genes have been identified (Xu and el-Gewely, 2001). It was reported that differential transregulation by p53, which means transactivation or transrepression of different subsets of p53 target genes, was important to determine what downstream event would be elicited (Szak et al., 2001; Khan and Bisen, 2013). Nonetheless, more details about the differential transregulation of p53 remain to be elucidated.

Many chemotherapeutic agents have been used for tumor therapy by inducing apoptosis of tumor cells. It was reported that cells treated with some antitumor agents had different susceptibility to apoptosis in a cell cycle-related manner (Gorczyca et al., 1993). This cell cycle stage-dependent susceptibility to apoptosis was also found in cells treated with staurosporine (Bernard et al., 2001) and arsenite (McCabe et al., 2000). This different susceptibility may be explained by the variety of the lesion severity and the repairability to the lesion. It is supported by the evidence that p53 can modulate the activity of base excision repair in a cell cycle-specific manner after genotoxic stress induced by γ-irradiation (Offer et al., 2001). However, the mechanism controlling the cell cycle-dependent susceptibility to apoptosis is still unknown.

Wild-type p53 is supposed to repress survivin expression at the mRNA level (Mirza et al., 2002). The repression mechanism of survivin was understood using an adenovirus vector-based transfection for wild-type p53, human ovarian cancer cell line 2774qwl (which expresses mutant p53). The downregulation of survivin mRNA levels was well analyzed by real-time quantitative PCR (RT-PCR). The results of the study showed time-dependent downregulation of when the cells were infected with wild-type p53. A 3.6-fold decrease of survivin mRNA level was observed 16 hours after infection initiation and decreased 6.7-fold 24 hours after infection. Western blot results do show that indeed the p53 from the adenoviral vector was being expressed in the cells using an antibody specific for p53. The expression of p53 levels indicates its role in survivin repression. It was also reported that p53 expressed 6 hours after the onset of infection and had its highest level at 16 to 24 hours. Further confirmation of repression of survivin was established by the authors, after induction of A549 (human lung cancer cell line with wild-type p53) and T47D (human breast cancer cell line with mutant p53) cells with DNA-damaging agent adriamycin. This induction was done to trigger the physiological p53 apoptotic response in these cancer cells and compare the survivin levels measured to the same cells without DNA damage induction. The A549 line, which intrinsically has functioning wild-type p53, showed a significant reduction in survivin levels compared to noninduced cells (Mirza et al., 2002). This same effect was not seen in T47D cells that carry mutant inactive p53.

P53's normal function is to regulate genes regulating apoptosis. Survivin is a known inhibitor of apoptosis, and the p53 repression of survivin is one mechanism by which cells can undergo apoptosis upon induction by any apoptotic stimuli or signals. Survivin is overexpressed in the cell lines mentioned in the previous paragraph, the apoptotic response from the DNA-damaging agent. The downregulation of survivin by p53 is important for p53-mediated apoptosis. Most tumors showed overexpression of survivin with the complete loss of wild-type p53. The evidence put forth by Mirza et al. (2002) shows that there exists a link between survivin and p53 that can explain a critical event that contributes to cancer progression.

In order to see whether p53 reexpression in cancer cells (that have lost p53 expression) has the suppressive effect on the promoter of the survivin gene, a luciferase reporter construct was made. The isolated survivin promoter was placed upstream of the luciferase reporter gene. In a luciferase reporter assay, if the promoter is active, the luciferase gene is transcribed and translated into a product that gives off light that can be measured quantitatively, and thus represents the activity of the promoter. This construct was transfected into cancer cells that had either wild-type or mutant p53. High luciferase activity was measured in the cells with mutant p53, and significantly lower luciferase levels were measured for cells with wild-type p53 (Mirza et al., 2002).

Transfection of different cell types with wild-type p53 was associated with a strong repression of the survivin promoter. Transfection with mutant p53 was not shown to strongly repress the survivin promoter. More luciferase constructs were prepared with varying degrees of deletion from the 5' end of the survivin promoter region. At one point, there was the deletion that caused the survivin levels to be indifferent to the presence of the p53 overexpression plasmid, indicating that there is a specific region proximal to the transcription start site that is needed for p53 suppression of survivin. Although it has been found that two p53 binding sites are located on the survivin gene promoter, analysis using deletions and mutations has shown that these sites are not essential to transcriptional inactivation. Instead, it is observed that modification of the chromatin inside of the promoter region may be responsible for the transcriptional repression of the survivin gene (Mirza et al., 2002).

7.4.2 Regulation of Survivin by c-Myc in BCR/ABL-Transformed Cells

BCR/ABL can cause chronic myelogenous leukemia (CML), in part by altering the transcription of specific genes with growth- or survival-promoting functions. Recently, BCR/ABL has been shown to activate survivin, an important regulator of cell growth and survival, but the precise molecular mechanisms behind its expression and consequences thereof, in CML cells, remain unclear. The increase of survivin was largely controlled at the transcriptional level through a mechanism mediated by JAK2/PI3K signal pathways that activated c-Myc, leading to transactivation of the survivin promoter. Dynamic downregulation of survivin was a key event involved in imatinib-induced cell death, while forced expression of survivin partially counteracted imatinib's effect on cell survival. Additionally, shRNA-mediated silencing of survivin or c-Myc eradicated colony formation of K562 cells in a semisolid culture system, imply an essential role for this transcriptional network in BCR/ABL-mediated cell transformation and survival. Finally, interruption of c-Myc activity by 10058-F4 exerted an antileukemia effect with a synergistic interaction with imatinib and overcame the anti-apoptosis rescued by the IL-3 supplement (Fang et al., 2009).

7.4.3 Regulation of Survivin by Hsp90

Pathways controlling cell proliferation and cell survival require flexible adaptation to environmental stresses. These mechanisms are frequently exploited in cancer, allowing tumor cells to thrive in unfavorable milieus. Hsp90, a molecular chaperone that is central to the cellular stress response, associates with survivin, an apoptosis inhibitor and essential regulator of mitosis. This interaction involves the ATPase domain of Hsp90 and the survivin baculovirus inhibitor of apoptosis repeat. Global suppression of the Hsp90 chaperone function or targeted Ab-mediated disruption of the survivin-Hsp90 complex results in proteasomal degradation of survivin, mitochondrial-dependent apoptosis, and cell cycle arrest with mitotic defects. These data link the cellular stress response to an anti-apoptotic and mitotic checkpoint maintained by survivin. Targeting the survivin-Hsp90 complex may provide a rational approach for cancer therapy.

Tumor cells exhibit extraordinary plasticity to adapt to noxious stimuli and thrive in unfavorable environments. This typically involves increased resistance to apoptosis (Hanahan and Weinberg, 2000) by deregulated overexpression of cell death antagonists of the Bcl-2 or inhibitor of apoptosis (IAP) gene family or loss of cell death activators/effectors (Shin et al., 2002). Another cancer-promoting factor is enhanced adaptation to environmental challenges, which leads to upregulation of the cellular stress response. This preserves folding of nascent polypeptides, prevents protein aggregation (Schiene and Fischer, 2000), and ensures specialized intracellular trafficking of client proteins. The protein-folding quality control machinery is orchestrated by heat shock proteins (Hsps), a family of evolutionary conserved ATPase-directed molecular chaperones (Lindquist and Craig, 1988).

In particular, Hsp90 (Prodromou et al., 2000) controls the balance between folding/maturation and proteasomal destruction of a restricted number of client proteins that are typically involved in signal transduction and cell proliferation (Figure 7.6).

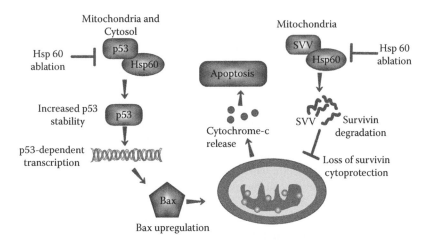

FIGURE 7.6   Model of Hsp60-regulated cytoprotection in cancer.

This pathway is exploited in cancer where Hsp90 is upregulated and may be linked to resistance to apoptosis by inhibition of caspase-9 activation (Pandey et al., 2000), induction of anti-apoptotic Bcl-2, or stabilization of survival kinase RIP-1 (Lewis et al., 2000) or Akt (Sato et al., 2000). Hsp90 antagonists are being explored as novel cancer therapeutics (Isaacs et al., 2003).

Survivin is found to be overexpressed in many tumor types. However, the exact mechanism of abnormal expression of survivin is unknown. p53 is downregulated in almost all cancers, so it is tempting to suggest that survivin overexpression is due to p53 inactivity. Wagner et al. (2008) investigated the possible molecular mechanism involved with the overexpression of survivin in acute myeloid leukemia (AML) by performing an epigenetic and genetic analysis of the survivin gene promoter region in AML patients and compared the observations to what was seen in peripheral blood mononuclear cells (PBMCs) that have been shown to express no survivin, and found that the methylation status is not an important regulator of survivin reexpression during leukemogenesis. Unmethylated survivin promotes in AML and PBMCs treated with bisulfite, and the survivin promoter region sequence. Three single-nucleotide polymorphisms (SNPs) were identified and were all present, both in AML patients and in healthy donors, suggesting that the occurrence of these SNPs in the promoter region of the survivin gene also appears to be of no importance to survivin expression (Wagner et al., 2008). However, it has not been ruled out yet that there may be other possible epigenetic mechanisms that may be responsible for a high level of survivin expression observed in cancer cells and not in normal cells; for example, the acetylation profile of the survivin promoter region can also be looked at. Different cancer and tissue types may have slight or significant differences in the way survivin expression is regulated in the cell, and thus the methylation status or genetic differences in the survivin promoter may be observed to be different in different tissues. Thus, further experiments assessing the epigenetic and genetic profile of different tumor types must be investigated.

7.5   ANTI-APOPTOTIC NETWORK OF SURVIVIN

The role of survivin in the inhibition of apoptosis has a similar degree of complexity, connecting to multiple parallel pathways that regulate gene expression, protein-protein interactions, and mitochondrial functions.

7.5.1   PROVIDING A HEIGHTENED CELL SURVIVAL THRESHOLD

In addition to a stable and protracted mitotic arrest, acute lowering of survivin expression (for instance, using antisense, small interfering RNA, ribozymes, or dominant negative mutants) has often been associated with spontaneous apoptosis, depending on the cell type and its complement of checkpoints. Accordingly, pathways that regulate gene expression and control protein stability extensively intersect with the survivin cytoprotection network. Many prototype tumor suppressor genes result in efficient silencing of the *BIRC5* promoter.

These include the adenomatous polyposis coli protein (Zhang et al., 2001), which is often deleted or mutated in colorectal cancer; p53; fragile histidine triad (FHIT) gene (Semba et al., 2006), which is a pro-apoptotic molecule that binds and hydrolyzes diadenosine polyphosphates; and PML4, a pro-apoptotic promyelocytic leukemia protein (Xu et al., 2004). By contrast, oncogenic factors have been shown to promote *BIRC5* transcription. This is the case for TCF4-β-catenin (Kim et al., 2003), a developmentally regulated transcriptional activator complex participating in colon cancer, signal transduction, and activator of transcription 3 (Stat3) (Gritsko et al., 2006), which is an oncogenic transcription factor involved in cytokine signaling, and a group of E2F transcription factors (Jiang et al., 2004), which function in the G1-S transition of the cell cycle. Of these regulators, discrete binding sites on the *BIRC5* promoter have been identified for TCF4, p53, and Stat3 (Gu et al., 2007), suggesting that these molecules might directly control *BIRC5* expression. A second posttranscriptional network that controls survivin mRNA or protein stability has also been characterized. This involves several factors: the mammalian target of rapamycin (mTOR) (also known as FRAP1), which is required for stability and translation of a cytosolic pool of *BIRC5* mRNA (Vaira et al., 2007); intermediaries of growth factor receptor signaling, especially the phosphatidylinositol-3-kinase–Akt axis, which has been frequently implicated in the modulation of survivin levels (Asanuma et al., 2005); CDK1 phosphorylation, which promotes increased survivin stability at mitosis, and binding of survivin to molecular chaperones, including the aryl hydrocarbon receptor-interacting protein (AIP) (Kang and Altieri, 2006); and Hsp90 (Fortugno et al., 2003), which participates in survivin stability and subcellular trafficking pathways. Pharmacological antagonists of some of these pathways are being tested for cancer therapy, and their ability to lower survivin levels may contribute to their anticancer activity. In addition, changes in survivin expression could provide an accessible biomarker of target validation for patients treated with inhibitors of Hsp90 (17-AAG), the EGFR family (lapatinib), or CDK1 (flavopiridol).

7.5.2 Intermolecular Cooperation

One of the crucial features of this cytoprotective network is that it relies on physical interactions between survivin and other adaptor or cofactor molecules. This may explain why earlier studies with isolated recombinant survivin in a cell-free system did not show anti-apoptotic effects. In the cytosol, survivin associates with the hepatitis B X-interacting protein (HBXIP), and this complex, but not either protein alone, binds caspase-9 and inhibits mitochondrial apoptosis (Marusawa et al., 2003). Survivin exhibits parallel interactions with other members of the IAP gene family. One interaction involves XIAP, which binds the pool of survivin released from mitochondria in response to cell death stimuli (Dohi et al., 2007), resulting in increased XIAP stability against proteasomal degradation and inhibition of apoptosis in vivo. This pathway can be recapitulated in vitro with recombinant proteins with synergistic inhibition of caspase-3 and -9 activity. Assembly of the survivin-XIAP complex in vivo is regulated in subcellular compartments, and phosphorylation of survivin on Ser20 by protein kinase A in the cytosol, but not in mitochondria, disassembles the complex, and abolishes its anti-apoptotic function (Dohi et al., 2007) (Figure 7.5).

The interaction between survivin and the RAN-GTP effector XPO1 may also bridge cell division and cytoprotective networks, as it may be required to localize survivin for apoptosis inhibition in the cytosol. Conversely, a complex of survivin and cIAP1 (also known as *BIRC2*) might not participate in apoptosis inhibition, but seems to feed back on the regulation of survivin during cell division (Samuel et al., 2005). Cells overexpressing cIAP1 displayed extensive mitotic defects, cytokinesis failure, and a propensity for chromosomal instability, suggesting that a survivin-cIAP1 complex might antagonize the function of survivin in late-stage cell division (Samuel et al., 2005). Finally, survivin has been implicated in heterodimeric interaction with at least some of its alternatively spliced forms. With the caveats that these results were obtained using overexpression approaches, and that the balance of survivin dimers versus monomers in vivo is far from understood, it has been suggested that these complexes may also participate in cytoprotection (Caldas et al., 2005b).

7.5.3 Mitochondrial Dynamics

Recent evidence suggests that survivin cytoprotection hinges on a pool of the molecule compartmentalized in mitochondria, and released in the cytosol in response to cell death stimuli. Accordingly, there are multiple signaling pathways of mitochondrial homeostasis that connect to survivin cytoprotection. First, although it is as yet unclear how survivin is transported to mitochondria, its regulated association with molecular chaperones, AIP56 or Hsp90, might contribute to this process, potentially through the import receptor complexes at the outer mitochondrial membrane, TOM20 (also known as TOMM20) and TOM70 (also known as TOMM70A). Second, mitochondrial survivin is posttranslationally modified, and this step is required for its anti-apoptotic function.

Only survivin that is kept unphosphorylated on Ser20 retains the ability to bind XIAP and antagonize cell death, and this process may involve compartmentalized proximity in mitochondria between survivin and the broad-spectrum phosphatase,

PP2A, which dephosphorylates survivin on Ser20. Third, once transported in mitochondria and properly processed, survivin binds Smac (also known as DIABLO) (Sun et al., 2005), a molecule that relieves the inhibitory effect of XIAP on caspases and thus promotes cell death. The actual physiological relevance of a survivin-Smac complex in vivo has not been fully established, a caveat that might apply to other reported interactions involving survivin, for instance, when using supraphysiological overexpression approaches.

However, there have been reports that this interaction may regulate apoptosis directly, by sequestering Smac away from XIAP (Song et al., 2003), or indirectly, by preventing its release from mitochondria (Ceballos-Cancino et al., 2007). In a parallel pathway, an alternatively spliced survivin variant, called survivin-Ex-3, containing a novel carboxy terminus sequence due to a frameshift, has been shown to localize to mitochondria, where it interacts with Bcl-2 and inhibits caspase-3 activity (Wang et al., 2002). Because anti-apoptotic Bcl-2 proteins function as inhibitors of mitochondrial permeability transition, this recognition would position survivin, or at least one of its spliced variants, in the regulation of mitochondrial membrane integrity.

Variations of this pathway have been suggested, involving hyperphosphorylation of Bcl-2, and reduced activation of pro-apoptotic Bcl-2-associated X protein (Bax) by the survivin–aurora kinase B complex, potentially upstream of caspase activation (Vogel et al., 2007), thus further dampening mitochondrial permeability. It is too soon to tell whether a broader basis exists for a role of survivin in mitochondrial homeostasis, but it is intriguing that survivin-Ex-3 was recently shown to maintain mitochondrial membrane potential and control the production of reactive oxygen species in response to cell death stimuli (You et al., 2006). It should be kept in mind that the actual abundance of survivin splice variants in tumor cells appears to be quite low. Although this does not negate the a priori role of these molecules in survivin regulation, definitive validation of this model awaits the availability of specific reagents capable of faithfully discriminating the various endogenous survivin isoforms in different cells and tissues.

Therefore, the survivin networks in cell division and cell death emerge as highly flexible signaling hubs, connecting to multiple independent pathways of cellular homeostasis. It seems plausible to hypothesize that the cytoprotective and mitotic functions of survivin intersect at cell division. There is compelling experimental evidence to support this model, as interference with survivin expression or function in synchronized cultures often culminates in a form of apoptosis of dividing cells called mitotic catastrophe. Probably, this is not the whole story, and survivin cytoprotection is probably operative in interphase as well. This is consistent with the fact that tumor cells have constitutively high levels of survivin in interphase in vivo, and that cell cycle-regulated transcription of *BIRC5* at mitosis cannot account for the expression of endogenous survivin in transgenic mice (Xia and Altieri, 2006), and with the dynamics of mitochondrial survivin, which are uncoupled from cell cycle progression. This is consistent with the formation of a survivin-caspase-9 complex in vivo (O'Connor et al., 2000a, 2000b), the requirement of apoptosome components Apaf-1 and caspase-9 in the survivin pathway, and the ability of survivin to prevent incorporation of caspase-9 into a functional apoptosome.

7.6   EXPRESSION OF SURVIVIN IN NORMAL TISSUES

Apoptosis plays a major role during embryonic development, especially in the maintenance of tissue homeostasis (Vaux et al., 1994). Dysregulation of apoptosis leads to the survival of unwanted cells or death of required cells, resulting in cancer (Thompson, 1995). The inhibitor of apoptosis (IAP) protein family is intracellular proteins that suppress the apoptosis induced by a variety of stimuli (Deveraux et al., 1998). Survivin is a member of the IAP family. Its levels remain undetectable in most adult normal tissues, whereas its expression in human common tumors has been associated with increased aggressiveness and decreased patient survival (Kawasaki et al., 1998; Lu et al., 1998; Monzo et al., 1999; Swana et al., 1999; Tanaka et al., 2000), which suggests that apoptosis inhibition by survivin could be an important predictive/prognostic parameter of poor outcome in human cancers and can help in developing a diagnostic/therapeutic target in the case of malignant tumors. However, the mechanisms of survivin involved in oral squamous cell carcinomas (OSCCs) remain unknown.

Survivin, an inhibitor of apoptosis protein, is highly expressed in most cancers and is associated with chemotherapy resistance, increased tumor recurrence, and shorter patient survival, making antisurvivin therapy an attractive cancer treatment strategy. However, growing evidence indicates that survivin is expressed in normal adult cells, particularly primitive hematopoietic cells, T-lymphocytes, polymorphonuclear neutrophils, and vascular endothelial cells, and may regulate their proliferation or survival. In preclinical animal models, targeted antisurvivin therapies show efficacy without overt toxicity. However, consequences of prolonged survivin disruption in normal cells, particularly those associated with continuous renewal, have not been clearly determined.

Menstruation has been recognized as ischemic necrosis in the endometrium (Speroff and Vande Wiele, 1971). However, some authors (Hopwood and Levison, 1975) reported apoptotic bodies in the human endometrium through light and electron microscopy. Recently, it has been reported that apoptosis appears in the human endometrium during the menstrual cycle by DNA gel fragmentation and in situ apoptosis analysis techniques (Kokawa et al., 1996). The cyclic Bcl-2 expression in the human endometrium has been demonstrated by several investigators (Gompel et al., 1994; Otsuki et al., 1994; Tabibzadeh et al., 1995; Koh et al., 1995; Jones et al., 1998). Endometrial glandular cells express Bcl-2 during the proliferative phase, but not during the late secretory phase. The disappearance of Bcl-2 expression during the late secretory phase is consistent with the appearance of apoptotic cells during this same phase. Bcl-2 expression in glandular cells may also be regulated by steroid hormones (Gompel et al., 1994; Otsuki et al., 1994; Koh et al., 1995; Tabibzadeh et al., 1995).

7.6.1   SURVIVIN IN EMBRYOGENESIS

A high expression of survivin has been found during embryonic development, while the gene is quiescent in most terminally differentiated tissues. IAP family members, especially survivin, showed restricted expression in adult tissues compared to several apoptosis-regulated fetal tissues (Adida et al., 1998a; 2000a,b). In contrast, survivin transcripts were undetectable in terminally differentiated adult tissues, including peripheral blood leukocytes, lymph node, spleen, pancreas, kidney, skeletal muscle, liver, lung, brain, and heart (Ambrosini et al., 1997), reminiscent of the expression of another apoptosis inhibitor, Bcl-2, in these adult organs (LeBurn et al., 1993). Normal adult human tissues that express survivin include thymus (Ambrosini et al., 1997), CD341 bone marrow stem cells (Carter et al., 2001), and basal colonic epithelium (Gianani et al., 2001).

The regulation of programmed cell death and normal morphogenesis of embryonic tissues and their developmental homeostasis are due to the presence of survivin. Gene targeting experiments indicate that survivin is both a pro-survival and an anti-apoptotic factor (Ambrosini et al., 1998). Survivin has also been shown to be essential for mitosis during development since null embryos exhibit disrupted microtubule formation, become polyploid, and fail to survive beyond day 4.5 (Uren et al., 2000). Survivin, expressed in a cell cycle-dependent manner, has been shown, in cell lines, to translocate into the nucleus where it competitively binds to Cdk4/p16^{INK4a}. The resultant Cdk4-survivin complex directly or indirectly activates Cdk2/cyclin E for S phase entry (Suzuki et al., 2000a, 2000b). The survivin-Cdk4 complex formation also causes the release of p21, which translocates to mitochondria to form a complex with procaspase-3, which inhibits cell death.

Survivin is also expressed in the gastric mucosa of adult humans and rats (Chiou et al., 2003) and also in human thymocyte subsets; double positive thymocytes express the highest levels of survivin, demonstrating upregulation in this subset of T-lymphocytes (Kobayashi et al., 2002). There is downregulation of survivin, however, in mature single positive (CD41 or CD81) T-cells, and peripheral T-cells are negative for survivin. Thus, survivin may be important in T-cell development. Survivin is also normally expressed in CD341 hematopoietic stem cells, which raises the concern that survivin-targeted therapy for cancer cells could disturb normal hematopoiesis (Fukuda and Pelus, 2001). CD341 cells expressed survivin in all phases of the cell cycle.

7.7 EXPRESSION OF SURVIVIN IN HUMAN CANCER

One of the clinically significant features of survivin is its differential distribution in many cancers compared with its limited expression in normal, terminally differentiated tissues. In human medicine, detection of survivin expression by immunohistochemical staining is becoming an important prognostic parameter in a variety of cancers, including carcinomas, sarcomas, and hematologic neoplasms. High survivin expression in tumors correlates with a more aggressive and invasive clinical phenotype. Subsequently, a poorer prognosis and a decreased responsiveness to chemotherapeutic agents can be expected (Adida et al., 1998a, 2000a,b; Islam et al., 2000; Kato et al., 2001; Kawasaki et al., 1998; Monzo et al., 1999; Serela et al., 2000; Swana et al., 1999; Tanaka et al., 2000; Zhao et al., 2000). Several human carcinomas have been shown to express high levels of survivin: lung (Monzo et al., 1999), breast (Tanaka et al., 2000), colon (Kawasaki et al., 1998), stomach (Lu et al., 1998), esophagus (Kato et al., 2001), pancreas (Satoh et al., 2001), bladder (Swana et al., 1999), uterus (Saitoh et al., 1999), ovary (Yoshida et al., 2001), liver (Ito et al., 2000), and nonmelanoma skin cancer (Grossman and Leffell, 1997, 2008; Grossman et al., 1999).

The overexpression of survivin is consistently associated with more aggressive tumor types and a poorer prognosis than tumors that are negative for survivin. For example, normal oral mucosa and skin, including adnexal structures, are negative for survivin. In contrast, one study found that 56 and 64% of oral and cutaneous squamous cell carcinomas, respectively, were strongly positive for survivin (Muzio et al., 2001). Likewise, survivin expression correlated with larger, more poorly differentiated tumors, including those that had metastasized to lymph nodes. Detection of survivin in urine may also prove to be a diagnostic marker of bladder cancer (Smith et al., 2001). Its presence in urine has been used to diagnose bladder cancer and differentiate neoplastic lesions from inflammatory conditions with a sensitivity of 100% and specificity of 95%.

Overexpression of survivin has also been documented in leukemias (Adida et al., 2000a; Kamihira et al., 2001), neuroblastoma (Adida et al., 1998a; Islam et al., 2000), melanoma (Grossman et al., 1999), soft tissue sarcoma (Kappler et al., 2001), and high-grade non-Hodgkin's lymphoma. Low-grade non-Hodgkin's lymphomas, however, do not express survivin (Ambrosini et al., 1997). The overexpression of survivin is not related to carcinomas; hence, survivin could be a poor prognostic marker (Adida et al., 2000a,b). Survivin expression is also associated with viral-induced neoplasms, and its expression may be an early indicator of malignancy (Frost et al., 2002). Immunohistochemical evaluation of normal and abnormal human cervical squamous epithelia, including low- and high-grade squamous intraepithelial lesions (LSILs, HSILs) and squamous cell carcinomas, reveals nuclear staining in the normal epithelium, LSILs, and HSILs. Staining intensity is variable, however, with the greatest intensity of staining in the tissues containing human papillomavirus (HPV). In situ hybridization of these HPV positive tissues reveals colocalization of HPV DNA and survivin, which reveals that survivin plays a role in HPV-mediated cervical dysplasia. Survivin expression is also present in juvenile-onset recurrent respiratory papillomatous (Poetker et al., 2002). In comparison with normal laryngeal tissue that is negative for survivin, viral-induced papillomas have increased mRNA and protein expression of survivin. Likewise, the strongest survivin expression is present in papillomas that have undergone malignant transformation.

Survivin is expressed in the G_2-M phase of the cell cycle in a cell cycle-regulated manner and associates with microtubules of the mitotic spindle. Disruption of survivin-microtubule interactions results in loss of survivin's anti-apoptosis function and increased caspase-3 activity during mitosis. The overexpression of survivin in cancer may obliterate this apoptotic checkpoint and allow aberrant progression of transformed cells through mitosis. In breast carcinoma and in many of the most common human cancers, inhibition of apoptosis may be a general feature, and expression of *survivin* alone or *survivin* plus other anti-apoptosis genes like *bcl-2* may cause more pronounced anti-apoptotic effects. Identification of prognostic significance of the apoptotic index should be clarified with additional use of these prognostic markers. *bcl-2* has been reported to be frequently expressed in breast cancer and to be associated with positivity for estrogen receptors (ERs) in both node negative and node positive breast cancers (Leek et al., 1994). In clinical studies, the *bcl-2* expression is inversely correlated with S phase fraction and tumor size in breast cancer (Lipponen et al., 1995). Furthermore, *bcl-2* has proved to have the potential to lead to prolongation of the cell cycle as well as a decrease in in vitro breast cancer growth. It is well known that there are multiple genetic pathways that control apoptosis, a part of which is probably regulated by survivin or *bcl-2*. Therefore, changing the level of expression of these proteins may not necessarily have an effect on the outcome of therapy.

Another potential novel molecular progression marker is survivin, an apoptosis inhibitor related to the *bcl-2* family with the particular characteristic of being selectively expressed in all of the most common human cancers, although not in normal tissue counterparts. Furthermore, as occurs with the sonic hedgehog gene (*Shh*), survivin is expressed in the embryonic lung as well as in other organs in developmental stages. The reexpression of this novel class of genes, which are involved in lung morphogenesis, but not in normal adult tissue, prompted us to speculate that survivin expression might be present in non-small cell lung cancer (NSCLC) tumor samples and influence the prognosis of the disease. Apoptosis resistance is a fundamental property of many solid tumors and may play a critical role in determining tumor response phenotype following treatment with cytotoxic therapy. The low level of spontaneous apoptosis and general resistance of NSCLC cells to a diversity of cytotoxic modalities, including radiotherapy, suggest a defect in intrinsic apoptosis signaling.

The recent growth in understanding of the core apoptosis machinery of the cell has provided a framework for understanding how apoptosis is regulated in NSCLC cells. A critical event during apoptosis involves proteolytic activation of a constitutively expressed family of cytoplasmic zymogens with aspartate-specific cysteine protease activity, termed caspases. These enzymes play a critical role in executing the final common pathway of apoptosis. It may be hypothesized that caspases are targets for suppression in NSCLC, and that this may, in part, account for the apoptosis-resistant phenotype.

7.8   CONCLUSION

Survivin is a chromosomal passenger protein that mediates the spindle assembly checkpoint and cytokinesis. In the nucleus, survivin binds to microtubules and assists in chromosomal segregation and cytokinesis during mitosis. Survivin expression is also highly regulated by the cell cycle and is only expressed in the G2-M phase. It is known that survivin localizes to the mitotic spindle by interaction with tubulin during mitosis, and may play a contributing role in regulating mitosis, but regulation of survivin seems to be linked to the p53 protein. Survivin is observed uniquely in tumor and developmental cells, which undergo either inappropriate or programmed cell growth. Survivin not only controls apoptosis but also participates in cell cycle entry and progression. Survivin and its alternative splice forms are involved in cellular processes, such as cell division and programmed cell death. Expression levels of survivin correlate with tumor aggressiveness and resistance to therapy. This fact therefore makes survivin an ideal target for cancer therapy, as cancer cells are targeted while normal cells are left alone.

REFERENCES

Adida, C., D. Berrebi, M. Peuchmaur, et al. 1998a. Anti-apoptosis gene, survivin, and prognosis of neuroblastoma. *Lancet* 351: 882–883.

Adida, C., P.L. Crotty, J. McGrath, et al. 1998b. Developmentally regulated expression of the novel cancer anti-apoptosis gene survivin in human and mouse differentiation. *Am J Pathol* 152: 43–49.

Adida, C., C. Haioun, P. Gaulard, et al. 2000a. Prognostic significance of survivin expression in diffuse large B-cell lymphomas. *Blood* 96: 1921–1925.

Adida, C., C. Recher, E. Raffoux, et al. 2000b. Expression and prognostic significance of survivin in de novo acute myeloid leukaemia. *Br J Haematol* 111: 196–203.

Altieri, D.C. 2003a. Survivin in apoptosis control and cell cycle regulation in cancer. *Prog Cell Cycle Res* 5: 447–452.

Altieri, D.C. 2003b. Survivin, versatile modulation of cell division and apoptosis in cancer. *Oncogene* 22: 8581–8589.

Altieri, D.C. 2003c. Validating survivin as a cancer therapeutic target. *Nat Rev Cancer* 3: 46–54.

Ambrosini, G., C. Adida, and D.C. Altieri. 1997. A novel anti-apoptosis gene, survivin, expressed in cancer and lymphoma. *Nat Med* 3: 917–921.

Ambrosini, G., C. Adida, G. Sirugo, et al. 1998. Induction of apoptosis and inhibition of cell proliferation by survivin gene targeting. *J Biol Chem* 273: 11177–11182.

Asanuma, H., T. Torigoe, K. Kamiguchi, et al. 2005. Survivin expression is regulated by coexpression of human epidermal growth factor receptor 2 and epidermal growth factor receptor via phosphatidylinositol 3-kinase/AKT signaling pathway in breast cancer cells. *Cancer Res* 65: 11018–11025.

Badran, A., A. Yoshida, K. Ishikawa, et al. 2004. Identification of a novel splice variant of the human anti-apoptosis gene survivin. *Biochem Biophys Res Commun* 314: 902–907.

Barrett R.M., T.P. Osborne, and S.P. Wheatley. 2009. Phosphorylation of survivin at threonine 34 inhibits its mitotic function and enhances its cytoprotective activity. *Cell Cycle* 8: 278–283.

Bernard, B., T. Fest, J.L. Pretet, et al. 2001. Staurosporine-induced apoptosis of HPV positive and negative human cervical cancer cells from different points in the cell cycle. *Cell Death Differ* 8: 234–244.

Bolton, M.A., W. Lan, S.E. Powers, et al. 2002. Aurora B kinase exists in a complex with survivin and INCENP and its kinase activity is stimulated by survivin binding and phosphorylation. *Mol Biol Cell* 13: 3064–3077.

Caldas, H., L.E. Honsey, and R.A. Altura. 2005a. Survivin 2alpha: A novel survivin splice variant expressed in human malignancies. *Mol Cancer* 4: 11.

Caldas, H., Y. Jiang, M.P. Holloway, et al. 2005b. Survivin splice variants regulate the balance between proliferation and cell death. *Oncogene* 24: 1994–2007.

Carmena M., S. Ruchaud, W.C. Earnshaw. 2009. Making the auroras glow: Regulation of aurora A and B kinase function by interacting proteins. *Curr Opin Cell Biol* 21: 796–805.

Carter, B.Z., M. Milella, D.C. Altieri, et al. 2001. Cytokine-regulated expression of survivin in myeloid leukemia. *Blood* 97: 2784–2790.

Carvalho, A., M. Carmena, C. Sambade, W.C. Earnshaw, and S.P. Wheatley. 2003. Survivin is required for stable checkpoint activation in taxol-treated HeLa cells. *J Cell Sci* 116: 2987–2998.

Ceballos-Cancino, G., M. Espinosa, V. Maldonado, et al. 2007. Regulation of mitochondrial Smac/DIABLO-selective release by survivin. *Oncogene* 26: 7569–7575.

Chen, J., W. Wu, S.K. Tahir, et al. 2000. Down-regulation of survivin by antisense oligonucleotides increases apoptosis, inhibits cytokinesis and anchorage-independent growth. *Neoplasia* 2: 235–241.

Chiou, S., W.S. Moon, M.K. Jones, et al. 2003. Survivin expression in the stomach: Implications for mucosal integrity and protection. *Biochem Biophys Res Commun* 305: 374–379.

Conway, E.M., S. Pollefeyt, J. Cornelissen, et al. 2000. Three differentially expressed survivin cDNA variants encode proteins with distinct antiapoptotic functions. *Blood* 95: 1435–1442.

Cory, S., and J.M. Adams. 2002. The Bcl2 family: Regulators of the cellular life-or-death switch. *Nat Rev Cancer* 2: 647–656.

Deveraux, Q.L., and J.C. Reed. 1999. IAP family proteins-suppressors of apoptosis. *Genes Dev* 13: 239–252.

Deveraux, Q.L., N. Roy, H.R. Stennicke, et al. 1998. IAPs block apoptotic events induced by caspase-8 and cytochrome c by direct inhibition of distinct caspases. *EMBO J* 17: 2215–2223.

Dohi, T., F. Xia, and D.C. Altieri. 2007. Compartmentalized phosphorylation of IAP by protein kinase A regulates cytoprotection. *Mol Cell* 27: 17–28.

Du Pasquier, D., A.C. Phung, Q. Ymlahi-Ouazzani, et al. 2006. Survivin increased vascular development during *Xenopus* ontogenesis. *Differentiation* 74: 244–253.

Evan, G.I., and K.H. Vousden. 2001. Proliferation, cell cycle and apoptosis in cancer. *Nature (London)* 411: 342–348.

Fang, Z.H., C.L. Dong, Z. Chen, et al. 2009. Transcriptional regulation of survivin by c-Myc in BCR/ABL-transformed cells: Implications in anti-leukaemic strategy. *J Cell Mol Med* 13: 2039–2052.

Fortugno, P., E. Beltrami, J. Plescia, et al. 2003. Regulation of survivin function by Hsp90. *Proc Natl Acad Sci USA* 100: 13791–13796.

Fortugno, P., N.R. Wall, A. Giodini, et al. 2002. Survivin exists in immunochemically distinct subcellular pools and is involved in spindle microtubule function. *J Cell Sci* 115: 575–585.

Frost, M., E.A. Jarboe, D. Orlicky, et al. 2002. Immunohistochemical localization of survivin in benign cervical mucosa, cervical dysplasia, and invasive squamous cell carcinoma. *Am J Clin Pathol* 117: 738–744.

Fukuda, S., and L.M. Pelus. 2001. Regulation of the inhibitor-of-apoptosis family member survivin in normal cord blood and bone marrow CD34(+) cells by hematopoietic growth factors: Implication of survivin expression in normal hematopoiesis. *Blood* 98: 2091–2100.

Gianani, R., E. Jarboe, D. Orlicky, et al. 2001. Expression of survivin in normal, hyperplastic, and neoplastic colonic mucosa. *Hum Pathol* 32: 119–125.

Giodini, A., M.J. Kallio, N.R. Wall, et al. 2002. Regulation of microtubule stability and mitotic progression by survivin. *Cancer Res* 62: 2462–2467.

Gompel, A., J.C. Sabourin, A. Martin, *et al.* 1994. Bcl-2 expression in normal endometrium during the menstrual cycle. *Am J Pathol* 144: 1195–1202.

Gorczyca, W., J. Gong, B. Ardelt, et al. 1993. The cell cycle related differences in susceptibility of HL-60 cells to apoptosis induced by various antitumor agents. *Cancer Res* 53: 3186–3192.

Gritsko, T., A. Williams, J. Turkson, et al. 2006. Persistent activation of stat3 signaling induces survivin gene expression and confers resistance to apoptosis in human breast cancer cells. *Clin. Cancer Res* 12: 11–19.

Grossman, D., and D.J. Leffell. 1997. The molecular basis of nonmelanoma skin cancer: New understanding. *Arch Dermatol* 133: 1263–1270.

Grossman, D., and D.J. Leffell. 2008. Squamous cell carcinoma. In *Fitzpatrick's dermatology in general medicine*, ed. K. Wolff, et al., 1028–1036. 7th ed., vol. 1. McGraw-Hill Medical, New York.

Grossman, D., J.M. McNiff, F. Li, et al. 1999. Expression of the apoptosis inhibitor, survivin, in nonmelanoma skin cancer and gene targeting in a keratinocyte cell line. *Lab Invest* 79: 1121–1126.

Gu, L., K.Y. Chiang, N. Zhu, et al. 2007. Contribution of STAT3 to the activation of survivin by GM-CSF in CD34+ cell lines. *Exp Hematol* 35: 957–966.

Hanahan, D., and R.A. Weinberg. 2000. The hallmarks of cancer. *Cell* 100: 57–70.

Hengartner, M.O. 2000. The biochemistry of apoptosis. *Nature (London)* 407: 770–776.

Hopwood, D., and D. Levison. 1975. Atrophy and apoptosis in the cyclical human endometrium. *J Pathol* 119: 159–166.

Isaacs, J.S., W. Xu, and L. Neckers. 2003. Heat shock protein 90 as a molecular target for cancer therapeutics. *Cancer Cells* 3: 213–217.

Islam, A., H. Kageyama, K.Y. Hashizume, et al. 2000. Role of survivin, whose gene is mapped to 17q25, in human neuroblastoma and identification of a novel dominant-negative isoform, survivin-beta/2B. *Med Pediatr Oncol* 35: 550–553.

Ito, Y., K. Suzuki, N. Ichino, et al. 2000. The risk of *Helicobacter pylori* infection and atrophic gastritis from food and drink intake: A cross-sectional study in Hokkaido, Japan. *Asia Pac J Cancer Prev* 1: 147–156.

Jeyaprakash, A.A., U.R. Klein, D. Lindner, et al. 2007. Structure of a survivin- borealin-INCENP core complex reveals how chromosomal passengers travel together. *Cell* 131: 271–285.

Jiang, Y., H.I. Saavedra, M.P. Holloway, et al. 2004. Aberrant regulation of survivin by the RB/E2F family of proteins. *J Biol Chem* 279: 40511–40520.

Jones, R.K., R.F. Searle, J.A. Stewart, et al. 1998. Apoptosis, bcl-2 expression, and proliferative activity in human endometrial stroma and endometrial granulated lymphocytes. *Biol Reprod* 58: 995–1002.

Kamal, A., L. Thao, J. Sensintaffar, et al. 2003. A high-affinity conformation of Hsp90 confers tumour selectivity on Hsp90 inhibitors. *Nature (London)* 425: 407–410.

Kamihira, S., Y. Yamada, Y. Hirakata, et al. 2001. Aberrant expression of caspase cascade regulatory genes in adult T-cell leukaemia: Survivin is an important determinant for prognosis. *Br J Haematol* 114: 63–69.

Kang, B.H., and D.C. Altieri. 2006. Regulation of survivin stability by the aryl hydrocarbon receptor-interacting protein. *J Biol Chem* 281: 24721–24727.

Kappler, M., T. Kohler, C. Kampf, et al. 2001. Increased survivin transcript levels: An independent negative predictor of survival in soft tissue sarcoma patients. *Int J Cancer* 95: 360–363.

Kato, J., Y. Kuwabara, M. Mitani, et al. 2001. Expression of survivin in oesophageal cancer: Correlation with the prognosis and response to chemotherapy. *Int J Cancer* 95: 92–95.

Kawasaki, H., D.C. Altieri, C.D. Lu, et al. 1998. Inhibition of apoptosis by survivin predicts shorter survival rates in colorectal cancer. *Cancer Res* 58: 5071–5074.

Khan, Z., N. Khan, A.K. Varma, et al. 2010. Oxaliplatin-mediated inhibition of survivin increases sensitivity of head and neck squamous cell carcinoma cell lines to paclitaxel. *Curr Cancer Drug Targets* 10: 660–669.

Khan, Z., R.P. Tiwari, N. Khan, et al. 2012. Induction of apoptosis and sensitization of head and neck squamous carcinoma cells to cisplatin by targeting survivin gene expression. *Curr Gene Ther* 12: 444–453.

Khan, Z., P. Bhadouria, R. Gupta, et al. 2006. Tumor control by manipulation of the human anti-apoptotic survivin gene. *Curr Cancer Ther Rev* 2: 73–79.

Khan, Z., and Bisen, P.S. 2013. Oncoapoptotic signaling and deregulated target genes in cancers: Special reference to oral cancer. *Biochim Biophy Acta Reviews on Cancer*, dx.doi.org/10.1016/j.bbcan.2013.04.002.

Kim, P.J., J. Plescia, and H. Clevers. 2003. Survivin and molecular pathogenesis of colorectal cancer. *Lancet* 362: 205–209.

Kobayashi, Y., H. Yukiue, H. Sasaki, et al. 2002. Developmentally regulated expression of survivin in the human thymus. *Hum Immun* 63: 101–107.

Koh, E.A., P.J. Illingworth, W.C. Duncan, et al. 1995. Immunolocalization of bcl-2 protein in human endometrium in the menstrual cycle and simulated early pregnancy. *Hum Reprod* 10: 1557–1562.

Kokawa, J., T. Shikone, and R. Nakano. 1996. Apoptosis in the human uterine endometrium during the menstrual cycle. *J Clin Endocrinol Metab* 81: 4111–4147.

Kroemer, G., and J.C. Reed. 2000. Mitochondrial control of cell death. *Nat Med* 6: 513–519.

LaCasse, E.C., S. Baird, R.G. Korneluk, et al. 1998. The inhibitors of apoptosis (IAPs) and their emerging role in cancer. *Oncogene* 17: 3247–3259.

LeBurn, D.P., R.A. Warnke, and M.L. Cleary. 1993. Expression of Bcl-2 in fetal tissue suggests a role in morphogenesis. *Am J Pathol* 142: 743–753.

Leek, R.D., L. Kaklamanis, F. Pezzella, et al. 1994. bcl-2 in normal human breast and carcinoma, association with oestrogen receptor-positive, epidermal growth factor receptor-negative tumours and in situ cancer. *Br J Cancer* 69: 135–139.

Lewis, J., A. Devin, A. Miller, et al. 2000. Disruption of hsp90 function results in degradation of the death domain kinase, receptor-interacting protein (RIP), and blockage of tumor necrosis factor-induced nuclear factor-kappaB activation. *J Biol Chem* 275: 10519–10526.

Li, F., E.J. Ackermann, C.F. Bennett, et al. 1999. Pleiotropic cell-division defects and apoptosis induced by interference with survivin function. *Nat Cell Biol* 1: 461–466.

Li, F., G. Ambrosini, E.Y. Chu, et al. 1998. Control of apoptosis and mitotic spindle checkpoint by survivin. *Nature (London)* 396: 580–544.

Li, F., and M.G. Brattain. 2006. Role of the survivin gene in pathophysiology. *Am J Pathol* 169: 1–11.

Lindquist, S., and E.A. Craig. 1988. The heat shock protein. *Annu Rev Genet* 22: 631–677.

Lipponen, P., T. Pietilainen, V.M. Kosma, et al. 1995. Apoptosis suppressing protein bcl-2 is expressed in well-differentiated breast carcinomas with favourable prognosis. *J Pathol* 177: 49–55.

Lu, C.D., D.C. Altieri, N. Taniqawa, et al. 1998. Expression of a novel antiapoptosis gene, survivin, correlated with tumor cell apoptosis and p53 accumulation in gastric carcinomas. *Cancer Res* 58: 1808–1812.

Mahotka, C., J. Liebmann, M. Wenzel, et al. 2002. Differential subcellular localization of functionally divergent survivin splice variants. *Cell Death Differ* 9: 1334–1342.

Mahotka, C., M. Wenzel, E. Springer, et al. 1999. Survivin-deltaEx3 and survivin-2B: Two novel splice variants of the apoptosis inhibitor survivin with different antiapoptotic properties. *Cancer Res* 59: 6097–6102.

Martinou, J.C., and D.R. Green. 2001. Breaking the mitochondrial barrier. *Nat Rev Mol Cell Biol* 2: 63–67.

Marusawa, H., S. Matsuzawa, K. Welsh, et al. 2003. HBXIP functions as a cofactor of survivin in apoptosis suppression. *EMBO J* 22: 2729–2740.

Matsumura, S., Toyoshima, F., Nishida, E. 2007. Polo-like kinase 1 facilitates chromosome alignment during prometaphase through BubR1. *J. Biol Chem* 282: 15217–15227.

McCabe, M.J., Jr., K.P. Singh, S.A. Reddy, et al. 2000. Sensitivity of myelomonocytic leukemia cells to arsenite-induced cell cycle disruption, apoptosis, and enhanced differentiation is dependent on the inter-relationship between arsenic concentration, duration of treatment, and cell cycle phase. *J Pharmacol Exp Ther* 295: 724–733.

Mirza, A., M. McGuirk, T.N. Hockenberry, et al. 2002. Human survivin is negatively regulated by wild-type p53 and participates in p53-dependent apoptotic pathway. *Oncogene* 21: 2613–2622.

Mollinari, C., C. Reynaud, S. Martineau-Thuillier, et al. 2003. The mammalian passenger protein TD-60 is an RCC1 family member with an essential role in prometaphase to metaphase progression. *Dev Cell* 5: 295–307.

Monzo, M., R. Rosell, E. Felip, *et al.* 1999. A novel anti-apoptosis gene: Re-expression of survivin messenger RNA as a prognostic marker in non-small-cell lung cancers. *J Clin Oncol* 17: 2100–2104.

Muzio, L., S. Staibano, G. Pannone, et al. 2001. Expression of the apoptosis inhibitor survivin in aggressive squamous cell carcinoma. *Exp Mol Pathol* 70: 249–254.

O'Connor, D.S., D. Grossman, J. Plescia, et al. 2000a. Regulation of apoptosis at cell division by p34cdc2 phosphorylation of survivin. *Proc Natl Acad Sci USA* 97: 13103–13107.

O'Connor, D.S., J.S. Schechner, C. Adida, et al. 2000b. Control of apoptosis during angiogenesis by survivin expression in endothelial cells. *Am J Pathol* 156: 393–398.

Offer, H., I. Zurer, G. Banfalvi, et al. 2001. p53 modulates base excision repair activity in a cell cycle-specific manner after genotoxic stress. *Cancer Res* 61: 88–96.

Otsuki, Y., O. Misaki, O. Sugimoto, et al. 1994. Cyclic *bcl-2* gene expression in human uterine endometrium during menstrual cycle. *Lancet* 344: 28–29.

Pandey, P., A. Saleh, A. Nakazawa, et al. 2000. Negative regulation of cytochrome c-mediated oligomerization of Apaf-1 and activation of procaspase-9 by heat shock protein 90. *EMBO J* 19: 4310–4322.

Poetker, D.M., A.D. Sandler, D.L. Scott, et al. 2002. Survivin expression in juvenile-onset recurrent respiratory papillomatosis. *Ann Otol Rhinol Laryngol* 111: 957–961.

Prodromou, C., B. Panaretou, S. Chohan, et al. 2000. The ATPase cycle of Hsp90 drives a molecular "clamp" via transient dimerization of the N-terminal domains. *EMBO J* 19: 4383–4392.

Sah, N.K., Z. Khan, G.J. Khan, et al. 2006. Structural, functional and therapeutic biology of survivin. *Cancer Lett* 244: 164–171.

Saitoh, Y., Y. Yaginuma, and M. Ishikawa. 1999. Analysis of Bcl-2, Bax and survivin genes in uterine cancer. *Int J Oncol* 15: 137–141.

Salvesen, G.S., and C.S. Duckett. 2002. IAP proteins: Blocking the road to death's door. *Nat Rev Mol Cell Biol* 3: 401–410.

Sampath, S.C., R. Ohi, O. Leismann, et al. 2004. The chromosomal passenger complex is required for chromatin-induced microtubule stabilization and spindle assembly. *Cell* 118: 187–202.

Samuel, T., K. Okada, M. Hyer, et al. 2005. cIAP1 localizes to the nuclear compartment and modulates the cell cycle. *Cancer Res* 65: 210–218.

Sandall, S., F. Severin, I.X. McLeod, et al. 2006. A Birl–Sli15 complex connects centromeres to microtubules and is required to sense kinetochore tension. *Cell* 127: 1179–1191.

Sato, S., N. Fujita, and T. Tsuruo. 2000. Modulation of Akt kinase activity by binding to Hsp90. *Proc Natl Acad Sci USA* 97: 10832–10837.

Satoh, K., K. Kaneko, M. Hirota, et al. 2001. Expression of survivin is correlated with cancer cell apoptosis and is involved in the development of human pancreatic duct cell tumors. *Cancer* 92: 271–278.

Scaffidi, C., S. Fulda, A. Srinivasan, et al. 1998. Two CD95 (APO-1/Fas) signaling pathways. *EMBO J* 17: 1675–1687.

Schiene, C., and G. Fischer. 2000. Enzymes that catalyse the restructuring of proteins. *Curr Opin Struct Biol* 10: 40–45.

Semba, S., F. Trapasso, M. Fabbri, et al. 2006. Fhit modulation of the Akt-survivin pathway in lung cancer cells: Fhit-tyrosine 114 (Y114) is essential. *Oncogene* 25: 2860–2872.

Serela, A.I., R.C. Macadam, S.M. Farmery, et al. 2000. Expression of the antiapoptosis gene, survivin, predicts death from recurrent colorectal carcinoma. *Gut* 46: 645–650.

Shin, M.S., H.S. Kim, C.S. Kang, et al. 2002. Inactivating mutations of CASP10 gene in non-Hodgkin lymphomas. *Blood* 99: 4094–4099.

Skoufias, D.A., C. Mollinari, F.B. Lacroix, et al. 2000. Human survivin is a kinetochore-associated passenger protein. *J Cell Biol* 151: 1575–1582.

Smith, S.D., M.A. Wheeler, J. Plescia, et al. 2001. Urine detection of survivin and diagnosis of bladder cancer. *JAMA* 285: 324–328.

Song, Z., X. Yao, and M. Wu. 2003. Direct interaction between survivin and Smac/DIABLO is essential for the antiapoptotic activity of survivin during taxol-induced apoptosis. *J Biol Chem* 278: 23130–23140.

Speroff, L., and R. Vande Wiele. 1971. Regulation of the human menstrual cycle. *Am J Obstet Gynecol* 109: 234–247.

Strasser, A., A.W. Harris, M.L. Bath, et al. 1990. Novel primitive lymphoid tumours induced in transgenic mice by cooperation between myc and bcl-2. *Nature (London)* 348: 331–333.

Sun, C., M. Cai, A.H. Gunasekera, et al. 1999. NMR structure and mutagenesis of the inhibitor of apoptosis protein XIAP. *Nature (London)* 40: 818–821.

Sun, C., D. Nettesheim, Z. Liu, et al. 2005. Solution structure of human survivin and its binding interface with Smac/Diablo. *Biochemistry* 44: 11–17.

Susin, S.A., H.K. Lorenzo, N. Zamzami, et al. 1999. Mitochondrial release of caspase-2 and -9 during the apoptotic process. *J Exp Med* 189: 381–394.

Suzuki, A., M. Hayashida, T. Ito, et al. 2000a. Survivin initiates cell cycle entry by the competitive interaction with Cdk4/p16(INK4a) and Cdk2/cyclin E complex activation. *Oncogene* 19: 3225–3234.

Suzuki, A., T. Ito, H. Kawano, et al. 2000b. Survivin initiates procase3/p21 complex formation as a result of interaction with Ckd4 to resist Fas-mediated cell death. *Oncogene* 19: 1346–1353.

Swana, H.S., D. Grossman, J.N. Anthony, et al. 1999. Tumor content of the anti-apoptotic molecule survivin and recurrence of bladder cancer. *N Engl J Med* 341: 4522–4530.

Szak, S.T., D. Mays, and J.A. Pietenpol. 2001. Kinetics of p53 binding to promoter sites in vivo. *Mol Cell Biol* 21: 3375–3386.

Tabibzadeh, S., E. Zupi, A. Babaknia, *et al.* 1995. Site and menstrual cycle-dependent expression of proteins of the tumour necrosis factor (TNF) receptor family, and BCL-2 oncoprotein and phase-specific production of TNF-α in human endometrium. *Hum Reprod* 10: 277–286.

Tanaka, K., S. Iwamoto, G. Gon, et al. 2000. Expression of survivin and its relationship to loss of apoptosis in breast carcinomas. *Clin Cancer Res* 6: 127–134.

Thompson, C.B. 1995. Apoptosis in pathogenesis and treatment of disease. *Science* 267: 1456–1462.

Uren, A.G., L. Wong, M. Pakusch, et al. 2000. Survivin and the inner centromere protein INCENP show similar cell-cycle localization and gene knockout phenotype. *Curr Biol* 10: 1319–1328.

Vagnarelli, P., and W.C. Earnshaw. 2004. Chromosomal passengers: The four-dimensional regulation of mitotic events. *Chromosoma* 113: 211–222.

Vaira, V., C.W. Lee, H.L. Goel, et al. 2007. Regulation of survivin expression by IGF-1/mTOR signaling. *Oncogene* 26: 2678–2684.

van de Wetering, M., E. Sancho, C. Verweij, et al. 2002. The β-catenin/TCF-4 complex imposes a crypt progenitor phenotype on colorectal cancer cells. *Cell* 111: 241–250.

Vanoosthuyse, V., S. Prykhozhij, and K.G. Hardwick. 2007. Shugoshin 2 regulates localization of the chromosomal passenger proteins in fission yeast mitosis. *Mol Biol Cell* 18: 1657–1669.

van't Veer, L.J., H. Dai, M.J. van de Vijver, et al. 2002. Gene expression profiling predicts clinical outcome of breast cancer. *Nature* 415: 530–536.

van Vugt, M.A., B.C. van de Weerdt, G. Vader, et al. 2004. Polo-like kinase-1 is required for bipolar spindle formation but is dispensable for anaphase promoting complex/Cdc20 activation and initiation of cytokinesis. *J Biol Chem* 279: 36841–36854.

Vaux, D.L., G. Haecker, and A. Strasser. 1994. An evolutionary perspective on apoptosis. *Cell* 76: 777–779.

Velculescu, V.E., S.L. Madden, L. Zhang, et al. 1999. Analysis of human transcriptomes. *Nat Genet* 23: 387–388.

Vermeulen, K., Z.N. Berneman, and D.R. Van Bockstaele. 2003. Cell cycle and apoptosis. *Cell Prolif* 36: 165–175.

Vogel, C., C. Hager, and H. Bastians. 2007. Mechanisms of mitotic cell death induced by chemotherapy-mediated G2 checkpoint abrogation. *Cancer Res* 67: 339–345.

Vogelstein, B., D. Lane, and A.J. Levine. 2000. Surfing the p53 network. *Nature (London)* 408: 307–310.

Wagner, M., K. Schmelz, B. Dörken, et al. 2008. Epigenetic and genetic analysis of the survivin promoter in acute myeloid leukemia. *Leuk Res* 32: 1054–1060.

Wang, H.W., T.V. Sharp, A. Koumi, et al. 2002. Characterization of an anti-apoptotic glycoprotein encoded by Kaposi's sarcoma-associated herpesvirus which resembles a spliced variant of human survivin. *EMBO J* 21: 2602–2615.

Wheatley, S.P., A. Carvalho, P. Vagnarelli, et al. 2001. INCENP is required for proper targeting of survivin to the centromeres and the anaphase spindle during mitosis. *Curr Biol* 11: 886–890.

Wheatley, S.P., and I.A. McNeish. 2005. Survivin: A protein with dual roles in mitosis and apoptosis. *Int Rev Cytol* 247: 35–88.

Xia, F., and D.C. Altieri. 2006. Mitosis-independent survivin gene expression in vivo and regulation by p53. *Cancer Res* 66: 3392–3395.

Xu, H., and M.R. el-Gewely. 2001. P53-responsive genes and the potential for cancer diagnostics and therapeutics development. *Biotechnol Annu Rev* 7: 131–164.

Xu, Z., H. Ogawa, P. Vagnarelli, et al. 2009. INCENP–aurora B interactions modulate kinase activity and chromosome passenger complex localization. *J Cell Biol* 187: 637–653.

Xu, Z.X., R.X. Zhao, T. Ding, et al. 2004. Promyelocytic leukemia protein 4 induces apoptosis by inhibition of survivin expression. *J Biol Chem* 279: 1838–1844.

Yoshida, H., O. Ishiko, T. Sumi, et al. 2001. Survivin, bcl-2 and matrix metalloproteinase-2 enhance progression of clear cell- and serous-type ovarian carcinomas. *Int J Oncol* 19: 537–542.

You, R.I., M.C. Chen, H.W. Wang, et al. 2006. Inhibition of lymphotoxin-β receptor-mediated cell death by survivin-DeltaEx3. *Cancer Res* 66: 3051–3061.

Young, J.C., N.J. Hoogenraad, and F.U. Hartl. 2003. Molecular chaperones Hsp90 and Hsp70 deliver preproteins to the mitochondrial import receptor Tom70. *Cell* 112: 41–50.

Zamzami, N., and G. Kroemer. 2001. The mitochondrion in apoptosis: How Pandora's box opens. *Nat Rev Mol Cell Biol* 2: 67–71.

Zhang, T., T. Otevrel, Z. Gao, et al. 2001. Evidence that APC regulates survivin expression: A possible mechanism contributing to the stem cell origin of colon cancer. *Cancer Res* 61: 8664–8667.

Zhao, J., T. Tenev, L.M. Martins, et al. 2000. The ubiquitin-proteasome pathway regulates survivin degradation in a cell cycle-dependent manner. *J Cell Sci* 113: 4363–4371.

8 Therapeutics of Survivin

KEY WORDS

Cancer immunotherapy
Diagnosis
Head and neck squamous carcinoma
Marker
Prognosis
Survivin
Therapeutics

8.1 INTRODUCTION

Survivin is a unique member of the inhibitor of apoptosis (IAP) protein family that interferes with postmitochondrial events, including activation of caspases. Survivin regulates the cell cycle also. It is expressed in most of the human tumors, but it is barely detectable in the terminally differentiated normal cells/tissues. Molecular mechanisms of regulation of survivin in cancer are not clearly understood. Nevertheless, the functional loss of wild-type (wt) p53 is often associated with upregulation of survivin. Tumors that overexpress survivin generally bear a poor prognosis and are associated with resistance to therapy. The differential expression of survivin in cancer versus normal tissues makes it a useful tool in cancer diagnosis and a promising therapeutic target. A growing body of literature suggests nuclear expression of survivin as a good prognostic marker. Disruption of the survivin induction pathway has resulted in an increase in apoptosis and inhibition of tumor growth.

Survivin is expressed in a cell cycle-dependent manner at mitosis. The promoter of the survivin gene has a cell cycle-dependent element and cell cycle protein homology regions. These regions are typically present in genes expressed in the G2/M phase of the cell cycle, such as cyclin A and cyclin B. Survivin expression is very high in most cancers, particularly colon, lung, breast, brain, and melanoma. The mechanisms of survivin overexpression in cancer are partially understood, but its widespread expression in cancer gives certain clues that multiple pathways may govern this process. Several mechanisms involved in survivin expression have been elucidated. In the normal ovaries, survivin exon 1 is silenced by methylation, but it becomes demethylated and transcriptionally active in ovarian cancer, thereby imparting survival benefits to cancer cells (Altieri, 2001).

On the other hand, p53 protein is a transcription factor that can induce apoptosis by regulating the apoptotic genes. Survivin is a target of p53 for downregulation. p53 may induce apoptosis by antagonizing the anti-apoptotic activity of survivin. However, survivin may also influence p53 activity through regulation of MDM2 and proteosome (Wang et al., 2004). The negative regulation of survivin by p53 is poorly understood. Survivin promoter has a p53 binding element. It may be possible that p53 directly binds survivin promoter alone or in combination with other protein(s) to repress survivin. E2F (a transcriptional activator) may also bind survivin promoter (Jiang et al., 2004).

8.2   MANIPULATION OF THE SURVIVIN GENE

Survivin is a critical regulator of multiple processes, including proliferation and apoptosis, and its expression appears to be a consistent feature of hyperproliferative lesions contributing to the formation of hyperplasia. It is expressed in the G2/M phase of the cell cycle to support rapidly dividing cell machinery (Li et al., 1998), and helps in proper segregation of chromosomes during cell division (Kaitan et al., 2000; Lens et al., 2003).

Survivin is abundantly expressed in embryonic tissues and in a wide range of cancer tissues, but undetectable in normal differentiated tissues (Ambrosini et al., 1997). Overexpression of survivin in cancer may overcome cell cycle checkpoints to facilitate aberrant progression of transformed cells through mitosis. Several chemotherapeutic agents kill tumor cells through apoptosis (Zaffaroni et al., 2002). It has been found that survivin expression is high when tumor cells are treated with anticancer agents by counteracting the effect of chemotherapeutic agents, and provides resistance to apoptosis (Zaffaroni and Daidone, 2002; Taran et al., 2002). Downregulation of survivin sensitizes tumor cells to apoptosis induced by chemotherapy and radiotherapy in many types of cancer (Wall et al., 2003). Exploitation of the survivin signaling pathway may offer new therapeutic alternatives for cancer treatment. Differential expression of survivin in cancer, compared to most normal tissues, makes survivin a candidate for a molecular marker of cancer. There is good evidence that survivin may provide a quick prognostic indicator for identifying patients at risk of recurrent disease (Tetsuhisa and Nobuhiko, 2001).

Survivin or its autoantibody present in biological fluids of cancer patients could provide them a potential diagnostic tool. Molecular mechanisms of survivin action are not fully elucidated and are, at least in some aspects, controversial. Nevertheless, it is well accepted that survivin is an inhibitor of apoptosis and interferes with cell cycle progression and microtubule stability. In general, mammalian IAPs block apoptosis by direct or indirect inhibition of initiator caspase-9 or terminal effector caspases-3 and -7 (Figures 8.1 and 8.2).

Studies suggested that survivin inhibits the intrinsic pathway of apoptosis by interacting with postmitochondrial events (Tamm et al., 1998; Altieri, 2003a, 2003b; Song et al., 2003). Several hypotheses exist to explain the mechanisms of antiapoptotic activity of survivin.

They might directly bind and inhibit caspases like any another IAPs, such as X chromosome-linked inhibitor of apoptosis protein (XIAP) (Tamm et al., 1998). Smac/DIABLO may act as a pro-apoptotic protein through its participation in the activation of caspase-9 (apoptosome formation). It may inhibit apoptosis through antagonizing the pro-apoptotic ability due to its affinity with Smac/DIABLO (Song et al., 2003).

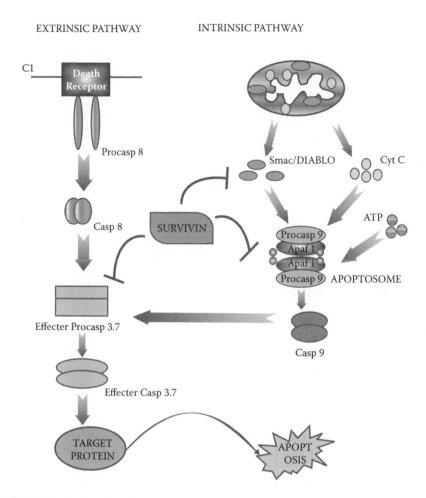

FIGURE 8.1   Apoptotic pathways and the site of survivin anti-apoptotic action.

In addition to anti-apoptotic function, survivin also regulates cell division because of its presence on the mitotic machinery of dividing cells. Targeted survivin has resulted in aberrant mitotic progression, leading to failed cytokinesis and multi-nucleation (Kaitan et al., 2000; Altieri, 2003; Skaufias et al., 2000; Uren et al., 2000; Honda et al., 2003). Survivin is indispensable during embryonic development. The homozygous deletion in mice leads to inevitable lethality at day 4 or 5 due to defects in mitotic spindles formation (Reed, 2001). The apparent requirement of survivin in normal cell division suggests that overexpression of survivin in tumors could perturb normal cell cycle control.

8.3   FACTORS INDUCING SURVIVIN EXPRESSION

The expression of survivin is undetectable in normal differentiated cells, but slowly expressed in fast-dividing normal cells, such as CD34+ bone marrow-derived stem

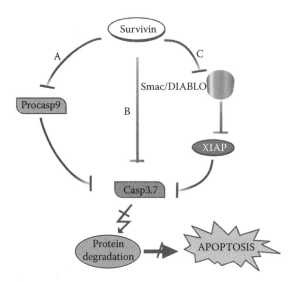

FIGURE 8.2 Models of survivin in the inhibition of apoptosis.

cells, basal epithelial cells, thymocytes, and basal epithelial cells of the normal uterine cervix (Altieri, 2003a; Konno et al., 2000; Endoh et al., 2001). Survivin expression is very high in most cancers, particularly colon, lung, breast, brain, and melanoma. The molecular mechanisms of survivin overexpression in cancer seem to be complex and are only partially understood. It is very likely that multiple pathways are involved in the reactivation of the survivin gene.

Several molecular mechanisms in survivin overexpression in tumor cells have been elucidated. In the normal ovaries, survivin exon 1 is silenced by methylation, but it becomes demethylated and transcriptionally active in ovarian cancer. In neuroblastoma, amplification of 17q25, comprising the survivin locus, has been reported (Konno et al., 2000; Endoh et al., 2001). A transcriptional factor such as p53 has been reported to regulate survivin expression in various human cancer cell lines (Hoffman et al., 2002; Mirza et al., 2002). In many cancers, such as gastric, pancreatic, prostate, lung, and epidermoid carcinomas, a correlation between p53 accumulation and survivin expression has been demonstrated (Lu et al., 1998; Sarela et al., 2002).

8.4 DOWNREGULATION OF SURVIVIN

8.4.1 ANTISENSE TECHNOLOGY

Antisense technology, with its potential to selectively control gene expression and cellular phenotype, is proving useful as a therapeutic application. Several types of antisense approaches (viz., antisense oligonucleotides, antisense RNA, and small interfering RNA) can be used to inhibit expression of target genes (Figure 8.3). Downregulation of survivin by a targeted antisense oligonucleotide could be potential gene therapy to combat several cancers. Transfection of cancer cells with antisense of survivin enhances sensitivity of tumor cells to chemotherapy and radiotherapy.

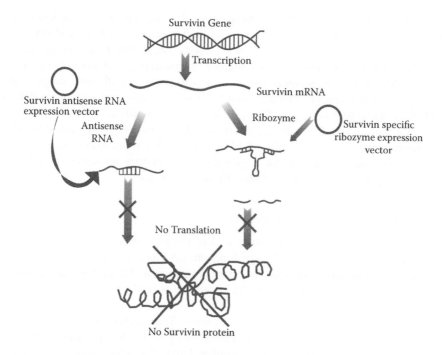

FIGURE 8.3 Schematic representation of inhibition of survivin synthesis by ribozyme-mediated degradation and antisense technology.

Several antisurvivin oligonucleotides have been tested for their ability to block survivin expression in tumor cell lines. Antisense oligonucleotide 4003 was found to be the most effective in growth inhibition and apoptosis in lung carcinoma cell lines (Olie et al., 2000).

Most of the mesothelioma cancer cells underwent apoptosis and were more sensitive to chemotherapy and radiotherapy when treated with 20-mer phosphorothioate antisense oligonucleotide targeting nucleotides 232 to 251 of survivin mRNA (Chuunyao et al., 2002). Growth of lymphoma cell lines was significantly inhibited by using antisense oligonucleotide (ASOND) (Ansell et al., 2004). Treatment of colon cancer by adenoviral antisense vectors (pAd-CMV-SAS) resulted in an increase of the Go/G1 phase population in the cell cycle and increased their sensitivity to chemotherapeutic drugs in vitro (Yamamoto et al., 2003). In PC-3 prostate cancer cell treatment with antisense survivin cDNA caused nuclear fragmentation, hypodiploidy, cleavage of a 32 kDa proform caspase-3 to active caspase-3, and proteolysis of the caspase substrate poly(ADP-ribose) polymerase (Mara et al., 2003).

In another study, a similar observation has been made with human neuroblastoma cell line SK-N-MC (Guan et al., 2002). Expression of survivin gene can be checked by interference RNA, including both interfering RNA (siRNA) and short-hairpin RNA (shRNA). These interfering RNAs can be chemically synthesized and transfected into cells or directly expressed intracellularly from a plasmid DNA or an adenoviral vector (Zhao et al., 2003).

Transfection of cancer cell lines by adenoviral vector harboring a tandem type siRNA expression unit targeting survivin resulted in gene knockdown and induced apoptosis (Ling and Li, 2004; Uchida et al., 2004). The transfection of cancer cells with Adv-siSurv showed significantly attenuated growth potential, both in vitro and in vivo. Moreover, intratumoral injection of Adv-siSurv significantly suppressed tumor growth in a xenograft model using U251 glioma cells. A short-hairpin RNA (shRNA), containing two 20 to 21 bp reverse repeat motifs of a survivin target sequence with a 4 to 8 bp spacer, effectively downregulates expression survivin in liver cancer cell lines Hep G2 and SMMC-7721, after transfection (Yang et al., 2003). These findings suggest that targeting of survivin using antisense technology may have a potential role in the selective therapy of cancer.

8.4.2 Dominant Negative Construct

In this technology, an essential amino acid of the protein is replaced by another amino acid, which leads to the loss of function, but this nonfunctional protein has the same target as normal protein. Dominant negative mutants compete with normal proteins for their target and suppress the function of normal proteins.

Several dominant negative constructs have been discovered for survivin, from which the T34A mutant of survivin is best known. Transduction of breast, cervical, prostate, lung, and colorectal cancer cell lines with pAd-T34A (replication-deficient adenovirus vector encoding a nonphosphorylated The34Ala mutant survivin) increased cyt-c release from mitochondria, processing caspase-3 to the active sub-units of approximately 17 and 19 kDa increased caspase-3 catalytic activity, and facilitated tumor cell apoptosis induced by taxol and adriamycin anticancer drugs (Mesri et al., 2001). In malignant HeLa cells, transfection with survivin mutant (sur-vivin-N and survivin-T34A) could partially reverse the malignancy of HeLa cells (Zhu et al., 2003). The efficacy of negation of anti-apoptotic activity of survivin was also demonstrated in gastric cancer cell lines (BCG-823 and MKN-45) through decreased cell growth and increased rate of spontaneous apoptosis, by introducing a plasmid construct expressing survivin antisense and a dominant negative (DN) mutant with a cysteine residue at amino acid 84 replaced with alanine (Cys84Ala) (Shui et al., 2003). In the PC-3 prostate cancer cell line, transfection with C84A mutant was sufficient to visualize all biochemical hallmarks of apoptosis, including hypoploid DNA content, caspase-3 activities, and cleavage of caspase substrates.

8.4.3 Acetylation and Deacetylation Pathway

Histone acetylation/deacetylation plays an important role in epigenetic transcriptional regulation in eukaryotic cells (Grunstein, 1997). Histone acetyltransferases (HATs) are recruited by transcription factors and are associated with activation of transcription, whereas histone deacetylases (HDACs) are involved in transcriptional silencing. Histone acetylation is tightly controlled by the dynamic equilibrium between competing HATs and HDACs (Grunstein, 1997; Hassing and Schreiber, 1999). The HATs bring about transcriptional activation by the addition of acetyl groups on the amino group of lysine residues in the NH2-terminal tails of core histones, and HDACs reverse this reaction by removing the acetyl groups from the acetylated lysines in histones, and thereby bringing transcriptional silencing (Hassing and Schreiber, 1999; Kouzarides, 1999; Jenuwein and Allis, 2001).

Abnormality in these enzymes can disturb the expression of several apoptotic and cell cycle regulatory genes, including survivin. Inhibition of HDACs is emerging as a new strategy in human cancer therapy. Several drugs have been discovered, which specifically inhibit the activity of HDAC. An important drug, clamydocin, used as a potent inhibitor of cell proliferation, is a potent inhibitor of HDAC (De Schepper et al., 2003).

Like other HDAC inhibitors, clamydocin induces the expression of hyperacetylated histones H3 and H4 in cancer cells, increases the expression of p21 (cip/waf1), and causes an accumulation of cells in the G2/M phase of the cell cycle. In addition, clamydicin induces apoptosis by activating caspase-3, which in turn leads to the cleavage of p21 (cip/waf1) and drives cells from growth arrest into apoptosis. Clamydocin decreases the level of survivin, which is mediated by proteasome-mediated degradation. LAQ824, another inhibitor of HDAC, also downregulates the levels of survivin and induces apoptosis of cancer cells (Fei et al., 2004).

8.5 RIBOZYME-MEDIATED DEGRADATION OF SURVIVIN

Survivin expression can be inhibited by ribozymes. Survivin-specific mRNA undergoes cleavage due to ribozyme treatment resulting in inhibition of translation and consequent retardation in tumor growth. Two hammerhead ribozymes, viz., RZ1 and RZ2, targeting human survivin mRNA have been designed (Choi et al., 2003). Transduction of MCF-7 (Michigan Cancer Foundation-7) cancer cell lines with adenovirus encoding the ribozyme resulted in a significant reduction of survivin mRNA and protein as well and sensitized the cells to apoptosis.

Infected PC-3 cells by the adenoviral vector that encodes a ribozyme targeting the 3' end of the survivin mRNA increased susceptibility of PC-3 and DU145 cancer cells to apoptosis. These treatments increased susceptibility of PC-3 and DU145 cancer cells to apoptosis (Pennati et al., 2004). In another study, transduction of melanoma cell lines JR8 and M14 with a vector-carrying ribozyme sequence led to a decrease of the survivin level and sensitized cancer cells to radiotherapy and chemotherapy (Pennati et al., 2003, 2004).

8.6 IMMUNOTHERAPY

The antigens expressed by tumor cells can be recognized by the host immune system. The immune system recognizes antigens in the form of short peptides binding to the major histocompatibility complex (MHC) molecules. It was recently discovered that the immune response against survivin-derived epitopes can be induced (Reker et al., 2004). Several survivin-derived epitopes have been tested to induce cytotoxic T-lymphocyte activity against tumor cells. Dendritic cells were transduced successfully with an adenoviral vector containing a full-length dominant negative survivin gene. Immunization with these dendritic cells induced a T-cell immune response against three different survivin-derived HLA-A2 matching peptides. The significant CTL activity was found against HLA-A2 positive MCF-7 tumor cells that express survivin (Pisarev et al., 2003).

Recently, an HLA-24-restricted immunogenic peptide-based vaccination (survivin-2B 80–88, AYACNTSTL) to neutralize survivin-2B splice variant, which is abnormally expressed in various types of tumor tissues and tumor cell lines, was subjected to phase I trial in patients with advanced or recurrent colorectal cancer (Tetsuhiro et al., 2004). Immunotherapy of cancer based on survivin may give promising hope.

Survivin belongs to an IAP family of proteins that play an important role in cell cycle progression and cancer cell viability. It is one of the proteins that are specially expressed in most cancer cells. It may therefore be a good marker for cancer detection and may serve as an attractive therapeutic target. Survivin is expressed in the majority of various cancers studied. Various strategies to target survivin in cancer cells are currently under investigation with promising results in both in vivo and in vitro models. Moreover, some currently explored anticancer agents might mediate their anticancer effects by inhibiting the survivin pathway.

8.7 THERAPEUTIC BIOLOGY OF SURVIVIN

Along with the apoptosis inhibitory role, survivin is also known to play a regulatory role in the cell cycle. It plays a critical role during development; however, its expression is practically absent in differentiated cells. Its expression has been detected in various cancerous cells. Its aberrant expression in different cancers has drawn a lot of attention from the research community, for looking at the possibility to use it as a prognostic or diagnostic marker, and also to devise targeted therapies against survivin. Many approaches have been attempted at downregulating survivin in vivo and in vitro. Expression of survivin can be blocked by antisense techniques. Many companies synthesize antisense drugs targeting survivin (Taran et al., 2002; Li, 2005; Ling and Li, 2004).

Isis Pharmaceutical made an antisense drug that inhibits tumor growth and induces apoptosis in tumor cells. Survivin expression is also inhibited by ribozyme. Survivin-specific RNA undergoes cleavage at the 3' end due to ribozyme treatment, resulting in inhibition of translation and consequent retardation in tumor growth (Choi et al., 2003; Pennati et al., 2003). Recent observations also suggest that survivin may be targeted through P38, and transforming growth factor beta (TGF-β) pathways to derive therapeutic benefits (Li et al., 2005). These approaches in combination with chemotherapy and radiotherapy may be highly promising. Gene therapy using negative dominant constructs of the survivin gene may yield very good clinical results. Experiments using a nonphosphorylated Thr34-Ala mutant of survivin caused spontaneous apoptosis in cell lines of cervical, prostate, lung, and colorectal cancer and in breast xenograft models.

Wild-type survivin is phosphorylated at Thr34 by cdc2 cyclin B1 kinase. This phosphorylated form of survivin is required for cancer cell viability and possibly to help survivin stability. Survivin phosphorylation can be blocked by several drugs, like flavopiridol, which functions as an antagonist of cyclin-dependent kinases including cdc2. Treatment of cancer by anticancer drugs, including adriamycin, taxol, and UV-B, enhances expression of survivin four- to fivefold (Kappler et al., 2003). Sequential ablation of survivin phosphorylation could enhance tumor cell apoptosis induced by anticancer agents independent of p53. Recently, the immune response against a survivin-derived epitope has been demonstrated. Several survivin-derived epitopes have been tested, which can induce cytotoxic T-lymphocyte activity against tumor cells.

An HLA-A24-restricted immunogenic peptide-based vaccination (survivin-2B 80–83) to neutralize survivin-2B splice variant, which is abundantly expressed in various types of tumor tissues and tumor cell lines, has been subjected to phase I trial on patients with advanced and colorectal cancer (Tsuruma et al., 2004). More recently, it has been demonstrated that vaccination of four out of five stage IV melanoma patients with HLA-II-restricted survivin (96–104) epitope presented by autologous dendritic cells resulted in strong T-cell responses as measured by ELISPOT assay (Otto et al., 2005). This therapy, however, needs more rigorous testing to avoid the possibility of occurrence of other apoptosis-related pathological complications. On the other hand, there are sound rationales for designing extra-immunological drugs that would target survivin to achieve success in cancer therapy. Notwithstanding this, it is advisable to combine it with other means, such as radiation, to achieve better success. In addition to cancer, expression of survivin has been observed to play a significant role in the origin of rheumatoid arthritis (RA), particularly the RA on the erosive course. The nonerosive RA patients produce more antisurvivin antibodies to afford protection (Bokarewa et al., 2005).

Survivin is one of the few proteins that is differentially expressed in tumor cells compared to most normal tissues. This characteristic enables it to be a potential marker of cancer. Overexpression of survivin is related to overall shortened human survival, and an increased rate of recurrence of tumors that are resistant to chemotherapy and radiotherapy (Takai et al., 2002; Kawasaki et al., 2001). The dual function of survivin indicates that it is expressed in tumor cells associated with growth and survival advantages, resulting in tumor onset and progression.

A positive correlation has been established between survivin expression and the high rate of proliferation and diminished rate of apoptosis. An increasing number of original publications suggest that nuclear expression of survivin may be a better and more reliable prognostic marker (Li et al., 2005). An increased level of survivin is a powerful negative prognostic marker for tumors. It can be detected in tumor specimens by immunohistochemical methods, which are quick prognostic indicators for identifying patients at risk of recurrence of the disease, and those who would benefit from more aggressive follow-up and alternative protocols (Trieb et al., 2003; Kappler et al., 2003).

Survivin can be detected in biological fluids of cancer patients. In case of bladder cancer, survivin has been detected in the urine of the patients, which provides a specific and sensitive diagnostic marker. In addition, antisurvivin antibodies circulate in cancer patients, which may be used as a diagnostic tool (Altieri, 2001). A few companies have come up with diagnostic kits for detection of cancer, which are largely based on survivin. For example, Biocore Pvt. Ltd. has recently developed a kit for quantification of human survivin from cell lysates by colorimetric immunoassay. However, the kit may be less useful in cases where certain factors, including chemotherapeutic agents, such as DNA-damaging agents, may mediate suppression of expression of survivin through induction of p53 (Smith and Seo, 2002). The drugs that enhance expression of survivin may also vitiate efficiency of the diagnostic kit (Wall et al., 2003).

Survivin is a 16.5 kDa large and is the smallest member of the inhibitor of apoptosis (IAP) gene family that contains one to three zinc finger folds. It is an evolutionary conserved suppressor of caspases (Salvesen and Duckett, 2002). The cross-pathway role and differential expression of survival make it a unique cancer target. Survivin plays an important role in mitosis, in which it is responsible for the completion of various phases of cell division, and it is believed to regulate cell division by regulating microtubule dynamics (Figure 8.4) (Altieri, 2003a). Survivin's apoptosis inhibitory activity can be attributed to its role in interfering with processing of caspase-9 (Dohi et al., 2004). These physiological activities of survivin help cells to deal with harsh microenvironments, such as cellular stress, in which survivin interacts with the molecular chaperone Hsp90 to preserve cell proliferation and viability (Fortugno et al., 2003).

Survivin is often implicated as a cancer-causing gene, which has an "oncofetal" expression pattern, i.e., it is significantly expressed during developmental stages, but not in differentiated tissues, and is reexpressed in the host of cancerous tissues. Its aberrant expression in various cancers is mediated by oncogenes, including E2F (Jiang et al., 2004), Ras (Sommer et al., 2003), Stat3 (Kanda et al., 2004), and by loss of tumor suppressor genes like p53 (Hoffman et al., 2002) or the adenomatous polyposis coli protein (Zhang et al., 2001). The aberrant expression of survivin is often associated with clinical conditions like advanced-stage cancer, poor survival, resistance to therapy, and increased recurrence. Survivin holds potential to be used as a diagnostic marker; some studies have demonstrated its predictive potential in breast (Paik et al., 2004) and bladder (Smith et al., 2001) cancers.

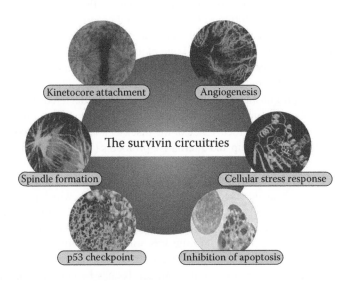

FIGURE 8.4 Molecular circuitries of survivin.

8.8   SURVIVIN-DIRECTED CANCER THERAPY

Survivin occupies a very critical position in the molecular network involved in various cancer pathways. Its cancer-specific expression along with involvement in various critical pathways makes it an attractive target for cancer therapy. Therapeutic targeting of survivin offers various advantages (Reed, 2003), as listed below:

1. Targeting *hub* of tumor biomolecular network: Survivin occupies a central or controlling position (*hub*) of various survival pathways necessary for the growth and maintenance of a tumor. Targeting such *hubs* produces a cascading effect and has an added advantage against normal apoptotic targets that are notorious for developing resistance over time due to mutation at the drug-protein interaction site (Sawyers, 2004; Mesri et al., 2001; Sakao et al., 2005).
2. Possibility of regulating cancer stem cells through survivin antagonism (Reya et al., 2001).
3. Survivin is a flexible target amenable to various therapeutic strategies, like molecular antagonism, vaccination, inhibition by small molecules, and gene therapy.
4. The expression of survivin in normal cells is practically absent, which is very desirable in targeted therapies.

The tumor-specific expression of survivin was hypothesized to elicit an immune response because it will be recognized by the immune system as a foreign protein (Andersen and Thor, 2002). This hypothesis has been corroborated by the presence of antibodies and cytolytic T-cells against survivin in cancer patients (Rohayem et al., 2000; Schmidt et al., 2003; Casati et al., 2003; Idenoue et al., 2005; Pisarev et al., 2003; Siegel et al., 2003). The feasibility of a survivin-based vaccine has been demonstrated in preclinical studies, in which a survivin-induced immune response efficiently controlled pulmonary metastasis by targeting tumor cells and tumor-associated angiogenesis (Xiang et al., 2005). Motivated by such preclinical studies, several survivin-based immunotherapies are underway in various stages of clinical trials (Hirschowitz et al., 2004; Otto et al., 2005; Tsuruma et al., 2004). Survivin-based vaccination was reported to be safe, without any significant side effects (Hirschowitz et al., 2004; Otto et al., 2005).

Many gene therapies against survivin have shown promising results during proof of concept preclinical studies (Wall et al., 2003; Mesri et al., 2001; Kanwar et al., 2001; Tu et al., 2005; Blanc-Brude et al., 2003). The gene therapy strategies involve the introduction of a dominant negative mutant of survivin, which dimerizes with endogenous survivin present in tumor cells, and thereby brings about degradation of survivin dimers and affects various survival pathways dependent on survivin. The "suicide" gene therapy strategies are yet another method to kill tumor cells by induced expression of cytotoxic gene mediated by survivin (Zhu et al., 2004; Chen et al., 2004).

Cancer cells are characterized by the high rate of cell division. Survivin is known to play an important role in cell division and evasion of apoptosis. The survivin is known to be overexpressed in various cancer cells, which is responsible for development of resistance to anticancer drugs like paclitaxel (microtubule stabilizer) and cisplatin (DNA-damaging agent). However, it does affect sensitivity of microtubule destabilizers such as *Vinca* alkaloids and comberstatin A-4 (CA-4)-related compounds. The resistance to *Vinca* alkaloids is caused by development of multidrug resistance (MDR) (Dumontet and Sikic, 1999; Cornelissen et al., 1994). MDR is a multifactorial phenomenon of developed overexpression of transmembrane efflux pumps like P-glycoprotein (P-gp170/MDR) and multidrug resistance protein (MRP) (Deng et al., 2002). MDR and survivin overexpression have been implicated for development of resistance against anticancer drugs (Pennati et al., 2007; Kojima et al., 2006).

Survivin is known to promote and facilitate cell division, especially in cancer cells. It was observed that survivin binds to centromeres during prophase or metaphase, relocates to spindle midzone during anaphase or telophase, and disappears toward the end to cell division (Uren et al., 2000). Inhibition of survivin in malignant cells resulted in cell cycle defects like a failure of cytokinesis, multipolar mitotic spindles, and formation of multinucleated cells (Li et al., 1999; Fortugno et al., 2002). Survivin plays an important role in microtubule dynamics (Rosa et al., 2006). Microtubules are cytoskeletal protein filaments that consist of α- and β-tubulin molecules (Sawada and Cabral, 1989; Wade and Hyman, 1997). Microtubules are responsible for regulation of mitotic machinery, and therefore are recognized as potential anticancer drug targets (Kiselyov et al., 2007; Shi et al., 1998), which leads to development of antimitotic compounds such as vincristine, vinblastine (microtubule-destabilizing *Vinca* alkaloid), and paclitaxel (microtubule-stabilizing taxane) for clinical use (Arrieta et al., 2006; Tanaka et al., 2004; Raitanen et al., 2002).

8.9   REGULATION OF SURVIVIN IN HEAD AND NECK SQUAMOUS CARCINOMA

Head and neck cancers are primarily squamous carcinomas (Vokes et al., 1993). Head and neck cancers are managed by radiation therapy, surgery, and chemotherapy (Veenita and Singh, 2005). Because of complexities in treating advanced cancers, a single treatment modality is not sufficient, and therefore often requires a combination of more than one treatment option, like chemotherapy along with radiation therapy. The development of radioresistance makes treatment management quite complex with poor outcome (Raybaud-Diogene et al., 1997; Murphy et al., 2007). Survivin, which is associated with carcinogenesis, cancer progression, and drug resistance, is regarded as a predictor of radioresistance, and inhibition of survivin expression is reported to trigger apoptosis (Khan et al., 2010).

The development of head and neck cancer is a multistep process that involves genetic and epigenetic alterations in tumor suppressor genes (TSGs) and oncogenes (Forastiere et al., 2005; Ha and Califano, 2006). However, the exact course of events leading to development of head and neck cancer is not yet fully understood. The current knowledge suggests that most of the alterations at the molecular level converge to apoptotic pathways (Gastman, 2001; Okada and Mak, 2004). Apoptosis is a regulated process in which damaged or worn-out cells with defective cellular components are marked to die in a well-controlled manner. Apoptotic pathway includes factors that promote (pro-apoptotic) and inhibit (anti-apoptotic) the process of apoptosis. During carcinogenesis an apoptotic mechanism is compromised by downregulation of pro-apoptotic factors and upregulation of anti-apoptotic factors; consequently, stress signals (e.g., DMA damage) from cell cycle checkpoints are overlooked and cancer cells survive. The defective apoptotic machinery makes cancer cells resistant to chemo- and radiotherapy (Sharma et al., 2004; Reed, 2002). The evasion of apoptosis is considered one of the major hallmarks in various cancers, including those of head and neck. Various signal transduction pathways (like epidermal growth factor receptor (EGFR), Stat3, and nuclear factor (NF)-κB) contribute toward the survival pathway in cancer cells, which essentially involves dysregulation of pro-survival proteins (Fesik, 2005; Khan and Bisen, 2013). The importance of survival pathways and involved factors has evoked much interest for their utilization as therapeutic and prognostic targets. Among various factors involved in the survival mechanism of cancer cells, the inhibitor of apoptosis proteins (IAPs) like XIAP, cIAP1/2, and survivin have been pursued by various research groups for therapeutic, prognostic, and diagnostic applications (Salvesen and Duckett, 2002; Altieri, 2003b). The expression of survivin is undetectable in normal cells, but it is specifically expressed in cancer cells, and its expression has been positively correlated with radio-/chemoresistance and poor survival of cancer patients (Engels et al., 2007; Freier et al., 2007).

Much is already known about the role of survivin in apoptosis, but there are still many gaps left in our mechanistic understanding of survivin (Lens et al., 2006). The survivin gene located on chromosome 17q is reported to give rise to four alternatively spliced transcripts (Li, 2005). These spliced transcripts are debated to inhibit or promote wt survivin's activity (Knauer et al., 2007).

Apart from its well-known role as an inhibitor of apoptosis, survivin is also known as a mitotic effector. The chromosomal passenger complex (CPC) consists of aurora B, borealin, INCENP, and survivin. The cell division is marked by equal distribution of chromosomes between resulting cells, and is broadly regulated by the mitotic spindle (consisting of microtubules). During cell division sister chromatids start moving toward opposite poles, through dynamic constriction of microtubules from opposite spindle poles attached to kinetochores present on chromatids. The attachment of microtubules to the kinetochore is a random event, in which pairing of the kinetochore with microtubes is erroneous at times. The CPC is responsible for correcting such errors and plays a role in detachment of incorrectly paired kinetochore-microtubules. This event is corroborated by peaks in expression of survivin at the G2/M transition of the cell cycle (Knauer et al., 2006). The involvement of survivin in the cell cycle can be regarded as a mechanism to tolerate increasing chromosomal abnormalities (Vagnarelli and Earnshaw, 2004).

8.10   PROGNOSTIC AND DIAGNOSTIC
SIGNIFICANCE OF SURVIVIN

The specificity of survivin expression in cancer cells makes it an ideal marker to be used for prognostic or diagnostic applications. The high level of survivin expression in cancer patients is often correlated with aggressive cancer subtype and poor survival. The survivin expression follows a well-defined pattern, temporally expressed in tissues during the developmental phase, expressed in cancer cells during peaks at the G2/M transition in the cell cycle (Reed and Bischoff, 2000), and spatially expressed in fetal lung, heart, liver, kidney, gastrointestinal tract, and other tissues where apoptosis occurs and tumor cells (Altieri, 2003b). The differential expression of survivin in primary oral squamous cell carcinoma and oral premalignant lesions was found to be significantly higher than in normal oral tissues (Khan et al., 2009). The survivin expression was found to be significantly correlated with oral carcinogenesis, lymph node metastasis, and poor survival (Kim et al., 2010). More such studies in multicenter setups should be taken to evaluate clinical use of survivin as a diagnostic or prognostic biomarker.

8.11 THERAPEUTIC APPROACH

Survivin has been actively pursued as a therapeutic target because (1) it's specifically expressed in malignant cells, (2) targeting survivin will sensitize cancer cells toward existing chemotherapeutic drugs, (3) targeting survivin will inhibit tumor growth through induced apoptosis, and (4) targeting survivin could control proliferation, because of its role in the cell cycle. Some of the therapeutic strategies to target survivin follows.

8.11.1 ANTISENSE OLIGONUCLEOTIDES (ASONDs)

Antisense oligonucleotides (ASONDs) are used to selectively target and inactivate mRNA of genes known to be critically involved in disease progression. Targeting survivin through survivin-antisense oligonucleotide is one of the promising strategies to inhibit survivin, and thereby reduce the concentration of survivin protein in cancer cells. The survivin ASONDs were found to significantly inhibit proliferation and induce apoptosis in high metastatic human mucoepidermoid carcinoma cell lines (Mc3) (Qi et al., 2010). The clinical efficacy of using antisense oligonucleotides in selectively targeting survivin has shown promising results in the proof-of-concept study conducted in human cancer patients (Talbot et al., 2010); in this study antisense oligonucleotide LY2181308 was used to downregulate survivin in tumor cells. The survivin-antisense oligonucleotides can also be used to increase the efficacy of chemotherapeutic and targeted therapeutic drugs.

8.11.2 SURVIVIN RIBOZYME APPROACH

Ribozymes are yet another tool to selectively silence target mRNA, and have potential for therapeutic application. The hammerhead ribozyme is a specialized class of ribozyme that has three parts: (1) stem loop structured catalytic domain with conserved nucleotide, and (2) two arms extending from stem loop structure, which is complementary to target RNA. Ribozyme selectively cleaves target mRNA after NUH triplets (N → any nucleotide, U → uridine, H → any nucleotide except G), and thus inactivates mRNA and brings down protein expression of the target (Zaffaroni et al., 2005). The ribozyme directed against survivin in prostate cancer cells was found to induce apoptosis and sensitize cancerous cells for chemo- and radiotherapy (Pennati et al., 2003, 2004). In an experiment, the multiple hammerhead ribozymes were used to target all major transcripts of survivin, resulting in significant inactivation of survivin and in vivo inhibition of tumor growth (Fei et al., 2008).

8.11.3 SURVIVIN RNA INTERFERENCE APPROACH

RNA interference (RNAi) has emerged as a powerful technology for selective silencing of target genes of interest (Izquierdo, 2005). RNA interference is a naturally evolved system in living cells that is responsible for posttranscriptional regulation of gene expression. The RNA interference system can broadly consist of two types of central biomolecules: (1) miRNA and (2) siRNA. The RNAi molecule base pairs with a complementary subsequence present in target mRNA and mark them to be systematically degraded. There has been considerable work on understanding the RNAi pathway during the last decade, and because of this it has become an indispensable research tool with various applications. This technology has also been successfully utilized in inhibition of various cancer-related genes, including survivin (Uchida et al., 2004; Beltrami et al., 2004; Caldas et al., 2006; Carvalho et al., 2003).

Once the causative gene/protein has been identified in target identification experiments, the computational methods are used to identify a unique subsequence in the target gene. Based on the computationally identified subsequence, RNAi molecules are synthesized in vitro as 21–15 nucleotide dsRNAs (siRNA) or 21–29 nucleotide short-hairpin dsRNAs (shRNA). The synthesized RNAi molecules are introduced in the cell to silence target mRNA through the RNAi pathway. The *knack-down* of survivin by siRNA was reported to sensitize tumor cells for radiotherapy, along with increased expression of caspase-3/-7, p53 (Kappler et al., 2005).

8.11.4 SURVIVIN-DERIVED CANCER IMMUNOTHERAPY

Survivin protein is deemed as an antigen by our immune system, because of the near-exclusive expression of survivin in tumor cells. There has been evidence of survivin being a strong T-cell-activating antigen, which can be utilized for designing effective therapy based on vaccination (Friedrichs et al., 2006). The antibodies against survivin were isolated in cancer patients, which supports the hypothesis that survivin elicits a full humoral immune response. The isolation of survivin-specific antibody is of great value for rational designing of epitope-based cancer vaccines. The structural study of survivin antibodies would help us in understanding features present in the binding groove in the antibody (paratope), to which the epitope binds to induce an immune response. The immunotherapeutic aspect of survivin has been actively pursued in the last couple of years and is now available as a cancer vaccine for clinical use.

8.11.5 SMALL-MOLECULE INHIBITORS (SMIs)

Inhibition by small molecules offers another attractive strategy for controlling tumorous growth by inhibition by survivin. YM155, a novel small-molecule inhibitor of survivin, was found to effectively inhibit survivin (Nakahara et al., 2007), and has demonstrated a promising safety and efficacy profile in phase I (Tolcher et al., 2008) and phase II (Giaccone et al., 2009) clinical trials. In another study 5-deazaflavin analog was found to inhibit interaction between survivin and Smac/DIABLO, consequently sensitizing cancer cells toward stress factors (like DNA damage) and activating apoptotic cell death (Oikawa et al., 2010).

8.12 SURVIVIN FOR CANCER IMMUNOTHERAPY

Survivin has generated much enthusiasm to be leveraged as an agent for cancer immunotherapy, because of its exclusive expression in cancer cells. Survivin evidently induces activation of T-cell antigen. Activating the body's immune system to deal with tumor cells has been a long-standing goal of cancer therapeutics (Gold et al., 1997).

Some of the desirable conditions to be present in candidate tumor-associated antigens (TAAs) to be used for cancer immunotherapy are as follows:

1. They should be presented on human leukocyte class I or II antigens (HLA-I or HLA-II), which can invoke immune response by activation of host T-cells.
2. TAAs should be specifically expressed on tumor cells.
3. TAAs should be a critical entity of cancer pathway, and their expression should be mandatory for proliferation and survival of cancer cells.

Survivin can be regarded as a potential candidate for immunotherapy, as it satisfies the above-mentioned conditions. It has been reported as an essential gene in gene ablation studies (Uren et al., 2000). The knockdown of survivin expression arrested cell division and increased apoptosis, thus halting tumor development.

The peptides derived from survivin protein can generate an immunogenic response, which lead to designing experiments in which autologous dendritic cells (DCs) are infected with survivin-expressing recombinant adenovirus. Such interaction was performed with an expectation that some of the endogenously processed peptides (Figure 8.5) from survivin could be displayed in immunogenic forms on HLA-A2. The dominant negative (DN) mutant of survivin was used in place of wt survivin in order to avoid any pro-oncogenic side effects (Pisarev et al., 2003).

8.12.1 ACTIVATION OF THE ADAPTIVE IMMUNE SYSTEM

8.12.1.1 Cellular T-Cell Response

Survivin-derived HLA class I restricted T-cell epitopes were reported to induce CTL response in cancer patients (Andersen and Thor, 2002; Hirohashi et al., 2002; Reker et al., 2004). Survivin-induced CTC was reported to kill HLA-matched tumors with origin from different tissue types (Schmidt et al., 2003; Siegel et al., 2003).

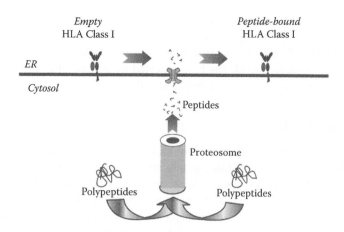

Empty
HLA Class I

Peptide-bound
HLA Class I

ER

Cytosol

Peptides

Proteosome

Polypeptides

Polypeptides

FIGURE 8.5 Antigen processing and presentation.

Survivin's potential to induce a cellular T-cell response was demonstrated in an experimental assay in which cytotoxic T-lymphocytes (CTLs) induced lysis of B-cells transfected to present survivin peptides on its surface (Friedrichs et al., 2006).

8.12.1.2 Humoral Antibody Response

Survivin-specific antibodies were isolated from blood samples from cancer patients (Friedrichs et al., 2006). In the same study, it was observed that survivin antibodies were absent in healthy subjects, which proves survivin's ability to elicit a full humoral immune response. The isolation of survivin-specific antibody has also opened the opportunity for rational computer-aided designing of epitopes, which would be consequently used in the development of effective cancer vaccines.

Survivin can be used as a universal antigen for diagnostic as well as therapeutic purposes. Its ubiquity across various cancers can be corroborated by the fact that it was found to be the fourth most highly expressed protein in human cancer tissue in comparison to normal tissue, as observed in large-scale analysis of human transcripts (Velculescu et al., 1999). Along with its ubiquity in various cancers, survivin is also reported to be positively associated with unfavorable prognosis and chemoresistance (Kawasaki et al., 1998; Olie et al., 2000). Such observations have motivated researchers across the world to look into the possibility of using survivin antigen as an agent for cancer immunotherapy. The processed survivin was presented on dendritic cells reported to induce specific CTL in vitro (Schmitz et al., 2000). Another approach relied on using epitopes present on survivin to detect a specific T-cell response in cancer patients by ELISPOT assay (Andersen et al., 2001a, 2001b). Moreover, the evidence of the presence of survivin-specific T-cells in the blood, and in the tumor lesions, should be regarded as encouraging proof that a survivin-induced T-cell response would be targeted toward tumorous tissues (Andersen et al., 2001b).

Survivin-based vaccines like SurVaxM have entered into clinical trials and are expected to be a promising cancer immunotherapeutic agent (National Cancer Institute trial NCT01250470). The vaccine SurVaxM is a survivin peptide mimetic based on a reverse immunology experiment (Ciesielski et al., 2010). The efficacy of cancer immunotherapeutic agents is often reduced by immunosuppressive factors such as regulatory T-cells (Tregs). Histone deacetylase (HDAC) inhibitor entinostat, in combination with SurVaxM, has been reported to increase efficacy and circumvent immunosuppression by Tregs (Shen et al., 2012).

8.13   CONCLUSION

Survivin is ubiquitously and exclusively expressed in cancerous cells of diverse origin, which makes it an ideal therapeutic and diagnostic/prognostic marker. It is significantly expressed in various cancers, including oral cancer. High expression of survivin has been associated with poor survival and chemo- and radioresistance among cancer patients. Functionally, it is known to promote tumorigenesis by inhibiting apoptosis and ensuring normal cell division among cancer cells. The critical role of survivin in cancer cell proliferation has been established through gene silencing experiments in which selective silencing of survivin has been shown to inhibit tumor growth and increase efficacy of other treatment options. Various therapeutic approaches like immunotherapy, small-molecule inhibition, and gene silencing are in different clinical/discovery phases; therefore, we can expect survivin-based therapy to be available in the near future. The success of survivin-based therapy will depend on efforts of the scientific community to design a cancer management plan by convergence of diverse approaches targeting survivin, for instance, by effective combination of immunotherapy, targeted drug, and radiotherapy.

REFERENCES

Altieri, D.C. 2001. The molecular basis and potential role of survivin in cancer diagnosis and therapy. *Trends Mol Med* 12: 542–547.

Altieri, D.C. 2003a. Survivin, versatile modulation of cell division and apoptosis in cancer. *Oncogene* 22: 8581–8589.

Altieri, D.C. 2003b. Validating survivin as a cancer therapeutic target. *Nat Rev Cancer* 3: 46–54.

Ambrosini, G., C. Addie, and D.C. Altieri. 1997. A noble anti-apoptotic gene, survivin, is expressed in cancer and lymphoma. *Nature Med* 3: 917–921.

Andersen, M.H., L.O. Pedersen, J.C. Becker, et al. 2001a. Identification of a cytotoxic T lymphocyte response to the apoptose inhibitor protein survivin in cancer patients. *Cancer Res* 61: 869–872.

Andersen, M.H., L.O. Pedersen, B. Capeller, et al. 2001b. Spontaneous cytotoxic T-cell responses against survivin-derived MHC class I-restricted T-cell epitopes in situ as well as ex vivo in cancer patients. *Cancer Res* 61: 5964–5968.

Andersen, M.H., and S.P. Thor. 2002. Survivin—A universal tumor antigen. *Histol Histopathol* 17: 669–675.

Ansell, S.M., B.K. Arendt, D.M. Grote, et al. 2004. Inhibition of survivin expression suppresses the growth of aggressive non-Hodgkin's lymphoma. *Leukemia* 18: 616–623.

Arrieta, O., J.L. Rodriguez-Diaz, V. Rosas-Camargo, et al. 2006. Colchicine delays the development of hepatocellular carcinoma in patients with hepatitis virus-related liver cirrhosis. *Cancer* 107: 1852–1858.

Beltrami, E., J. Plescia, J.C. Wilkinson, et al. 2004. Acute ablation of survivin uncovers p53-dependent mitotic checkpoint functions and control of mitochondrial apoptosis. *J Biol Chem* 279: 2077–2084.

Blanc-Brude, O.P., M. Mesri, N.R. Wall, et al. 2003. Therapeutic targeting of the survivin pathway in cancer: Initiation of mitochondrial apoptosis and suppression of tumor-associated angiogenesis. *Clin Cancer Res* 9: 2683–2692.

Bokarewa, M., S. Lindblad, D. Bokarew, et al. 2005. Balance between survivin, a key member of the apoptosis inhibitor family, and its specific antibodies determines erosivity in rheumatoid arthritis. *Arthritis Res Ther* 7: 349–358.

Caldas, H., M.P. Holloway, B.M. Hall, et al. 2006. Survivin-directed RNA interference cocktail is a potent suppressor of tumour growth in vivo. *J Med Genet* 43: 119–128.

Carvalho, A., M. Carmena, C. Sambade, et al. 2003. Survivin is required for stable checkpoint activation in taxol-treated HeLa cells. *J Cell Sci* 116: 2987–2998.

Casati, C., P. Dalerba, L. Rivoltini, et al. 2003. The apoptosis inhibitor protein survivin induces tumor-specific CD8+ and CD4+ T cells in colorectal cancer patients. *Cancer Res* 63: 4507–4515.

Chen, J.S., J.C. Liu, L. Shen, et al. 2004. Cancer-specific activation of the survivin promoter and its potential use in gene therapy. *Cancer Gene Ther* 11: 740–747.

Choi, K.S., T.H. Lee, and M.H. Jung. 2003. Ribozyme mediated cleavage of the human survivin mRNA and inhibition of anti-apoptotic function of survivin in MCF-7 cells. *Cancer Gene Ther* 10: 87–95.

Chuunyao, X., X. Zhidong, Y. Xiaocheng, et al. 2002. Induction of apoptosis in mesothelioma cells by antisurvivin oligonucleotides. *Mol Cancer Ther* 1: 687–694.

Ciesielski, M.J., M.S. Ahluwalia, S.A. Munich, et al. 2010. Antitumor cytotoxic T-cell response induced by a survivin peptide mimic. *Cancer Immunol Immunother* 59: 1211–1221.

Cornelissen, J.J., P. Sonneveld, M. Schoester, et al. 1994. MDR-1 expression and response to vincristine, doxorubicin, and dexamethasone chemotherapy in multiple myeloma refractory to alkylating agents. *J Clin Oncol* 12: 115–119.

Deng, L., S. Tatebe, Y.C. Lin-Lee, et al. 2002. MDR and MRP gene families as cellular determinant factors for resistance to clinical anticancer agents. *Cancer Treat Res* 112: 49–66.

De Schepper, S., H. Bruwiere, T. Verhult, et al. 2003. Inhibition of histone deacetylases by chlamydocin induces apoptosis and proteasome-mediated degradation of survivin. *J Pharmacol Exp Ther* 304: 881–888.

Dohi, T., E. Beltrami, N.R. Wall, et al. 2004. Mitochondrial survivin inhibits apoptosis and promotes tumorigenesis. *J Clin Invest* 114: 1117–1127.

Dumontet, C., and B.I. Sikic. 1999. Mechanisms of action of and resistance to antitubulin agents: Microtubule dynamics, drug transport, and cell death. *J Clin Oncol* 17: 1061–1070.

Endoh, A., K. Asanuma, R. Moriai, et al. 2001. Expression of survivin mRNA in CD34+ cell. *Clin Chim Acta* 306: 149–151.

Engels, K., S.K. Knauer, D. Metzler, et al. 2007. Dynamic intracellular survivin as a nearly prognostic marker in oral squamous cell carcinomas—Underlying molecular mechanism and potential. *J Pathol* 211: 532–540.

Fei, G., C. Sigua, J. Tao, et al. 2004. Cotreatment with histone deacetylase inhibitor LAQ824 enhances Apo-2L/tumor necrosis factor-related apoptosis inducing ligand-induced death inducing signaling complex activity and apoptosis of human acute leukemia cells. *Cancer Res* 64: 2580–2589.

Fei, Q., H. Zhang, L. Fu, et al. 2008. Experimental cancer gene therapy by multiple anti-survivin hammerhead ribozymes. *Acta Biochim Biophys Sin (Shanghai)* 40: 466–477.

Fesik, S.W. 2005. Promoting apoptosis as a strategy for cancer drug discovery. *Nat Rev Cancer* 5: 876–885.

Forastiere, A.A., K. Ang, D. Brizel, et al. 2005. Head and neck cancers. *J Natl Compr Canc Netw* 3: 316–391.

Fortugno, P., E. Beltrami, J. Plescia, et al. 2003. Regulation of survivin function by Hsp90. *Proc Natl Acad Sci USA* 100: 13791–13796.

Fortugno, P., N.R. Wall, A. Giodini, et al. 2002. Survivin exists in immunochemically distinct subcellular pools and is involved in spindle microtubule function. *J Cell Sci* 115: 575–585.

Freier, K., S. Pungs, C. Sticht, et al. 2007. High survivin expression is associated with favorable outcome in advanced primary oral squamous cell carcinoma after radiation therapy. *Int J Cancer* 120: 942–946.

Friedrichs, B., S. Siegel, M.H. Andersen, et al. 2006. Survivin-derived peptide epitopes and their role for induction of antitumor immunity in hematological malignancies. *Leuk Lymphoma* 47: 978–985.

Gastman, B.R. 2001. Apoptosis and its clinical impact. *Head Neck* 23: 409–425.

Giaccone, G., P. Zatloukal, J. Roubec, et al. 2009. Multicenter phase II trial of YM155, a small-molecule suppressor of survivin, in patients with advanced, refractory, non-small-cell lung cancer. *J Clin Oncol* 27: 4481–4486.

Gold, D.P., K. Shroeder, A. Golding, et al. 1997. T-cell receptor peptides as immunotherapy for autoimmune disease. *Crit Rev Immunol* 17: 507–510.

Grunstein, M. 1997. Histone acetylation in chromatin structure and transcription. *Nature (London)* 389: 349–352.

Guan, J., J. Chen, Y. Luo, et al. 2002. Effects of antisense bcl2 or survivin on the growth of neuroblastoma cell line SK-N-MC. *Zhonghua Yi Xue Za Zhi* 82: 1536–1540.

Ha, P.K., and J.A. Califano. 2006. Promoter methylation and inactivation of tumour-suppressor genes in oral squamous-cell carcinoma. *Lancet Oncol* 7: 77–82.

Hassing, C.A., and S.L. Schreiber. 1999. Nuclear histone acetylases and deacetylases and transcriptional regulation: HATs off to HDACs. *Curr Opin Chem Biol* 1: 300–308.

Hirohashi, Y., T. Torigoe, A. Maeda, et al. 2002. An HLA-A24-restricted cytotoxic T lymphocyte epitope of a tumor-associated protein, survivin. *Clin Cancer Res* 8: 1731–1739.

Hirschowitz, E.A., T. Foody, R. Kryscio, et al. 2004. Autologous dendritic cell vaccines for non-small-cell lung cancer. *J Clin Oncol* 22: 2808–2815.

Hoffman, W.H., S. Biade, J.T. Zilfou, et al. 2002. Transcriptional repression of the anti-apoptotic survivin gene wild type p53. *J Biol Chem* 277: 3247–3257.

Honda, R., K. Ronan, and A.N. Erich. 2003. Exploring the functional interactions between aurora B, INCENP and survivin in mitosis. *Mol Biol Cell* 14: 3325–3341.

Idenoue, S., Y. Hirohashi, T. Torigoe, et al. 2005. A potent immunogenic general cancer vaccine that targets survivin, an inhibitor of apoptosis proteins. *Clin Cancer Res* 11: 1474–1482.

Izquierdo, M. 2005. Short interfering RNAs as a tool for cancer gene therapy. *Cancer Gene Ther* 12: 217–227.

Jenuwein, T., and C.D. Allis. 2001. Translating the histone code. *Science* 293: 1074–1080.

Jiang, Y., H.I. Saavedra, M.P. Holloway, et al. 2004. Aberrant regulation of survivin by the RB/E2F family of proteins. *J Biol Chem* 279: 40511–40520.

Kaitan, S., M. Mendosa, V. Jantch-Plunger, et al. 2000. INCENP and aurora like kinase form a complex essential for chromosome segregation and efficient completion of cytokinesis. *Curr Biol* 10: 1072–1081.

Kanda, N., H. Seno, Y. Konda, et al. 2004. STAT3 is constitutively activated and supports cell survival in association with survivin expression in gastric cancer cells. *Oncogene* 23: 4921–4929.

Kanwar, J.R., W.P. Shen, R.K. Kanwar, et al. 2001. Effects of survivin antagonists on growth of established tumors and B7-1 immunogene therapy. *J Natl Cancer Inst* 93: 1541–1552.

Kappler, M., M. Kotzsch, F. Bartel, et al. 2003. Elevated expression level of survivin protein in soft tissue sarcoma is a strong independent predictor of survival. *Cancer Res* 9: 1098–1104.

Kappler, M., H. Taubert, F. Bartel, et al. 2005. Radiosensitization, after a combined treatment of survivin siRNA and irradiation, is correlated with the activation of caspases 3 and 7 in a wt-p53 sarcoma cell line, but not in a mt-p53 sarcoma cell line. *Oncol Rep* 13: 167–172.

Kawasaki, H., D.C. Altieri, C.D. Lu, et al. 1998. Inhibition of apoptosis by survivin predicts shorter survival rates in colorectal cancer. *Cancer Res* 58: 5071–5074.

Kawasaki, M., M. Toyoda, H. Shinohara, et al. 2001. Expression of survivin correlates with apoptosis, proliferation and angiogenesis during human colorectal tumorigenesis. *Cancer* 91: 2026–2032.

Khan, Z., N. Khan, R.P. Tiwari, et al. 2010. Down-regulation of survivin by oxaliplatin diminishes radioresistance of head and neck squamous carcinoma cells. *Radiother Oncol* 96: 267–273.

Khan, Z., R.P. Tiwari, R. Mulherkar, et al. 2009. Detection of survivin and p53 in human oral cancer: Correlation with clinicopathologic findings. *Head Neck* 31: 1039–1048.

Khan, Z., and Bisen, P.S. 2013. Oncoapoptotic signaling and deregulated target genes in cancers: Special reference to oral cancer. *Biochim Biophy Acta Reviews on Cancer*, dx.doi.org/10.1016/j.bbcan.2013.04.002.

Kim, Y.-H., S.-M. Kim, Y.-K. Kim, et al. 2010. Evaluation of survivin as a prognostic marker in oral squamous cell carcinoma. *J Oral Pathol Med* 39: 368–375.

Kiselyov, A., K.V. Balakin, S.E. Tkachenko, et al. 2007. Recent progress in discovery and development of antimitotic agents. *Anticancer Agents Med Chem* 7: 189–208.

Knauer, S.K., C. Bier, N. Habtemichael, et al. 2006. The survivin-Crm1 interaction is essential for chromosomal passenger complex localization and function. *EMBO Rep* 7: 1259–1265.

Knauer, S.K., C. Bier, P. Schlag, et al. 2007. The survivin isoform survivin-3B is cytoprotective and can function as a chromosomal passenger complex protein. *Cell Cycle* 6: 553–562.

Kojima, H., M. Iida, Y. Yaguchi, et al. 2006. Enhancement of cisplatin sensitivity in squamous cell carcinoma of the head and neck transfected with a survivin antisense gene. *Arch Otolaryngol Head Neck Surg* 132: 682–685.

Konno, R., H. Yamakawa, K. Ito, et al. 2000. Expression of survivin and bcl2 in the normal human endometrium. *Mol Hum Reported* 6: 529–534.

Kouzarides, T. 1999. Histone acetylases and deacetylases in cell proliferation. *Curr Opin Genet Dev* 9: 40–48.

Lens, S.M., G. Vader, and R.H. Medema. 2006. The case for survivin as mitotic regulator. *Curr Opin Cell Biol* 18: 616–622.

Lens, S.M., R.M. Wolthuis, R. Klompmaker, et al. 2003. Survivin is required for a sustained spindle checkpoint arrest in response to lack of tension. *EMBO J* 22: 2934–2947.

Li, F. 2005. Role of survivin and its splice variants in tumorigenesis. *Br J Cancer* 92: 212–216.

Li, F., E.J. Ackermann, C.F. Bennett, et al. 1999. Pleiotropic cell-division defects and apoptosis induced by interference with survivin function. *Nat Cell Biol* 1: 461–466.

Li, F., H. Huang, L. Brattain, et al. 2005. Differential expression of survivin and apoptosis by vitamin D3 in two isogenic MCF-7 breast cancer cell sublines. *Oncogene* 24: 1385–1395.

Li, F.Z., G. Ambrosin, E.Y. Chu, et al. 1998. Control of apoptosis and mitotic spindle checkpoint by survivin. *Nature (London)* 396: 580–583.

Ling, X., and F. Li. 2004. Silencing of anti-apoptotic survivin gene by multiple approaches of RNA interference technology. *Biotechniques* 36: 450–454.

Lu, C.-D., D.C. Altieri, and N. Tanugawa. 1998. Expression of novel antiapoptotic gene, survivin, correlated with tumor cell apoptosis and p53 accumulation in gastric carcinomas. *Cancer Res* 58: 1808–1812.

Mara, F., P. Jonet, C. Sophie, et al. 2003. Fibronectin protects prostate cancer cells from tumor necrosis factor-alpha-induced apoptosis via the AKT/survivin pathway. *J Biol Chem* 278: 50402–50411.

Mesri, M., N.R. Wall, J. Li, et al. 2001. Cancer gene therapy using a survivin mutant adenovirus. *J Clin Invest* 108: 981–990.

Mirza, A., M. McGuirk, T.N. Hockenberry, et al. 2002. Human survivin is negatively regulated by wild type p53 and participates in p53-dependent apoptotic pathway. *Oncogene* 2: 2613–2622.

Murphy, J.T., A. Beavis, M.B. Watson, et al. 2007. Development of novel radioresistant oral squamous cell carcinoma cell lines. *Clin Otolaryngol* 32: 510.

Nakahara, T., M. Takeuchi, I. Kinoyama, et al. 2007. YM155, a novel small-molecule survivin suppressant, induces regression of established human hormone-refractory prostate tumor xenografts. *Cancer Res* 67: 8014–8021.

Oikawa, T., Y. Unno, K. Matsuno, et al. 2010. Identification of a small-molecule inhibitor of the interaction between survivin and Smac/DIABLO. *Biochem Biophys Res Commun* 393: 253–258.

Okada, H., and T.W. Mak. 2004. Pathways of apoptotic and non-apoptotic death in tumour cells. *Nat Rev Cancer* 4: 592–603.

Olie, R.A., A.P. Simoes-Wust, B. Baumann, et al. 2000. A novel anti-sense oligonucleotide targeting survivin expression induces apoptosis and sensitizes lung cancer cells to chemotherapy. *Cancer Res* 60: 2805–2809.

Otto, K., M.H. Andersen, A. Eggert, et al. 2005. Lack of toxicity of therapy-induced T cell responses against the universal tumour antigen survivin. *Vaccine* 23: 884–889.

Paik, S., S. Shak, G. Tang, et al. 2004. A multigene assay to predict recurrence of tamoxifen-treated, node-negative breast cancer. *N Engl J Med* 351: 2817–2826.

Pennati, M., M. Binda, G. Colella, et al. 2003. Radiosesitisation of human melanoma cells by ribozyme mediated inhibition of survivin expression. *J Invest Dermatol* 120: 648–654.

Pennati, M., M. Binda, G. Colella, et al. 2004. Ribozyme mediated inhibition of survivin expression increases spontaneous and drug induced apoptosis and decreases the tumirigenic potential of human prostate cancer cells. *Oncogene* 23: 386–394.

Pennati, M., M. Folini, and N. Zaffaroni N. 2007. Targeting survivin in cancer therapy: Fulfilled promises and open questions. *Carcinogenesis* 28: 1133–1139.

Pisarev, V., B. Yu, R. Salup, et al. 2003. Full-length dominant-negative survivin for cancer immunotherapy. *Clin Cancer Res* 9: 6523–6533.

Qi, H., J. Guo, Y.C. Zahng, et al. 2010. Antisense oligonucleotide targeting survivin gene induces cell apoptosis in salivary mucoepidermoid carcinoma. *Zhonghua Kou Qiang Yi Xue Za Zhi* 45: 525–530.

Raitanen, M., V. Rantanen, J. Kulmala, et al. 2002. Paclitaxel combined with fractionated radiation in vitro: A study with vulvar squamous cell carcinoma cell lines. *Int J Cancer* 97: 853–857.

Raybaud-Diogene, H., A. Fortin, R. Morency, et al. 1997. Markers of radioresistance in squamous cell carcinomas of the head and neck: A clinicopathologic and immunohistochemical study. *J Clin Oncol* 15: 1030–1038.

Reed, J.C. 2001. The survivin saga goes in vivo. *J Clin Invest* 108: 965–969.

Reed, J.C. 2002. Apoptosis-based therapies. *Nat Rev Drug Discov* 1: 111–121.

Reed, J.C. 2003. Apoptosis-targeted therapies for cancer. *Cancer Cell* 3: 17–22.

Reed, J.C., and J.R. Bischoff. 2000. BIRinging chromosomes through cell division—and survivin' the experience. *Cell* 102: 545–548.

Reker, S, J.C. Becker, I.M. Svane, et al. 2004. HLA-B35-restricted immune responses against survivin in cancer patients. *Int J Cancer* 108: 937–941.

Reya, T., S.J. Morrison, M.F. Clarke, et al. 2001. Stem cells, cancer, and cancer stem cells. *Nature (London)* 414: 105–111.

Rohayem, J., P. Diestelkoetter, B. Weigle, et al. 2000. Antibody response to the tumor-associated inhibitor of apoptosis protein survivin in cancer patients. *Cancer Res* 60: 1815–1817.

Rosa, J., P. Canovas, A. Islam, et al. 2006. Survivin modulates microtubule dynamics and nucleation throughout the cell cycle. *Mol Biol Cell* 17: 1483–1493.

Sakao, S., L. Taraseviciene-Stewart, J.D. Lee, et al. 2005. Initial apoptosis is followed by increased proliferation of apoptosis-resistant endothelial cells. *FASEB J* 19: 1178–1180.

Salvesen, G.S., and C.S. Duckett. 2002. Apoptosis: IAP proteins: Blocking the road to death's door. *Nat Rev Mol Cell Biol* 3: 401–410.

Sarela, A.I., C.S. Verveke, J. Ramsdale, et al. 2002. Expression of survivin, a novel inhibitor of apoptosis and cell cycle regulatory protein, in pancreatic adenocarcinoma. *Br J Cancer* 86: 886–892.

Sawada, T., and F. Cabral. 1989. Expression and function of beta-tubulin isotypes in Chinese hamster ovary cells. *J Biol Chem* 264: 3013–3020.

Sawyers, C. 2004. Targeted cancer therapy. *Nature (London)* 432: 294–297.

Schmidt, S.M., K. Schag, M.R. Muller, et al. 2003. Survivin is a shared tumor-associated antigen expressed in a broad variety of malignancies and recognized by specific cytotoxic T cells. *Blood* 102: 571–576.

Schmitz, M., P. Diestelkoetter, B. Weigle, et al. 2000. Generation of survivin-specific CD8+T effector cells by dendritic cells pulsed with protein or selected peptides. *Cancer Res* 60: 4845–4849.

Sharma, H., S. Sen, M. Mathur, et al. 2004. Combined evaluation of expression of telomerase, survivin, and anti-apoptotic Bcl-2 family members in relation to loss of differentiation and apoptosis in human head and neck cancers. *Head Neck* 26: 733–740.

Shen, L., M. Ciesielski, S. Ramakrishnan, et al. 2012. Class I histone deacetylase inhibitor entinostat suppresses regulatory T cells and enhances immunotherapies in renal and prostate cancer models. *PLoS One* 7: e30815.

Shi, Q., K. Chen, S.L. Morris-Natschke, et al. 1998. Recent progress in the development of tubulin inhibitors as antimitotic antitumor agents. *Curr Pharm Des* 4: 219–248.

Shui, P.T., X.H. Jiang, M.C.M. Lin, et al. 2003. Suppression of survivin expression inhibits in vivo tumorigenicity and angiogenesis in gastric cancer. *Cancer Res* 63: 7724–7732.

Siegel, S., A. Wagner, and N. Schmitz, 2003. Induction of antitumour immunity using survivin peptide-pulsed dendritic cells in a murine lymphoma model. *Br J Haematol* 122: 911–914.

Skaufias, D.A., C. Mollinari, F.B. Lacroix, et al. 2000. human survivin is kinatochore-associated passenger protein. *J Cell Biol* 151: 1575–1582.

Smith, S.D., M.A. Wheeler, J. Plescia, et al. 2001. Urine detection of survivin and diagnosis of bladder cancer. *JAMA* 285: 324–328.

Smith, M.L., and Y.R. Seo. 2002. p53 regulation of DNA excision repair pathways. *Mutagenesis* 17: 149–156.

Sommer, K.W., C.J. Schamberger, G.E. Schmidt, et al. 2003. Inhibitor of apoptosis protein (IAP) survivin is upregulated by oncogenic c-H-Ras. *Oncogene* 22: 4266–4280.

Song, Z., X. Yao, and M. Wu. 2003. Direct interaction between Smac/DIABLO is essential for anti-apoptotic activity of survivin during taxol-induced apoptosis. *J Biol Chem* 278: 23130–23140.

Takai, N., T. Miyazaki, M. Nishida, et al. 2002. Survivin expression correlates with clinical stage, histological grade, invasive behavior, and survival rate in endometrial carcinoma. *Cancer Lett* 184: 105–116.

Talbot, D.C., M. Ranson, J. Davies, et al. 2010. Tumor survivin is downregulated by the antisense oligonucleotide LY2181308: A proof-of-concept, first-in-human dose study. *Clin Cancer Res* 16: 6150–6158.

Tamm, I., Y. Wang, E. Sausville, et al. 1998. IAP family protein survivin inhibits caspase activity and apoptosis induced by Fas (CD95), Bax, caspases, and anti-cancer drugs. *Cancer Res* 58: 5315–5320.

Tanaka, Y., K. Fujiwara, and H. Tanaka, 2004. Paclitaxel inhibits expression of heat shock protein 27 in ovarian and uterine cancer cells. *Int J Gynecol Cancer* 14: 616–620.

Taran, J., Z. Master, J.I. Yu, et al. 2002. A role of survivin in chemoresistance of endothelial cells mediated by VEGF. *Proc Natl Acad Sci USA* 99: 4345–4349.

Tetsuhiro, T., H. Fumitake, F. Toshihiko, et al. 2004. Phase I clinical study of anti-apoptosis protein, survivin derived peptide vaccine therapy for patients with advanced or recurrent colorectal cancer. *J Transl Med* 2: 19.

Tetsuhisa, Y., and T. Nobuhiko. 2001. The role of survivin as a new target of diagnosis and treatment in human cancer. *Med Electron Microsc* 34: 12–20.

Tolcher, A.W., A. Mita, L.D. Lewis, et al. 2008. Phase I and pharmacokinetic study of YM155, a small-molecule inhibitor of survivin. *J Clin Oncol* 26: 5198–5203.

Trieb, K., R. Lehnar, T. Stulning, et al. 2003. Survivin expression in human osteosarcoma is a marker for survival. *Eur J Surg Oncol* 29: 379–382.

Tsuruma, T., F. Hata, T. Torigoe, et al. 2004. Phase I clinical study of anti-apoptosis protein, survivin-derived peptide vaccine therapy for patients with advanced or recurrent colorectal cancer. *J Transl Med* 2: 19.

Tu, S.P., J.T. Cui, P. Liston, et al. 2005. Gene therapy for colon cancer by adeno-associated viral vector-mediated transfer of survivin Cys84Ala mutant. *Gastroenterology* 128: 361–375.

Uchida, H., T. Tanaka, K. Sasaki K, et al. 2004. Adenovirus-mediated transfer of siRNA against survivin induced apoptosis and attenuated tumor cell growth in vitro and in vivo. *Mol Ther* 10: 162–171.

Uren, A.G., L. Wong, M. Pakusch, et al. 2000. Survivin and the inner centromere protein INCENP show similar cell cycle localization and gene knockout phenotype. *Curr Biol* 10: 1319–1328.

Vagnarelli, P., and W.C. Earnshaw. 2004. Chromosomal passengers: The four-dimensional regulation of mitotic events. *Chromosoma* 113: 211–222.

Veenita, Y., and O.P. Singh. 2005. Induction followed with concurrent chemoradiotherapy in advanced head and neck cancer. *J Cancer Res Ther* 1: 198–203.

Velculescu, V.E., S.L. Madden, L. Zhang, et al. 1999. Analysis of human transcriptomes. *Nat Genet* 23: 387–388.

Vokes, E.E., R.R. Weichselbaum, S.M. Lippman, et al. 1993. Head and neck cancer. *N Engl J Med* 328: 184–194.

Wade, R.H., and A.A. Hyman. 1997. Microtubule structure and dynamics. *Curr Opin Cell Biol* 9: 12–17.

Wall, N.R., D.S. O'Connor, J. Plescia, et al. 2003. Suppression of survivin phosphorylation on Thr34 by flavopyridol enhances tumor cell apoptosis. *Cancer Res* 63: 230–235.

Wang, Z., S. Fukuda, and L.M. Pelus. 2004. Survivin regulates the p53 tumor suppressor gene family. *Oncogene* 23: 8146–8153.

Xiang, R., N. Mizutani, Y. Luo, et al. 2005. A DNA vaccine targeting survivin combines apoptosis with suppression of angiogenesis in lung tumor eradication. *Cancer Res* 65: 553–561.

Yamamoto, T., Y. Manome, A. Miyamoto, et al. 2003. Development of novel gene therapy using survivin anti-sense expressing adenoviral vectors [in Japanese]. *Gan To Kagaku Ryoho* 30: 1805–1808.

Yang, G., A.L. Huang, N. Tang, et al. 2003. Inhibition of survivin expression in liver cancer cells by shRNA. *Zhonghua Gan Zang Bing Za Zhi* 11: 712–715.

Zaffaroni, N., and M.G. Daidone. 2002. Survivin expression and resistance to anticancer treatments: Perspectives for new therapeutic interventions. *Drug Resist Updat* 5: 65–72.

Zaffaroni, N., M. Pennati, G. Collela, et al. 2002. Expression of antiapoptotic gene survivin correlates with taxol resistance in human ovarian cancer. *Cell Mol Life Sci* 59: 1406–1412.

Zaffaroni, N., M. Pennati, and M.G. Daidone. 2005. Survivin as a target for new anticancer
interventions. *J Cell Mol Med* 9: 360–372.

Zhang, T., T. Otevrel, Z. Gao, et al. 2001. Evidence that APC regulates survivin expression:
A possible mechanism contributing to the stem cell origin of colon cancer. *Cancer Res*
61: 8664–8667.

Zhao, L.J., H. Jian, and H. Zhu. 2003. Specific gene inhibition by adenovirus-mediated expression of small interfering RNA. *Gene* 16: 137–141.

Zhu, Z.B., S.K. Makhija, B. Lu, et al. 2004. Transcriptional targeting of tumors with a novel
tumor-specific survivin promoter. *Cancer Gene Ther* 11: 256–262.

Zhu, H.X., C.Q. Zhou, G. Zang, et al. 2003. Survivin mutants reverse the malignancy of HeLa
cells. *Ai Zheng* 22: 467–470.

9 Molecular Diagnosis of Oral Cancer

KEY WORDS

Angiogenesis
Apoptosis
Biopsy
HPLC
Immunoassay
Markers
Transcriptomics

9.1 INTRODUCTION

Oral and oropharynx cancer is the seventh most common form of cancer worldwide and is projected to increase by over 60% in the next two decades (Mathers and Loncar, 2006). Around half of the patients detected with oral cancer will die within 5 years of initial diagnosis. The high mortality rate can be attributed to technological limitations, which allows detection of oral cancer with confidence only in the advanced stage. A clear understanding of oral cancer-associated risk factors and early detection of oral cancer would be critical in improving survival. Treatment planning, patient stratification, and prognosis for patients with oral cancer are mainly based on tumor, node, and metastasis (TNM) classification (Patel and Shah, 2005), which is not perfect. Two-step carcinogenesis of the oral mucosa is well established, which postulates development of oral cancer from a precursor lesion. However, there is clinical evidence that contradicts the two-step carcinogenesis model for oral cancer (Reibel, 2003). These traditional clinical methods are subjective at the best, and are not sensitive for early detection of oral cancer, and have a very limited scope in predicting the aggressiveness of the tumor and identifying high-risk patients. Early detection of oral cancer by sensitive monitoring of tumor-specific markers should help in disease management and improving survival.

Oral cancer progression results from the complex interaction of various genetic and environmental factors. These genetic factors modulate various molecular changes, and monitoring such molecular changes offers a tremendous opportunity for early detection, and thereby management of oral cancer therapy. Tumor cells shed by primary cancers find their way to the circulatory system, which should allow early detection of cancer. In the progression of a tumor from a homogeneous proliferating clone to a group of heterogeneous subpopulations of cells, the cells generate an entire array of enzymes and surface molecules (Mendelsohn, 1987; Sah et al.,

2006; Khan et al., 2006, 2010a). These molecules, as well as those produced by the tumor's host in response to the presence of the tumor, are generally known as tumor markers. They can be used to detect the presence of malignancy when located in clinical samples like cytological smears and biopsied tissue samples or body fluids (Chan and Sell, 1994).

At present, immunoassay is the only readily available technique for detection of tumor markers (Maduskey and Ogden, 2000). Immuno-methods generally are designed for detection of a single marker at a time, and such assays often lead to false negative results because of various factors, which could prevent antibody-antigen interaction. Absence of a single molecular marker in the case of oral cancer makes utility of such immunoassay-based tests infeasible. It is highly probable that a combination of several markers, detected simultaneously, will definitely improve the chances of success in early detection of cancer (Venkatakrishna et al., 2003).

The oral cancer screening is based on conventional oral investigation, which is followed up with confirmatory biopsy and histopathological examination. The severity and grading of cancer are often dependent on results from imaging techniques like magnetic resonance imaging (MRI), computed tomography (CT), and dental panoramic tomography (DPT). Some other clinical techniques used are the toluidine blue test and light-based detection system.

Evolution of high-throughput *omics* technologies has enabled profiling of tumor cells to get insight into the complex mechanisms of cancer progression and related complications due to cancer. Holistic analysis of such omics data generates potential molecular markers that are significantly modulated under cancerous conditions. However, it's not always straightforward to translate such findings into a viable diagnostic or prognostic biomarker. High-throughput studies help in understanding the disease mechanism, and scrutiny of the resulting molecular markers under stringent criteria of sensitivity, specificity, and exclusivity should lay the foundation of discovery of potentially useful biomarkers. There has been improvement of prognosis of patients with oral cancer in recent years, which can be attributed to identification of new molecular targets for oral cancer.

Serum tumor markers have been accepted as valuable tools for prognosis and treatment monitoring over the last two decades (Matsubara et al., 1986; Doweck et al., 1995). Serum tumor markers like squamous cell carcinoma antigen (SCCAg), carcinoembryonic antigen (CEA), lipid-associated sialic acid, SCC marker (TA-4), and serum intercellular adhesion molecule-1 (S-ICAM-1) have been examined for their value in detecting head and neck cancers (Doweck et al., 1995; Silverman et al., 1976; Katopodis et al., 1982; Yoshimura et al., 1990; Walther et al., 1990; Calsen et al., 1990; Ropka et al., 1991; Wollenberg et al., 1997; Ogawa et al., 1999; Ayude et al., 2003). However, these serum tumor markers have not found their way into clinical practice for various reasons, like low sensitivity. The proteins from the intermediate filament (IF) family, in particular cytokeratins (CKs), have shown utility in cancer diagnostics (Schweizer et al., 2006). CK fragments released from proliferating or apoptotic cells have proved to be useful markers for epithelial malignancies. The most frequently used cytokeratins are tissue polypeptide antigen (TPA), tissue polypeptide specific antigen (TPS), and cytokeratin fragments 21-1 (Cyfra 21-1), which are being evaluated as serum markers for their clinical utility. Cytokines are involved in cell growth, invasion, and interruption of tumor suppression, immune status, and survival. Production of cytokines is dysregulated in oral squamous cell carcinoma (OSCC), and therefore offers the opportunity to be used for diagnosis of OSCC. Analytes present in saliva are a reflection of the health status of an individual. Saliva has already been established as a potential source of biomarkers for oral cancer, because it is simple to collect in a noninvasive manner (Wu et al., 2010).

9.2 MOLECULAR MARKERS FOR ORAL CANCER

Cancer progression has been monitored using traditional prognostic and diagnostic markers like tumor size, grade, age, steroid hormone receptor status or menopausal status, etc. Histological examination is still considered the gold standard for diagnosis of malignant oral lesions. The clinical behavior of OSCC as highly, moderately, and poorly differentiated cancer is traditionally predicted on the basis of the histological grading system developed by Broders. This system has been used for decades. Some precancerous lesions with mild dysplasia may go unrecognized by histology (Akrish et al., 2004), and hence molecular pathology using the potential biomarkers for disease diagnosis and prognosis is highly desired. Development of mature tumor from the primary lesion involves dysregulation of a host of biomolecules, which offers a great opportunity for early detection of oral cancer, and thereby controlling mortality. Mutations of a specific gene or overexpression or underexpression of its protein is frequently found in the onset or progression of OSCC. Generally, the markers under the high-risk category expressed in the oral squamous tissue at the earlier stage (Gx, G1, G2) suggest that the tumor will become more malignant (Wu et al., 2010).

9.3 MOLECULAR MECHANISM OF DEVELOPMENT OF ORAL CANCER

Chromosomal aberrations are common in cancer that induce the normal cellular mechanism toward malignancy by silencing or repressing expression of tumor suppressor genes (TSGs) or activation of oncogenes. The loss of alleles 3p14 and 9p21 has been reported as an early event in OSCC (Uzawa et al., 2001). A high frequency of loss of heterozygosity (LOH) at chromosomal locations 13q and 17p has been reported in premalignant oral lesions and early carcinomas (El-Naggar et al., 1995). Aberrations in chromosome 9 are quite often in early tumor development; allelic losses at 9p21 have been reported in premalignant oral lesions and early carcinomas (Ohta et al., 2009). Cell proliferation regulator genes like p16 and p14 are located in 9p21. Allelic losses at 5p21 to 22, 22q13, 4q, 11q, 18q, and 21q are often found in association with advanced tumor stages and poorly differentiated carcinomas (Figure 9.1) (Moles et al., 2008; Khan and Bisen, 2013).

Anomalies have been reported in oral cancer for tumor suppressor genes (TSGs) like CDKN2A, p14, and TP53 (Kozomara et al., 2007; Kresty et al., 2002; Shintani et al., 2001).

FIGURE 9.1   Molecular model of oral cancer.

Gene expression is regulated by epigenetic mechanisms like methylation, acetylation, etc., and such regulation is affected in carcinogenesis. DNMT3B is involved in epigenetic control along with DNMT1. SNPs in DNMT3B have been postulated to play a causative role in several cancers, including OSCC (Van Heerden et al., 2001). The expression of Ras association family genes (RASSF) in general and RASSF2 in particular has been found to be altered by methylation in OSCC (Imai et al., 2008).

High levels of cyclo-oxygenase 2 (COX-2) have been found in dysplastic lesions (Sudbo, 2004). The human trophoblast cell surface antigen (TROP2) has been associated with shorted survival (Fong et al., 2008). The epithelial adhesion molecule (EpCAM) has been associated with tumor size and invasiveness (Yanamoto et al., 2007). Matrix metalloproteinase (MMP-2 and MMP-9) has been associated with invasive potential of tumors and levels of alcohol, leading to inference that alcohol might play a role in oral carcinogenesis through the stimulation of these genes (Moles et al., 2008). Soluble fragments of cytokeratin 19 and Cyfra 21-1 has been found in patients with OSCC, which could be used as markers (Zhong et al., 2007a, 2007b). Proteins and oxidated DNA levels in saliva might indicate high levels of reactive oxygen species, which appears to be involved in the development of OSCC (Bahar et al., 2007). Expression of receptors of advanced glycation end products (RAGEs) decrease with an increase in OSCC differentiation (Landesberg et al., 2008).

9.4 THE HALLMARKS OF ORAL CANCER

Cancer is a multistep process that derives progressive transformation of normal cells into malignant counterparts; these steps can be grouped into six essential alterations in cell physiology that dictate malignant growth: (1) self-sufficiency in growth signals, (2) insensitivity to growth-inhibitory (antigrowth) signals, (3) evasion of programmed cell death (apoptosis), (4) limitless replicative potential, (5) sustained angiogenesis, and (6) tissue invasion and metastasis (Hanahan and Weinberg, 2000).

9.4.1 Self-Sufficiency in Growth Signals

Mitogenic growth signals are required by normal cells to move from the quiescent state into an active proliferative state. These signals are relayed into the cells by transmembrane receptors that bind to specific signaling molecules: extracellular matrix components, diffusible growth factors, and cell-to-cell interaction/adhesion molecules. Unlike normal cells, cancer cells can proliferate in the absence of such stimulatory signals by mimicking normal growth signaling through various mechanisms. Epithelial growth factor (EGF) (EGFR, c-erb1-4/Her-2/neu) belongs to the erbB family and is a class of cell surface receptors that transduce growth stimulatory signals necessary for proliferation. erbB-1 and erbB-2 are 20.2% overexpressed in oral carcinomas, which is a sign of the carcinogenic process, and such aberrations are very common in histologically nondysplasic premalignant oral lesions (Werkmeister et al., 2000). Overexpression of the EGFR gene has been reported in several human cancers, including oral squamous cell carcinoma (Whyte et al., 2002).

9.4.2 Insensitivity to Growth-Inhibitory (Antigrowth) Signals

Cellular quiescence and tissue homeostasis is maintained through the range of antiproliferative signals, which includes both soluble growth inhibitors and immobilized inhibitors, embedded in the extracellular matrix and on the surface of nearby cells. Antigrowth signals block cell proliferation by reversibly forcing cells in the active proliferative cycle into the quiescent (G_0) state or by irreversibly forcing cells into postmitotic states through cellular differentiation. These antiproliferative signals act through retinoblastoma protein (pRb), which in the hypophosphorylated state blocks proliferation by sequestering and altering the function of E2F transcription factors that control the expression of a host of genes essential for G1-to-S phase transition (Weinberg, 1995). Retinoblastoma protein (pRb) has been reported to be overexpressed in oral carcinomas (Koontongkaew et al., 2000). The cyclin:CDK complex is responsible for phosphorylation of pRb, which releases E2F transcription factors, thereby releasing a host of genes essential for the G1–to-S phase transition. Cyclin D_1 has been found to be overexpressed in oral squamous cell and pharyngeal carcinomas (Koontongkaew et al., 2000). Overexpression of CCND1 can induce overexpression of the cyclin D1 protein, which induces cell proliferation, and has been associated with poor prognosis in early-stage oral tumors (Zhou et al., 2009; Sathyam et al., 2008; Marsit et al., 2008). Nuclear cell proliferation antigens, which appear in the final phase of G1 and in the S phase, are indicative of cell proliferation (Schliephake, 2003). Some of the nuclear cell proliferation antigens associated with oral cancer are *P120* (Schliephake, 2003), *Ki-67/MIB* (Schliephake, 2003; Tumuluri et al., 2002), *AgNOR* (Schliephake, 2003), and *Skp2* (Schliephake, 2003). It has been shown that levels of proliferating cell nuclear antigen are increased in an experimental model (Schwartz et al., 2000). It has been suggested to use AgNOR and Ki-67 analyses to determine proliferative states of epithelial cells in oral cancer (Teresa et al., 2007; Costa et al., 1999).

9.4.3 EVASION OF PROGRAMMED CELL DEATH (APOPTOSIS)

The uncontrolled growth of tumor cells is determined not only by rate of cell proliferation, but also by apoptosis. The programmed cell death or apoptosis is the
mechanism that is triggered by a variety of physiologic signals, like DNA damage, signaling imbalance, survival factor insufficiency, and hypoxia (Evan and
Littlewood, 1998; Ishizaki et al., 1995; Giancotti and Ruoslahti, 1999), followed by
precise steps by which cellular components are degraded and removed from the system in less than 24 hours (Wyllie et al., 1980). Tumor cells have acquired the ability
to evade such apoptotic actions by promoting anti-apoptotic or survival mechanisms.
In oral cancer its survival is achieved by aberrant changes for pro-apoptotic genes
like p53 (Whyte et al., 2002), BAX (Schliephake, 2003; Cruz et al., 2002), and Fas
(Schliephake, 2003; Muraki et al., 2000), and survival genes like Bcl-2 (Schliephake,
2003; Whyte et al., 2002). Hsp27 and Hsp70 are associated with mutation of the p53
gene, and Hsp70 has been detected on OSCC (Schliephake, 2003).

9.4.4 LIMITLESS REPLICATIVE POTENTIAL OR IMMORTALIZATION

The replicative potential of the cell is limited by factors like senescence and crisis,
which eventually lead to massive cell death by end-to-end fusion of chromosomes,
with very rare variants having the ability to multiply without limit, the process termed
immortalization (Wright et al., 1989). Telomerase activity is critical for controlling
unlimited potential for division or immortalization (Schliephake, 2003; Ries et al.,
2001). Telomerase activity is not undetectable in normal somatic cells; however, it
can be evaluated in biopsied tissue from oral cancer (Liao et al., 2000a). Telomerase
activity is used as a marker in the diagnosis of preneoplasic or neoplasic oral mucosa
lesions, as 80 to 90% of such tumors have high levels of telomeric expression, particularly of the hTERT subunit (Ries et al., 2001; Lee et al., 2001; Epstein et al., 2003).

9.4.5 SUSTAINED ANGIOGENESIS

Cell function and survival critically depend on availability of oxygen and nutrients
supplied by the vasculature, obligating virtually all cells in a tissue to reside close to
the blood capillary system. Factors promoting growth of new blood vessels (angiogenesis) are imperative for tumorigenesis. Some of the pro-angiogenic genes implicated in OSCC are VEGF/VEGFR, NOS_2, PD-ECGF, FGF-2, PGF-3, and HIF1α
(Schliephake, 2003; Wakulich et al., 2002).

9.4.6 TISSUE INVASION AND METASTASIS

Tissue invasion and metastasis are among the adaptive mechanisms of tumor cells
by which they migrate and invade adjacent tissues, and thereby travel to distant
sites, to find a new source of nutrients necessary for growth. These distant settle-
ments of tumor cells (metastasis) are the cause of 90% of human cancer deaths
(Sporn, 1996). Some of the invasion and metastasis factors implicated in oral cancer are matrix metalloproteinases (MMPs), cathepsins, integrins, cadherins and

catenins, desmoplakin/placoglobin, and Ets-1 (Schliephake, 2003; Hamidi et al., 2000; Bankfalvi et al., 2002).

9.5 OTHER MARKERS

1. Enzymes: Glutathione S-transferase is an isoenzyme that acts in the second phase of cell metabolism. It plays an important role in protecting the cell against cytotoxic and carcinogenic agents. Three types of glutathione S-transferase (GST) are known: α, β, and π. Overexpression of GST-π has been reported in various cancer tissues, including premalignant oral lesions (Bautista and Santiago, 2001).

2. Intracellular markers: Cytokeratins (CKs) are epithelial cell proteins associated with malignization of oral lesions. There are over 19 cytokeratins, divided into two subfamilies. Some studies have suggested using the expression of CK 19 in the diagnosis of precancerous oral lesions and the early stages of oral carcinogenesis (Nie et al., 2002).

3. Cell surface markers: CD57 antigen is found in the membrane of lymphoid and neural cells. The percentage of CD57 lymphocytes is increased in oral leukoplakia with moderate or severe dysplasia compared with normal tissue (Bautista and Santiago, 2001). Histocompatibility antigen (HLA) is associated with poorly differentiated oral carcinomas (Chung-Ji et al., 2002).

4. Nuclear analysis: The importance of DNA analysis for the evaluation and prediction of oral cancer risk has been stressed in many studies (Sudbo et al., 2001; Lippman and Hong, 2001; Zhang et al., 2001). Polymerase chain reaction (PCR)-based analysis is sensitive and noninvasive, which can be used to detect nuclear changes in tumors and premalignant lesions. The parameters evaluated in nuclear analysis include:

   a. DNA ploidy state (of chromosomal pairing): Reflects the risk of oral cancer:
      - Anaploidy → high risk
      - Tetraploidy → intermediate risk
      - Diploidy → low risk

   b. Chromosomal polysomy: Determines genetic instability (Kim et al., 2001; Zhang et al., 2001). Extensive chromosomal polysomy is reported in areas classified as high risk, such as dysplasic epithelia.

Oral cancer research in the past decade has attempted to leverage technological advancements made in the field of genomics for utilization of genomic tools in various areas of oral cancer research to find reliable diagnostic and prognostic markers for clinical use (Zimmermann et al., 2007). Modern high-throughput genomic and proteomic approaches have been extensively used to monitor altered expressions of genes and proteins in a variety of cancers, including OSCC, which may facilitate the identification of potential biomarkers for OSCC. Evaluation of molecular expressions associated with histopathological changes during OSCC development revealed diagnostic significance of the following biomarkers in all grades of high-risk patients: urokinase-type plasminogen activator receptor, Ki-67, VEGFA, active matrix metalloproteinase-7 (MMP-7) expression, tissue inhibitor of metal-loproteinase-2, trophoblast 2, RhoC, keratin 8, serum midkine (S-MK), human centromere protein H (CENP-H), p75 neurotrophin receptor, and RHAMM. The increased expressions of biomarkers such as galectin-1, trophoblast 2, cyclin B1, CXCR4, p63, Rho C, keratin 8, p21 (WAF1/CIP1), and phosphorylated Akt, and loss of E-cadherin, were correlated with the metastasis of OSCC (Wu et al., 2010). The elevation of the following biomarkers, including human telomerase reverse transcriptase, tissue inhibitor of metalloproteinase-2, IL-6, c-Met, and stromal syndecan-1, was associated with the recurrence of OSCC. High levels of CD109, Akt activation, and p63, a protein encoded by the p63 gene, which is a P53 gene homolog and the adverse expression of E-cadherin, could be significant prognostic indicators for OSCC. Studies have shown that OSCC patients with activation of Akt, switching from E-cadherin to N-cadherin, had decreased cellular syndecan-1, but increased stromal syndecan-1, and overexpression of p63 had a poorer survival rate. The inhibition of Akt appears to be one of the molecular approaches for the treatment of OSCC (Lim et al., 2005; Khan and Bisen, 2013). Various salivary biomarkers (p53, CD44, EGF) have been identified in candidates with cancer in the oral cavity and are proposed to be used as diagnostic and prognostic mark-ers (Warnakulasuriya et al., 2000; Franzmann, 2005; Balicki et al., 2005). The saliva counts of three oral bacteria species were found to be diagnostic indicators of OSCC (Mager et al., 2005). Various genetic alterations (like epigenetic gene silencing, loss of heterozygosity, mutations, etc.) have been detected in tumor tis-sue and have been associated with oral carcinogenesis (Rosas et al., 2001; Jiang et al., 2005; Zimmermann and Wong, 2008; Spivavack et al., 2004).

It will be interesting to investigate the expression of the biomarkers correlated with the invasion and metastasis of OSCC and the markers with poor prognosis and survival rate in body fluids of patients with OSCC. Survivin, a novel anti-apoptotic protein, overexpressed in most human cancers, including OSCC, is reported to carry diagnostic significance (Khan et al., 2010b).

In fact, autoantibodies against p53, which is aberrantly expressed in patients with OSCC, have been identified in both saliva and serum. In addition, IL-6, which was correlated with the recurrence of OSCC, was also detected in saliva. Thus, the autoantibodies against p53 and IL-6 may be the potential biomark-ers for OSCC. Furthermore, cell surface markers such as CD109, N-cadherin, and the reduced expression of cellular syndecan-1 may be the other group of potential biomarkers to identify high-risk patients with OSCC. It should be noted that some of the cytokines or enzymes such as MMP that were associated with tumor metastasis may not be detectable in saliva since the concentration is too low to be identified using the conventional analytic method. A sensitive analytic method such as mass spectrometry technology may be able to solve the detec-tion problem. Screening with sensitive biomarkers in saliva in combination with molecular pathology analysis may improve the accuracy and staging diagnosis for OSCC.

9.6 DIAGNOSIS OF ORAL SQUAMOUS CELL CARCINOMA

The high mortality rate among oral cancer patients is primarily due to lack of early detection of malignancies; therefore, technologies that allow identification of high-risk oral premalignant lesions and intervention at premalignant stages are very much desirable. Histological examination of tissue is a well-established method for the detection of malignant oral lesions. However, such a method has a limited scope, as it is a noninvasive technique requiring subjective expertise of the surgeon and has psychological implications for most patients. Also, these methods are less sensitive because they rely on the subjective interpretation of the pathologist/surgeon (Sciubba, 1999; Mehrotra et al., 2006).

These limitations with histological tests emphasize the development of new diagnostic methods, improving the existing ones and discovering new therapeutic targets for oral neoplastic diseases (Epstein et al., 2003; Ogden et al., 1991; El-Naggar et al., 2001; Spafford et al., 2001). Changes at the molecular level happen much before their clinical manifestations can be observed through traditional diagnostics, and such molecular changes can be diagnosed by noninvasive cytological and other molecular techniques.

9.6.1 CYTOLOGICAL TECHNIQUES

9.6.1.1 Oral Brush Biopsy

Oral cells are collected from subjects for pathological investigation; some of the common practices of collecting oral cells are rinsing the oral cavity, collecting saliva, and scraping the surface of the mucosa (Ogden et al., 1992; Jones et al., 1994). Some of the salient features of making a good cytological smear are (1) simple and easy to use in any location, (2) it causes minimum trauma, and (3) it should provide an adequate number of epithelial cells for study (Jones et al., 1994). A brush is generally considered an appropriate device to collect oral cytological samples because of its operational simplicity, and other add-on features, like high sensitivity and risk-free sampling (Svirsky et al., 2001). Brush biopsy allows us to collect a full-thickness sample, which is critical in giving a complete histomorphological picture, which is necessary to make correct identification in the case of the absence of cancer-related histomorphological features in the keratin layer (Walling et al., 2003; Ahmed et al., 2003). The clinical importance of brush biopsy has been demonstrated in various studies in which around 5% of normal/benign-appearing mucosal lesions were sampled by brush biopsy and later confirmed as precancerous or cancerous by scalpel biopsy (Ahmed et al., 2003; Banoczy, 1976). However, clinical use of brush biopsy has been doubted because of its relatively high false positive rate; its sensitivity is around 90%, but it has a low specificity of 3% (Rick and Slater, 2003; Frist, 2003). Moreover, the brush-based approach reported generating delayed and wrong diagnoses at times (Potter et al., 2003; Nichols et al., 1991; Mehrotra et al., 2004). There is a need for large multicenter studies to determine the effectiveness of brush cytology.

9.6.1.2   Liquid-Based Cytology

Liquid-based cytology, which was initiated in the 1990s, offers significant advantages over conventional exfoliative cytology. Several comparative studies have been conducted to establish its edge over conventional exfoliative cytology. The common experimental errors in sampling, transfer, and fixation of cellular samples can be reduced by adapting liquid-based preparations, as demonstrated in results from uterine cervix examination, and can also reduce false negatives in comparison to conventional smears (Bishop et al., 1998; McGoogan and Reith, 1996; Vassilakos et al., 1996; Howell et al., 1998; Grohs et al., 1994; Sprenger et al., 1996). The reports from clinical studies were quite positive about the diagnostic utility of liquid-based cytology (Fabia et al., 2005).

9.6.2   Molecular Analyses

Oral cytological examination is quite labor-intensive and requires a high degree of expertise for identifying cells with suspicious morphology. Molecular analyses are meant to identify genetic anomalies (Spafford et al., 2001). The feasibility and viability of oral premalignant and cancerous lesions diagnostics by extracting RNA from cells obtained by scraping have recently been demonstrated (Patel et al., 2004).

9.6.2.1   Gene Alterations

Cancer is a complex disease caused by accumulation of multiple genetic alterations, which affects normal cellular processes, and which in turn affects a wide range of functions, like cell signaling, apoptosis, differentiation, survival, angiogenesis, invasion, and metastasis. Most of these genetic alterations are acquired (somatic), although some of them may be inherited. These genetic alterations mainly act by activating proto-oncogenes or inactivating tumor suppressor genes, thereby causing malignant transformation.

The most important causal factors of oral cancer are tobacco, alcohol, radiation, and HPV infection, which initiates and aggravates cancer growth through direct or indirect alterations of the molecular network responsible for normal cellular growth. These changes at the molecular level may occur spontaneously; such genetic alterations induce development of oral cancer and can form the basis of genetic marker-based detection of cancerous cells in clinical samples (Ogden et al., 1991; Spafford et al., 2001; Boyle et al., 1994). Mutations in the tumor suppressor gene p53 are the most frequent genetic alterations in human cancer and show a variable frequency in oral cancer (Boyle et al., 1994; Khan et al., 2009; Khan and Bisen, 2013). The mutations in tumor suppressor gene p53 have been implicated in oral carcinogenesis, and this fact has been utilized in the prediction or screening of neoplastic samples in OSCC (Boyle et al., 1994; Lopez et al., 2004; Liao et al., 2000b; Scheifele et al., 2002; Khan et al., 2009). Many point mutations have been reported in p53, and it's difficult to ascertain a particular point mutation that is specific for oral cancer, therefore limiting its application as a cost-effective marker in early detection of oral cancer (Rosas et al., 2001).

9.6.2.1.1 Epigenetic Alterations, Loss of Heterozygosity, and Microsatellite Instability

The feasibility of molecular markers in oral cancer diagnostics has been studied by various research groups, wherein they have tried to utilize important molecular events such as epigenetic alterations (hypermethylation of the promoter region) and genomic instabilities (loss of heterozygosity (LOH), microsatellite instability (MSI)) in the context of oral carcinogenesis (Rosas et al., 2001; Lopez et al., 2003; Debnath et al., 2010). Such epigenetic and genetic alterations invariably affect the normal functioning of protective genes like the tumor suppressor gene by various events like gene silencing through epigenetic changes in the promoter region or a mutation in the functional motif of genes. The methylation specific polymerase chain reaction (PCR) was used by Rosas et al. (2001) to study the methylation pattern of p16, DAP-K, and MGMT genes in smears of head and neck cancer patients, and they were able to detect an abnormal hypermethylation pattern in cancerous samples. The methylation-specific PCR was found to be sensitive and efficient in detecting tumoral DNA and can be used as a screening method for monitoring cancer recurrence.

Exfoliated cells can be used, as a source, to detect molecular alterations like loss of heterogeneity suggestive of oral carcinogenesis (Rossin et al., 2000; Mao et al., 1996; Partridge et al., 2000). Restriction fragment length polymorphism (RFLP) analysis was done on PCR-amplified samples collected from exfoliated cells (Huang et al., 1999). In this study, they reported the presence of LOH at one location in the p53 gene sequence in 66% of tumor samples, and also at other locations in 55% of tumor samples.

OSCC is characterized by the presence of microsatellite mutations (LOH or instability), and therefore offers the opportunity to be used as a molecular marker for oral carcinogenesis (Debnath et al., 2010). The association of a certain genomic region of chromosomes 3p, 9p, 17p, and 18q with oral carcinogenesis has been established in various studies by using microsatellite markers. Such microsatellite-based markers can be clinically used in early diagnosis of oral carcinogenesis (Nunes et al., 2000; Spafford et al., 2001).

9.6.2.1.2 Viral Genome Studies

There is growing evidence that points toward a causal role of human papillomavirus (HPV) in oral tumorigenesis. The presence of HPV DNA was detected in metastatic lesions from head and neck squamous cell carcinomas, and was reported to be positively correlated with metastasis (Umudum et al., 2005).

9.6.2.1.3 Proliferation Index and AgNOR Analysis

The expression of Ki-67 in oral cytological smears was found to be a potential diagnostic and prognostic marker (Sharma et al., 2005; Remmerbach et al., 2003). However, its prognostic significance has been debated in a recent study: Gonzalez-Moles et al. (2010) suggested Ki-67 to be a marker of the total fraction of proliferating cells contributed by proliferating cancerous or proliferating cells destined for terminal differentiation. The validity of oral cytology for analyzing the number of keratinized cells and the nucleolar activity (AgNORs) in smoking patients has been demonstrated (Orellana-Bustos et al., 2004).

9.6.2.1.4   Immunohistochemical Identification of Tumor Markers

The development of the tumor is a process realized by the occurrence and disappearance of various biomolecules (including proteins) at different stages of tumor growth. The study and identification of such tumoral markers, particularly cytokeratins, offers a great opportunity to detect the presence of tumor caused by oral cancer. The expression profile of cytokeratin gives invaluable information about the differentiation status of a cell; however, its utility as a marker for early diagnosis of oral cancer is limited (Remmerbach et al., 2003). The predictive power of the cytokeratins (particularly K8 and K19) can be increased when complemented with other genetic markers (Ogden et al., 1993, 1994).

9.6.2.1.5   HPLC-LIF for Early Detection of Oral Cancer

High-performance liquid chromatography–laser-induced fluorescence (HPLC-LIF) is one of the most promising diagnostic technologies; it combines the strength of ultrasensitive optical methods like LIF with highly efficient separation techniques like HPLC or CE to detect ultra-traces of biomolecules in complex, multicomponent physiological samples. It is a well-established fact that the protein profile is significantly different in body fluids (like saliva and serum) from normal, premalignant, and malignant subjects. The HPLC-LIF-based detection method is sensitive enough to monitor such changes and sensitive enough to detect changes in magnitude as low as femtomoles.

Laser-induced fluorescence scores over traditional fluorescence in many ways, including (1) high monochromaticity ensures that there is limited damage by stray light, (2) greater focusing ability ensures the necessity of a minute amount of sample, and (3) low background due to use of dielectric laser mirrors. HPLC-LIF can record fluorescence spectra and chromatograms simultaneously, which gives it power to detect small changes in body fluids like saliva and serum. It can be used for early diagnosis of cancer, because of its sensitivity to detect differential behavior of markers between normal, premalignant, and malignant stages (Venkatakrishna et al., 2003). The features of HPLC-LIF that make it tenable to be used as a diagnostic method are:

1. Ability to detect samples in femtomolar or less quantities
2. Ability to detect multiple markers simultaneously
3. Requires very short time for complete testing

9.6.3   POTENTIAL BIOMARKERS FOR OSCC

9.6.3.1   Cytokeratin Fragments in the Serum

There is at present only one study available reporting the status of these markers in sera of patients with oral SCC. Nagler et al. (1999) have evaluated the levels of CK markers Cyfra 21-1 and TPS in the sera of oral SCC patients. This study was performed on 38 patients with oral SCC, out of which 15 samples were from the tongue and 2 from the cheek, and the rest were from the mandible, maxilla, and lower lip. Although this study includes a relatively small number of patients with high rates of sensitivity and specificity for Cyfra 21-1 (84 and 93%, respectively) and TPS (69 and 87%, respectively), they found elevated Cyfra 21-1 levels during follow-up correlating with the detection of recurrences and second primaries with a significant reduction in the level of each CK, in the serum of oral cancer patients, within approximately 2 to 3 weeks of resection of the tumor. However, these studies are limited in the number of samples analyzed, and well-organized follow-up studies with a large group of patients are needed to establish the utility of this marker for the early detection of recurrence.

Simple epithelial-specific CK 8, 18, and 19 are normally not expressed in oral tissues; however, they are aberrantly expressed in oral SCC (Maass et al., 1999; Vaidya et al., 1989, 1996). More recently, expression of CK 19 in oral SCC has been linked with poor prognosis of patients (Thomas et al., 2006). Further, there is also experimental evidence that CK 8 and 18 are aberrantly expressed in the uninvolved surrounding tissue of the oral buccal mucosal tumor (Schaafsma et al., 1993). This suggests that aberrantly expressed CK could be released from tumor mass or surrounding adjacent tissue.

9.6.3.2 Toluidine Blue (TB)

The use of toluidine blue (tolonium chloride) as a diagnostic aid for the detection of oral cancer has been evaluated in a large number of studies over many decades (Epstein et al., 2007). It has also been suggested that TB may provide information on lesion margins, accelerate the decision to biopsy, and guide biopsy site selection and treatment of oral potentially malignant and malignant lesions. Based on data available up to 1989, a meta-analysis assessing the effectiveness of TB for identifying oral squamous cell carcinomas revealed a sensitivity between 93 and 97% and specificity between 73 and 92%. However, a large proportion of these many studies presented suffer from serious methodological limitations. It has recently been suggested that, overall, the sensitivity of TB staining for the detection of oral cancer varies between 78 and 100%, but the specificity between 31 and 100%. Review of various studies suggests that sensitivity of toluidine blue (TB) for detecting carcinomas is significantly higher than that for detecting dysplasia (Epstein et al., 2007).

9.6.3.3 Light-Based Detection Systems

Carcinogenesis is associated with various structural and metabolic changes, which are presumed to generate a distinct profile of absorption and refraction when exposed to different types of light or energy; light-based detection systems are designed to utilize such profiles for the detection of any abnormality (Epstein and Guneri, 2009). There have been remarkably few reliable studies on many of these devices before their release for clinical use.

9.6.3.4 Chemiluminescence

ViziLite® (Zila Pharmaceuticals, Phoenix, Arizona) is the most well-known system and has been found to have a high sensitivity (100%) in a number of cross-sectional studies, as all patients presented mucosal lesions previously detected on naked eye examination. However, its specificity (0 to 14.2%) and positive predictive value are very low. This procedure appears to potentiate certain visual aspects of the lesion, such as brightness and sharpness of margins. ViziLite does not help over-all in the identification of malignant and premalignant oral lesions, and therefore the combination with TB (ViziLite Plus®) was subsequently proposed to reduce the number or false positives. Although the specificity and positive predictive value of this test may have improved, there is very little scientific evidence for this combination in the literature to date. Another marketed device consists of a battery-powered light-emitting diode (LED) transilluminator with an autoclavable light guide that produces diffused light (Microlux/DL®, AdDent, Inc., Danbury, Connecticut). In a prospective study, the sensitivity and specificity of Microlux/DL for the detection of oral cancerous and precancerous lesions were reported to be 77 and 70%, respectively (Rosenberg and Cretin, 1989; Lingen et al., 2008; Farah and McCullough, 2007).

9.6.3.5 Tissue Fluorescence Imaging

The VELscope® system (Visually Enhanced Lesion Scope; LED Dental, Inc., White Rock, British Columbia, Canada) uses direct fluorescence for any loss of fluores-cence in visible and nonvisible high-risk oral lesions, such as cancers and precancers. VELscope reported sensitivity values of 97 to 100% and specificities of 94 to 100%, which are encouraging. This procedure has shown to be helpful in obtaining safer surgical margins in tumor excision (Huber, 2009). Its utility in screening low-risk patients with malignant or premalignant lesions is not yet reported (Fedele, 2009).

9.6.3.6 Tissue Fluorescence Spectroscopy

This is based on an automated system that uses small optical fibers to produce exci-tation wavelengths, and a spectroscope that records the resulting spectral data on a computer (Fedele, 2009). This technique eliminates the subjective interpretation of changes in tissue fluorescence. However, its main application is limited to the study of small lesions previously diagnosed by visual examination. This system has been shown to have high sensitivity and specificity when differentiating healthy mucosa from malignant oral lesions (De Veld et al., 2005).

9.6.3.7 Exfoliative Cytology

The analytical study of cells chipped off naturally or artificially from oral mucosa is known as exfoliative oral cytology. Oral cytology is useful for monitoring sev-eral sites for a large lesion and can guide the choice of sites for incisional biopsies (Perez-Sayans et al., 2009; Freitas et al., 2004). OralCDx® brush biopsy (OralCDx Laboratories Inc., Suffern, New York) is an oral transepithelial "biopsy" system that uses computer-assisted brushing (Sciubba, 1999). This technique was designed to detect any abnormality in oral epithelia, which may point toward the presence of dysplasia or cancer (Bhoopathi et al., 2009). The sensitivity of this technique ranges between 52 and 100%, and the specificity ranges between 29 and 100% (Fedele, 2009). OralCDx seems to overestimate the presence of dysplastic lesions even in nor-mal samples (Bhoopathi et al., 2009), so its practical clinical use remains debatable

(Hohlweg-Majert et al., 2009). Cytology specimens can, however, be further studied using quantitative cytomorphology, nuclear DNA content analysis, immunohisto-chemical tumor marker identification, and molecular markers (Poate et al., 2004; Acha et al., 2005; Mehrotra et al., 2006). It has recently been demonstrated that RNA can be extracted from exfoliated cells, and the use of RNA thus obtained may suit-ably be exploited to determine susceptibility to cancer among healthy populations and detect early markers of carcinogenesis.

9.6.3.8 Biopsy and Histopathology

Suspected malignant lesions must be biopsied in order to establish a definitive diag-nosis (Larsen et al., 2009). The correctness of prediction based on biopsy depends on various factors like adequate clinical information, quality of the biopsy (Seoane et al., 2002), and correct interpretation of the biopsy results.

9.6.3.9 Blood Markers

Immunological and biochemical alterations in the serum help in the early diagno-sis of oral cancer (Khana and Karjodkar, 2006). For example, circulating immune complexes have been detected in 75% of patients with head and neck carcinoma. In patients with OSCC, significantly lower iron and selenium levels have been found than in healthy controls. In contrast, serum copper levels were higher in patients with OSCC or precancerous lesions than in healthy controls (Khana and Karjodkar, 2006). Serum tumor markers for OSCC have shown only a moderate degree of sen-sitivity for diagnosis. The serum concentrations of carcinoembryonic antigen (CEA), squamous cell carcinoma-associated antigen (SCCAA), inhibitor of apoptosis (IAP), and cytokeratin fragments (CYFRAs) have been shown to have a sensitivity of 81% for the detection of OSCC (Nagler et al., 1999). Annexin A1 (ANXA1) was recently identified in peripheral blood by real-time PCR and has been proposed as a potential diagnostic biomarker for OSCC (Faria et al., 2010).

9.6.3.10 Saliva Markers

In the future, saliva analysis could prove to be an efficient, noninvasive, patient-friendly tool for the diagnosis of OSCC. The diagnostic capacity is based on the permanent and intimate contact between saliva and the mucosa where this can-cer evolves. Patients with OSCC have a global alteration of salivary composition. Salivary levels of total sugar, protein-bound sialic acid, free sialic acid, sodium, calcium, immunoglobulin G, albumin, and lactate dehydrogenase are significantly raised compared to controls with healthy mucosa (Sanjay et al., 2008). A signifi-cant increase in the salivary concentrations of specific cancer-related cytokines, insulin-like growth factor (IGF), and matrix metalloproteinases (MMP-2 and MMP-9) has also been detected (Shpitzer et al., 2007). In addition, OSCC patients are reported to have significant alterations of epithelial tumor markers Cyfra 21-1, tissue polypeptide specific (TPS) antigen, various oxidative stress-related salivary parameters, and RNA transcripts of interleukin (IL) 8, IL-1B, dual-specificity phosphatase (DUSP1) (regulator of cell proliferation), HA3 (oncogene), OAZ1 (regulator of polyamine synthesis), S100P (calcium binding protein), and spermi-dine N1-acetyltransferase (SAT) (involved in polyamine metabolism). Salivary

DNA promoter hypermethylation analysis has also been found to be an efficient tool for the early diagnosis of oral cancer (Nagler, 2009). Finally, salivary micro-RNA (miRNA) can be used for the detection of oral cancer; two miRNAs (miR-125a and miR-200a) can discriminate oral cancer patients from control subjects (Park et al., 2009).

9.6.3.10.1   Saliva as Diagnostic Specimen for OSCC

The use of saliva as a noninvasive medium to collect markers related to oral cancer holds tremendous potential, because of its operational simplicity and acceptance by patients. Body fluids like serum, saliva, and sweat are believed to carry markers (DNA, RNA, protein, etc.), which can be used to ascertain the presence of tumor (Carvalho et al., 2008; Brinkman and Wong, 2006). Invasive methods like tissue biopsy have dominated cancer diagnostics for several decades; however, research efforts indicate the trend of the emergence of noninvasive techniques for diagnostics, in which body fluids like saliva would be used for effective diagnosis of cancer. In addition, body fluids such as blood remain the best choice for disease screening and diagnosis, because of the advantages in accessibility, low invasive procedure, low cost, and multiple sampling for monitoring disease development (Good et al., 2007). Examples of using body fluids for tumor detection include sputum for lung cancer diagnosis (Palmisano et al., 2000), urine for urologic tumors (Hoque et al., 2004), saliva for OSCC (Nunes et al., 2000), breast fluid (Lee et al., 2004), and serum or plasma for almost all types of cancers. Compared to blood, diagnostic application based on saliva holds a certain edge, like relatively less complexity in sample handling and processing, low level of background, and inhibitory substances (Zimmermann et al., 2007). Physical proximity of saliva to the oral cavity tissue makes it the first choice as a noninvasive diagnostic medium for OSCC (Wu et al., 2010).

9.6.3.10.2   Salivary Genomics

Genomics-based diagnostic techniques are designed to monitor genomic level changes in abnormal physiological conditions like OSCC. The presence of mutated p53 gene in saliva of subjects with oral cancer has been reported (Liao et al., 2000b). DNA study of salivary *microbiota* using checkerboard DNA-DNA hybridization was conducted, which revealed elevated counts of *C. gingivalis*, *P. melaninogenica*, and *S. mitis* in OSCC patients when compared with normal subjects (Mager et al., 2005). The analysis of salivary DNA can also give an overview of epigenetic changes associated with OSCC (like hypermethylation of the promoter region of the p16 gene), and can be monitored by methylation arrays (Viet and Schmidt, 2008; Franzmann et al., 2007).

9.6.3.10.3   Salivary Transcriptomics

The stability of endogenous cell-free mRNA in saliva is somewhat similar to plasma (Park et al., 2006), which could be possibly due to stabilizing interaction with a certain macromolecule. The transcriptomics methods use high-throughput technologies like microarrays and RNASeq to identify transcripts that are significantly changed in malignancies like cancer, when compared with normal conditions. These methods usually generate a list of candidate biomarkers, which can be verified with more sensitive technologies like RT-PCR. Salivary genomic biomarkers are generally obtained from the cellular component of the saliva. Since the cellular extract of saliva is comprised predominantly of normal oral epithelial cells, the biomarkers present in cell-free extract of saliva would be more apt for diagnostic clinical utilization. The RNA counterparts of protein biomarkers can be detected in saliva, through transcription studies (e.g., IL-8 mRNA was detected in a cell-free saliva sample from a subject with oral cancer) (St. John et al., 2004). This discovery sparkled generation of salivary RNA profiles of healthy subjects on gene expression arrays establishing the normal salivary core transcripts (NSCT), a set of 185 mRNAs that were detected in each saliva supernatant of the studied 10 healthy subjects (Li et al., 2004). Seven transcripts, viz., DUSP1, H3F3A, IL-1B, IL-8, OAZ1, SAT, and S100P, were found to be elevated significantly in OSCC. The expression profile of an individual transcript may not have strong discriminatory power, but can be very powerful when transcripts are analyzed together, as evident from the improvement of sensitivity and specificity up to 90% in distinguishing OSCC patients from normal subjects, by the method based on the combination of seven transcripts for sample classification (Zimmermann and Wong, 2008). Salivary transcriptomics studies of OSCC subjects have detected an abnormal expression profile of various markers like CD44, IL-1b, IL-8, and S100 calcium binding protein (S100P).

Over the past few years, several advances have been made in transcriptome-based technologies, to identify RNA biomarkers in saliva. Owing to technological advancements in amplification of salivary RNA fragments through the poly-A-independent method, a comprehensive exon level salivary expression profile was studied, which led to the establishment of salivary exon core transcriptome (SECT). Development of multiplex RT-PCR-based preamplification of multiple RNA sequences followed by real-time PCR offered an efficient and affordable solution to the limitations posed by the dependence of a sample volume. These technological developments coupled with microfluidics may facilitate the development of point-of-care diagnostic devices in the future for measurement of salivary analytes.

9.6.3.10.4 Salivary Proteomics

Saliva can be considered as a potent medium to give firsthand information about the health of the oral cavity (Wong, 2006). The first biomarker for cancer found in saliva is HER2/neu, a biomarker for breast cancer (Streckfus and Dubinsky, 2007). The pro-inflammatory, pro-angiogenic cytokines (TNFα, IL-1α, IL-6, and IL-8) were significantly elevated in the whole saliva of subjects with OSCC compared to patients with oral premalignant lesions and the control (Rhodus et al., 2005). Mass spectrometry-based systems are currently been pursued for development of a proteomics biomarker-based system for OSCC diagnosis. The proteomic approach is also used for identifying the serum markers in the blood sample. Some of the proteomics-based biomarkers for OSCC identified in both serum and saliva of patients are Cyfra 21-1 (CK 19), tissue polypeptide antigen (TPA) (CK 8, 18, 19), pro-inflammatory cytokines (interleukin-6 and TNFα), and anti-p53 antibodies (Wu et al., 2010). The sensitivity of serum markers to detect early stages of OSCC is not well established; however, it can be used for the diagnosis of recurrence and prognosis

(Sawant et al., 2008). For example, squamous cell carcinoma antigen (SCCAg) and carcinoembryonic antigen (CEA) in serum were examined for detecting OSCC (Yoshimura et al., 1990; Silverman et al., 1976).

9.6.3.11 Imaging

Dental panoramic tomography (DPT), computed tomography (CT), and magnetic resonance imaging (MRI) are frequently used to supplement the clinical evaluation and staging of the primary tumor and regional lymph nodes. CT is the technique of choice to evaluate bone invasion by the tumor. The introduction of cone beam computed tomography (CBCT) provides an alternative for the preoperative study of patients with oral cancer to determine the degree of invasion and extension of the lesion toward the jawbone (Closman and Schimidt, 2007), with the added advantage of lower cost and a lower radiation dose than CT. The CBCT can be useful in the preoperative staging of oral cancer and in the planning of the surgical resection (Closman and Schimidt, 2007). Optical coherence tomography (OCT) is yet another imaging technique that is based on low-coherence interferometry using broadband light. This technique provides cross-sectional, high-resolution subsurface tissue images, and is a noninvasive method to study macroscopic characteristics of epithelial and subepithelial structures, which can be used to detect and diagnose oral premalignant lesions (Wilder-Smith et al., 2009).

9.7   CONCLUSION

Advancements in molecular analysis techniques and development of derived detection systems have laid the foundation for early detection of oral cancer, thereby reducing its rate of mortality. Molecular analyses of oral cancer have been instrumental in understanding the mechanisms of carcinogenesis and have exposed diverse mechanisms that can be used for therapeutic and diagnostic purposes. The platform is ready for early detection of OSCC through techniques like DNA cytometry, toluidine blue staining, brush biopsy, imaging, and HPLC-LIF.

Omics technologies have gifted us with some potential genomic/proteomic biomarkers that can form the basis of designing more robust and sensitive detection kits. Analytes present in body fluids like saliva and blood have become ideal candidates for noninvasive detection of oral cancer. However, the road for early detection of oral cancer by biomarkers is full of challenges: (1) biomarkers should show their activity or expression in the case of oral cancer, and their expression should be absent in the absence of oral cancer; (2) biomarkers should be expressed much before clinical manifestation of oral cancer, which can be detected by physical examination; i.e., they should be able to detect early molecular changes leading to oral cancer; (3) biomarkers should be available in body fluids like blood and saliva for easy detection through noninvasive tests; and (4) biomarkers should be validated thoroughly with a large and diverse sample pool, possibly by an independent authority. It is very challenging to discover biomarkers that qualify with all essential features.

REFERENCES

Acha, A., M.T. Ruesga, M.J. Rodriguez, et al. 2005. Applications of the oral scraped (exfoliative) cytology in oral cancer and precancer. *Med Oral Patol Oral Cir Bucal* 10: 95–102.

Ahmed, H.G., A.M. Idris, and S.O. Ibrahim. 2003. Study of oral epithelial atypia among Sudanese tobacco users by exfoliative cytology. *Anticancer Res* 23: 1943–1949.

Akrish, S., A. Buchner, and D. Dayan. 2004. Oral cancer: Diagnostic options as an aid to histology in order to predict patients at high risk for malignant transformation. *Refuat Hapeh Vehashinayim* 21: 6–15.

Ayude, D., G. Gacio, M. Paez de la Cadena, et al. 2003. Combined use of established and novel tumour markers in the diagnosis of head and neck squamous cell carcinoma. *Oncol Rep* 10: 1345–1350.

Bahar, G., R. Feinmesser, T. Shpitzer, et al. 2007. Salivary analysis in oral cancer patients: DNA and protein oxidation, reactive nitrogen species, and antioxidant profile. *Cancer* 109: 54–59.

Balicki, R., S.Z. Grabowska, and A. Citko. 2005. Salivary epidermal growth factor in oral cavity cancer. *Oral Oncol* 41: 48–55.

Bankfalvi, A., M. Krabort, A. Vegh, et al. 2002. Deranged expression of the E-cadherin/b-catenin complex and the epidermal growth factor receptor in the clinical evolution and progression of oral squamous cell carcinomas. *J Oral Pathol Med* 31: 450–457.

Banoczy, J. 1976. Exfoliative cytologic examinations in the early diagnosis of oral cancer. *Int Dent J* 26: 398–404.

Bautista, A.M., and R. Santiago. 2001. Immunolocalization of p53, glutathione S-transferase pi and CD57 antigens in oral leukoplakia. *Anticancer Res* 21: 379–386.

Bhoopathi, V., S. Kabani, and A.K. Mascarenhas. 2009. Low positive predictive value of the oral brush biopsy in detecting dysplastic oral lesions. *Cancer* 115: 1036–1040.

Bishop, J.W., S.H. Bigner, T.H. Colgan, et al. 1998. Multicenter masked evaluation of AutoCyte PREP thin layers with matched conventional smears—including initial biopsy results. *Acta Cytol* 42: 189–197.

Boyle, J.O., L. Mao, J.A. Brennan, et al. 1994. Gene mutations in saliva as molecular markers for head and neck squamous cell carcinomas. *Am J Surg* 168: 429–432.

Brinkman, B.M., and D.T. Wong. 2006. Disease mechanism and biomarkers of oral squamous cell carcinoma. *Curr Opin Oncol* 18: 228–233.

Calsen, B., P. Pere, R. Senekowitsch, et al. 1990. SCC as a tumour marker in the initial diagnosis of carcinoma of the head and neck region. *Laryngorhinootologie* 65: 275–280.

Carvalho, A.L., C. Jeronimo, M.M. Kim, et al. 2008. Evaluation of promoter hypermethylation detection in body fluids as a screening/diagnosis tool for head and neck squamous cell carcinoma. *Clin Cancer Res* 14: 97–107.

Chan, D.W., and S. Sell. 1994. *Teitz textbook of clinical chemistry*, ed. C.A. Burtis and E.R. Ashwood. W.B. Saunders, Philadelphia.

Chung-Ji, L., L. Yann-Jinn, L. Hsin-Fu, et al. 2002. The increase in the frequency of MICA gene A6 allele in oral squamous cell carcinoma. *J Oral Pathol Med* 31: 323–328.

Closman, J.J., and B.L. Schimidt. 2007. The use of cone beam computed tomography as an aid in evaluating and treatment planning for mandibular cancer. *J Oral Maxillofac Surg* 65: 766–771.

Costa Ade, L., N.S. de Araujo, S. Pinto Ddos, et al. 1999. PCNA/AgNOR and Ki-67/AgNOR double staining in oral squamous cell carcinoma. *J Oral Pathol Med* 28: 438–441.

Cruz, I., S.S. Napier, I. van der Waal, et al. 2002. Suprabasal p53 immunoexpression is strongly associated with high grade dysplasia and risk of malignant transformation in potentially malignant oral lesions from Northern Ireland. *J Clin Pathol* 55: 98–104.

Debnath, M., G.B.K.S. Prasad, and P.S. Bisen. 2010. *Molecular diagnostics: Promises and possibilities.* Springer, Dordrecht.

De Veld, D.C., M.J. Witjes, H.J. Sterenborg, et al. 2005. The status of in vivo auto fluorescence spectroscopy and imaging for oral oncology. *Oral Oncol* 41: 117–131.

Doweck, I., M. Barak, E. Greenberg, et al. 1995. Cyfra 21-1: A new potential tumor marker for squamous cell carcinoma of head and neck. *Arch Otolaryngol Head Neck Surg* 121: 177–181.

El-Naggar, A.K., K. Hurr, J.G. Batsakis, et al. 1995. Sequential loss of heterozygosity at microsatellite motifs in preinvasive and invasive head and neck squamous carcinoma. *Cancer Res* 55: 2656–2659.

El-Naggar, A.K., L. Mao, G. Staerkel, et al. 2001. Genetic heterogeneity in saliva from patients with oral squamous carcinomas: Implications in molecular diagnosis and screening. *J Mol Diagn* 3: 164–170.

Epstein, J.B., and P. Guneri. 2009. The adjunctive role of toluidine blue in detection of oral premalignant and malignant lesions. *Curr Opin Otolaryngol Head Neck Surg* 17: 79–87.

Epstein, J.B., J. Sciubba, S. Silverman, et al. 2007. Utility of toluidine blue in oral premalignant lesion and squamous cell carcinoma: Continuing research and implications for clinical practice. *Head Neck* 29: 948–958.

Epstein, J.B., L. Zhang, C. Poh, et al. 2003. Increased allelic loss in toluidine blue-positive oral premalignant lesions. *Oral Surg Oral Med Oral Pathol Oral Radiol Endod* 95: 45–50.

Evan, G., and T. Littlewood. 1998. A matter of life and death. *Science* 281: 1317–1322.

Fabia, H.H., M.C.F. Ana, S.P.G. Antonio, et al. 2005. Liquid-based preparations versus conventional cytology: Specimen adequacy and diagnostic agreement in oral lesions. *Oral Med Pathol* 23: 1927–1933.

Farah, C.S., and M.J. McCullough. 2007. A pilot case control study on the efficacy of acetic acid wash and chemiluminescence illumination (ViziLite) in the visualization of oral mucosa white lesions. *Oral Oncol* 43: 820–824.

Faria, P.C.B., A.A. Servino-Sena, R. Nascimento, et al. 2010. Expression of annexin A1 in peripheral blood from oral squamous cell carcinoma patients. *Oral Oncol* 46: 25–30.

Fedele, S. 2009. Diagnostic aids in the screening of oral cancer. *Head Neck Oncol* 1: 5.

Fong, D., G. Spizzo, J.M. Gostner, et al. 2008. TROP2: A novel prognostic marker in squamous cell carcinoma of the oral cavity. *Mod Pathol* 21: 186–191.

Franzmann, E.J. 2005. Salivary soluble CD44: A potential molecular marker for head and neck cancer. *Cancer Epidemiol Biomarkers Prev* 14: 735–739.

Franzmann, E.J., E.P. Reategui, and F. Pedroso. 2007. Soluble CD44 is a potential marker for the early detection of head and neck cancer. *Cancer Epidemiol Biomarkers Prev* 16: 1348–1355.

Freitas, M.D., A.G. Garcia, A.C. Abelleira, et al. 2004. Aplicaciones de la citología exfoliativa en el diagnóstico del cáncer oral. *Med Oral Patol Oral Cir Bucal* 9: 355–361.

Frist, S. 2003. The oral brush biopsy: Separating fact from fiction. *Oral Surg Oral Med Oral Pathol Oral Radiol Endod* 96: 654–656.

Giancotti, F.G., and Ruoslahti, E. 1999. Integrin signaling. *Science* 285: 1309–1312.

Gonzalez-Moles, M.A., I. Ruiz-Avila, J.A. Gil-Montoya, et al. 2010. Analysis of Ki-67 expression in oral squamous cell carcinoma: Why Ki-67 is not a prognostic indicator. *Oral Oncol* 46: 525–530.

Good, D.M., V. Thongboonkerd, J. Novak, et al. 2007. Body fluid proteomics for biomarker discovery: Lessons from the past hold the key to success in the future. *J Proteome Res* 6: 4549–4555.

Grohs, H.K., D.J. Zahniser, and J.W. Geyer. 1994. Standardization of specimen preparation through mono/thin-layer technology. In *Automated cervical cancer screening*, ed. H.K. Grohs and O.A.N. Husain, 176–185. Igaku Shoin, New York.

Hamidi, S., T. Salo, T. Kainulainen, et al. 2000. Expression of alpha(v)ß6 integrin in oral leukoplakia. *Br J Cancer* 82: 1433–1440.

Hanahan, D., and R.A. Weinberg. 2000. The hallmarks of cancer. *Cell* 100: 57–70.

Hohlweg-Majert, B., H. Deppe, M.C. Metzger, et al. 2009. Sensitivity and specificity of oral brush biopsy. *Cancer Invest* 27: 293–297.

Hoque, M.O., S. Begum, O. Topaloglu, et al. 2004. Quantitative detection of promoter hypermethylation of multiple genes in the tumor, urine, and serum DNA of patients with renal cancer. *Cancer Res* 64: 5511–5517.

Howell, L.P., R.L. Davis, T.I. Belk, et al. 1998. The AutoCyte preparation system for gynecologic cytology. *Acta Cytol* 42: 171–177.

Huang, M.F., Y.C. Chang, P.S. Liao, et al. 1999. Loss of heterozygosity of p53 gene of oral cancer detected by exfoliative cytology. *Oral Oncol* 35: 296–301.

Huber, M.A. 2009. Assessment of the VELscope as an adjunctive examination tool. *Tex Dent* 126: 528–535.

Imai, T., M. Toyota, H. Suzuki, et al. 2008. Epigenetic inactivation of RASSF2 in oral squamous cell carcinoma. *Cancer Sci* 99: 958–966.

Ishizaki, Y., L. Cheng, A.W. Mudge, et al. 1995. Programmed cell death by default in embryonic cells, fibroblasts and cancer cells. *Mol Biol Cell* 6: 1443–1458.

Jiang, W.W., B. Masayesva, and M. Zahurak. 2005. Increased mitochondrial DNA content in saliva associated with head and neck cancer. *Clin Cancer Res* 11: 2486–2491.

Jones, A.C., F.E. Pink, P.L. Sandow, et al. 1994. The Cytobrush Plus cell collector in oral cytology. *Oral Surg Oral Med Oral Pathol* 77: 95–99.

Katopodis, N., Y. Hirshaut, N.L. Geller, et al. 1982. Lipid-associated sialic acid test for the detection of human cancer. *Cancer Res* 42: 5270–5275.

Khan, Z., P. Bhadouria, R. Gupta, et al. 2006. Tumor control by manipulation of the human anti-apoptotic survivin gene. *Current Cancer Ther Rev* 2: 73–79.

Khan, Z., N. Khan, R.P. Tiwari, et al. 2010a. Down-regulation of survivin by oxaliplatin diminishes radioresistance of head and neck squamous carcinoma cells. *Radiother Oncol* 96: 267–273.

Khan, Z., N. Khan, A.K. Varma, et al. 2010b. Oxaliplatin-mediated inhibition of survivin increases sensitivity of head and neck squamous cell carcinoma cell lines to paclitaxel. *Current Cancer Drug Targets* 10: 660–669.

Khan, Z., R.P. Tiwari, R. Mulherkar, et al. 2009. Detection of survivin and p53 in human oral cancer: Correlation with clinicopathologic findings. *Head Neck* 31: 1039–1048.

Khan, Z., and P.S. Bisen. 2013. Oncoapoptotic signaling and deregulated target genes in cancers: special reference to oral cancer. *Biochim Biophy Acta Reviews on Cancer* dx.doi.org/10.1016/j.bbcan. 2013.04.002.

Khana, S.S., and F.R. Karjodkar. 2006. Circulating immune complexes and trace elements (copper, iron and selenium) as markers in oral precancer and cancer: A randomised, controlled clinical trial. *Head Face Med* 2: 33.

Kim, J., D.M. Shin, A. El-Naggar, et al. 2001. Chromosome polysomy and histological characteristics in oral premalignant lesions. *Cancer Epidemiol Biomarkers Prev* 10: 319–325.

Koontongkaew, S., A. Chareonkitkajorn, A. Chavitan, et al. 2000. Alterations in p53, pBb, cyclin D1 and cdk4 in human oral and pharyngeal squamous cell carcinomas. *Oral Oncol* 36: 334–339.

Kozomara, R.J., M.V. Brankovic-Magic, N.R. Jovic, et al. 2007. Prognostic significance of TP53 mutations in oral squamous cell carcinoma with human papilloma virus infection. *Int J Biol Markers* 22: 252–257.

Kresty, L.A., S.R. Mallery, T.J. Knobloch, et al. 2002. Alterations of p16(INK4a) and p14(ARF) in patients with severe oral epithelial dysplasia. *Cancer Res* 62: 5295–5300.

Landesberg, R., V. Woo, L. Huang, et al. 2008. The expression of the receptor for glycation endproducts (RAGE) in oral squamous cell carcinomas. *Oral Surg Oral Patho Oral Radiol Endod* 105: 617–624.

Larsen, S.R., J. Johansen, J.A. Sorensen, et al. 2009. The prognostic significance of histological features in oral squamous cell carcinoma. *J Oral Pathol Med* 38: 657–662.

Lee, A., Y. Kim, K. Han, et al. 2004. Detection of tumor markers including carcinoembryonic
antigen, APC, and cyclin D2 in fine-needle aspiration fluid of breast. *Arch Pathol Lab
Med* 128: 1251–1256.

Lee, B.-K., E. Diebel, F.W. Neukam, et al. 2001. Diagnostic and prognostic relevance of
expression of human telomerase subunits in oral cancer. *Int J Oncol* 19: 1063–1068.

Li, Y., X. Zhou, M.A. St. John, et al. 2004. RNA profiling of cell-free saliva using microarray
technology. *J Dent Res* 83: 199–203.

Liao, J., T. Mitsuyasu, K. Yamane, et al. 2000a. Telomerase activity in oral and maxillofacial
tumors. *Oral Oncol* 36: 347–352.

Liao, P.H., Y.C. Chang, M.F. Huang, et al. 2000b. Mutation of p53 gene codon 63 in saliva as a
molecular marker for oral squamous cell carcinomas. *Oral Oncol* 36: 272–276.

Lim, J., J.H. Kim, J.Y. Paeng, et al. 2005. Prognostic value of activated Akt expression in oral
squamous cell carcinoma. *J Clin Pathol* 58: 1199–1205.

Lingen, M.W., J.R. Kalmar, T. Karrison, et al. 2008. Critical evaluation of diagnostic aids for
the detection of oral cancer. *Oral Oncol* 44: 10–22.

Lippman, S.M., and W.K. Hong. 2001. Molecular markers of the risk of oral cancer. *N Engl J
Med* 344: 1323–1326.

Lopez, M., J.M. Aguirre, N. Cuevas, et al. 2003. Gene promoter hypermethylation in oral
rinses of leukoplakia patients—A diagnostic and/or prognostic tool? *Eur J Cancer* 39:
2306–2309.

Lopez, M., J.M. Aguirre, N. Cuevas, et al. 2004. Use of cytological specimens for p53 gene
alterations detection in oral squamous cell carcinoma risk patients. *Clin Oncol* 16:
366–370.

Maass, J.D., A.M. Niemann, B.M. Lippert, et al. 1999. CYFRA 8/18 in head and neck cancer.
Anticancer Res 19: 2699–2701.

Maduskey, M., and G.R. Ogden. 2000. An overview of the prevention of oral cancer and diagnostic markers of malignant change. 2. Markers of value in tumour diagnosis. *Dental
Update* 27: 148–152.

Mager, D.L., A.D. Haffajee, P.M. Devlin, et al. 2005. The salivary microbiota as a diagnostic
indicator of oral cancer: A descriptive, nonrandomized study of cancer-free and oral
squamous cell carcinoma subjects. *J Trans Med* 3: 27.

Mao, L., J.S. Lee, Y.H. Fan, et al. 1996. Frequent microsatellite alterations at chromosome
9p21 and 3p14 in oral premalignant lesions and their value in cancer risk assessment.
Nat Med 2: 682–685.

Marsit C.J., C.C. Black, M.R. Posner, et al. 2008. A genotype phenotype examination of cyclin
D1 on risk and outcome of squamous cell carcinoma of the head and neck. *Clin Cancer
Res* 14: 2371–2377.

Mathers, C.D., and D. Loncar. 2006. Projections of global mortality and burden of disease
from 2002 to 2030. *PLoS Med* 3: e442.

Matsubara, Y., Y. Yasuda, T. Hanawa, et al. 1986. SCC-antigen in patients with lung cancer.
Nippon Gan Chiryo Gakkai Shi 21: 1036–1048.

McGoogan, E., and A. Reith. 1996. Would monolayers provide more representative samples
and improved preparations for cervical screening? Overview and evaluation of systems
available. *Acta Cytol* 40: 107–119.

Mehrotra, R., A. Gupta, and M. Singh. 2004. Brush biopsy in the early diagnosis of oral soft tissue lesions. In *Tobacco counters health*, ed. A. K Verma, 216–219. Macmillan, New Delhi.

Mehrotra, R., A. Gupta, M. Singh, et al. 2006. Application of cytology and molecular biology
in diagnosing premalignant or malignant oral lesions. *Mol Cancer* 5: 11.

Mendelsohn, J. 1987. In *Harrison's principles of internal medicine*, ed. E. Braunwald, et al.
McGraw-Hill, New York.

Moles, M.A.G., J.A.G. Montoya, and I.R. Avila. 2008. Base moleculares de la carizacion de cavidad oral. *Av Odontosetomatol* 24: 55–60.

Muraki, Y., A. Tateshi, C. Seta, et al. 2000. Fas antigen expression and outcome of oral squamous cell carcinoma. *Int J Oral Maxillofac Surg* 29: 360–365.

Nagler, R.M. 2009. Saliva as a tool for oral cancer diagnosis and prognosis. *Oral Oncol* 45: 1006–1010.

Nagler, R.M., M. Barak, H. Ben-Aryeh, et al. 1999. Early diagnostic and treatment monitoring role of Cyfra 21-1 and TPS in oral squamous cell carcinoma. *Cancer* 35: 118–125.

Nichols, M.L., F.B. Quinn Jr., V.J. Schnadig, et al. 1991. Interobserver variability in the interpretation of brush cytologic studies from head and neck lesions. *Arch Otolaryngol Head Neck Surg* 117: 1350–1355.

Nie, M., L. Zhong, G. Zheng, et al. 2002. The changes of cytokeratin 19 during carcinogenesis. *Zonghua Kou Qiang Yi Xue Za Zhi* 37: 187–190.

Nunes, D.N., L.P. Kowalski, and A.J. Simpson. 2000. Detection of oral and oropharyngeal cancer by microsatellite analysis in mouth washes and lesions brushings. *Oral Oncol* 36: 525–528.

Ogawa, T., Y. Tsurusako, N. Kimura, et al. 1999. Comparison of tumor markers in patients with squamous cell carcinoma of the head and neck. *Acta Otolaryngol Suppl* 540: 72–76.

Ogden, G.R., J.G. Cowpe, D.M. Chisholm, et al. 1994. DNA and keratin analysis of oral exfoliative cytology in the detection of oral cancer. *Oral Oncol Eur J Cancer* 30B: 405–408.

Ogden, G.R., J.G. Cowpe, and M. Green. 1992. Cytobrush and wooden spatula for oral exfoliative cytology. A comparison. *Acta Cytol* 36: 706–710.

Ogden, G.R., J.G. Cowpe, and M.W. Green. 1991. Detection of field change in oral cancer using oral exfoliative cytologic study. *Cancer* 68: 1611–1615.

Ogden, G.R., S. McQueen, D.M. Chisholm, et al. 1993. Keratin profiles of normal and malignant oral mucosa using exfoliative cytology. *J Clin Pathol* 46: 352–356.

Ohta, S., H. Uemura, Y. Matsui, et al. 2009. Alterations of p16 and p14ARG genes and 9p21 locus in oral squamous cell carcinoma. *Oral Surg Oral Med Oral Pathol Oral Radiol Endod* 107: 81–91.

Orellana-Bustos, A.I., I.L. Espinoza-Santander, M.E. Franco-Martínez, et al. 2004. Evaluation of keratinization and AgNORs count in exfoliative cytology of normal oral mucosa from smokers and non-smokers. *Med Oral* 9: 197–203.

Palmisano, W.A., K.K. Divine, G. Saccomanno, et al. 2000. Predicting lung cancer by detecting aberrant promoter methylation in sputum. *Cancer Res* 60: 5954–5958.

Park, N.J., Y. Li, T. Yu, et al. 2006. Characterization of RNA in saliva. *Clin Chem* 52: 988–994.

Park, N.J., H. Zhou, D. Elashoff, et al. 2009. Salivary microRNA: Discovery, characterization, and clinical utility for oral cancer detection. *Clin Cancer Res* 15: 5473–5477.

Partridge, M., S. Pateromchelakis, E. Phillips, et al. 2000. A case control-study confirms that microsatellite assay can identify patients at risk of developing squamous cell carcinoma within field cancerization. *Cancer Res* 60: 3893–3898.

Patel, K., N.L. Rhodus, P. Gaffney, et al. 2004. Extraction of RNA from oral biopsies in oral leukoplakia. In *82nd IADR Congress*, Honolulu, HI, p. 1240.

Patel, S.G., and J.P. Shah. 2005. TNM staging of cancers of head and neck: Striving for uniformity among diversity. *CA Cancer J Clin* 55: 242–258.

Perez-Sayans, M., J.M. Somoza-Martin, F. Barros-Angueira, et al. 2009. Exfoliative cytology for diagnosing oral cancer. *Biotech Histochem* 25: 1–11.

Poate, T.W., J.A. Buchanan, T.A. Hodgson, et al. 2004. An audit of the efficacy of the oral brush biopsy technique in a specialist oral medicine unit. *Oral Oncol* 40: 829–834.

Potter, T.J., D.J. Summerlin, and J.H. Campbell. 2003. Oral malignancies associated with negative transepithelial brush biopsy. *J Oral Maxillofac Surg* 61: 674–677.

Reibel, J. 2003. Prognosis of oral pre-malignant lesions: Significance of clinical, histopathological, and molecular biological characteristics. *Crit Rev Oral Biol Med* 14: 47–62.

Remmerbach, T.W., H. Weidenbach, C. Muller, et al. 2003. Diagnostic value of nucleolar organizer regions (AgNORs) in brush biopsies of suspicious lesions of the oral cavity. *Anal Cell Pathol* 25: 139–146.

Rhodus, N.L., V. Ho, C.S. Miller, et al. 2005. NF-kappaB dependent cytokine levels in saliva of patients with oral preneoplastic lesions and oral squamous cell carcinoma. *Cancer Detect Prev* 29: 42–45.

Rick, G.M., and L. Slater. 2003. Oral brush biopsy: The problem of false positives. *Oral Surg Oral Med Oral Pathol Oral Radiol Endod* 96: 252.

Ries, J.C., E. Hassfurther, H. Steininger, et al. 2001. Correlation of telomerase activity, clinical prognosis and therapy in oral carcinogenesis. *Anticancer Res* 21: 1057–1064.

Ropka, M.E., W.J. Goodwin, P.A. Levine, et al. 1991. Effective head and neck tumor markers. The continuing quest. *Arch Otolaryngol Head Neck Surg* 117: 1011–1014.

Rosas, S.L., W. Koch, M.G. da Costa Carvalho, et al. 2001. Promoter hypermethylation patterns of p16, O6-methylguanine-DNA methyltransferase, and death-associated protein kinase in tumors and saliva of head and neck cancer patients. *Cancer Res* 61: 939–942.

Rosenberg, D., and S. Cretin. 1989. Use of meta-analysis to evaluate tolonium chloride in oral cancer screening. *Oral Surg Oral Med Oral Pathol* 67: 621–627.

Rossin, M.P., X. Cheng, C. Poh, et al. 2000. Use of allelic loss to predict malignant risk for low-grade oral epithelial dysplasia. *Clin Cancer Res* 6: 357–362.

Sah, N.K., Z. Khan, G.J. Khan, et al. 2006. Structural, functional and therapeutic biology of survivin. *Cancer Lett* 244: 164–171.

Sanjay, P.R., K. Hallikeri, and A.R. Shivashankara. 2008. Evaluation of salivary sialic acid, total protein, and total sugar in oral cancer: A preliminary report. *Indian J Dent Res* 19: 288–291.

Sathyam, K.M., K.R. Nalinakumari, T. Abraham, et al. 2008. CCND1 polymorphisms (A870G and C1722G) modulate its protein expression and survival in oral carcinoma. *Oral Oncol* 44: 689–697.

Sawant, S.S., S.M. Zingde, and M.M. Vaidya. 2008. Cytokeratin fragments in the serum: Their utility for the management of oral cancer. *Oral Oncol* 44: 722–732.

Schaafsma, H.E., L.A. Van Der Velden, J.J. Manni, et al. 1993. Increased expression of cyto-keratins 8, 18 and vimentin in the invasion front of mucosal squamous cell carcinoma. *J Pathol* 170: 77–86.

Scheifele, C., H. Schlechte, G. Bethke, et al. 2002. Detection of TP53-mutations in brush biopsies from oral leukoplakias. *Mund Kiefer Gesichtschir* 6: 410–414.

Schliephake, H. 2003. Prognostic relevance of molecular markers of oral cancer—A review. *J Oral Maxillofac Surg* 32: 233–245.

Schwartz, J.L., X. Gu, R.A. Kittles, et al. 2000. Experimental oral carcinoma of the tongue and buccal mucosa: Possible biologic markers linked to cancers at two anatomic sites. *Oral Oncol* 36: 225–235.

Schweizer, J., P.E. Bowden, P.A. Coulombe, et al. 2006. New consensus nomenclature for mammalian keratins. *J Cell Biol* 174: 169–174.

Sciubba, J.J. (U.S. Collaborative Oral CDx Study Group). 1999. Improving detection of pre-cancerous and cancerous oral lesions: Computer assisted analysis of the oral brush biopsy. *JAM Dent Assoc* 130: 1145–1157.

Seoane, J., P. Varela-Centelles, J.R. Ramirez, et al. 2002. Artefacts produced by suture traction during incisional biopsy of oral lesions. *Clin Otolaryngol* 27: 549–553.

Sharma, P., N. Kumar, A.K. Bahadur, et al. 2005. Ki-67 expression in cytologic scrapes from oral squamous cell carcinoma before and after 24 Gray radiotherapy—A study on 43 patients. *Med Oral Patol Oral Cir Bucal* 1: E15–E17.

Shintani, S., Y. Nakahara, M. Mihara, et al. 2001. Inactivation of p14(ARF), p15(INK4B) and p16(INK4A) genes is a frequent event in human oral squamous cell carcinomas. *Oral Oncol* 37: 498–504.

Shpitzer, T., G. Bahar, R. Feeinmesser, et al. 2007. A comprehensive salivary analysis for oral cancer diagnosis. *J Cancer Res Clin Oncol* 133: 613–617.

Silverman, N.A., J.C. Alexander Jr., and P.B. Chretien. 1976. CEA levels in head and neck cancer. *Cancer* 37: 2204–2211.

Spafford, M.F., W.M. Koch, A.L. Reed, et al. 2001. Detection of head and neck squamous cell carcinoma among exfoliated oral mucosal cells by microsatellite analysis. *Clin Cancer Res* 7: 607.

Spivavack, S.D., G.J. Hurteau, R. Jain, et al. 2004. Gene–environment interaction signatures by quantitative mRNA profiling in exfoliated buccal mucosal cells. *Cancer Res* 64: 6805–6813.

Sporn, M.B. 1996. The war on cancer. *Lancet* 347: 1377–1381.

Sprenger, E., P. Schwarmann, M. Kirkpatrick, et al. 1996. The false negative rate in cervical cytology. *Acta Cytol* 40: 81–89.

St. John, M.A., Y. Li, X. Zhou, et al. 2004. Interleukin 6 and interleukin 8 as potential biomarkers for oral cavity and oropharyngeal squamous cell carcinoma. *Arch Otolaryngol Head Neck Surg* 130: 929–935.

Streckfus, C.F., and W.P. Dubinsky. 2007. Proteomic analysis of saliva for cancer diagnosis. *Expert Rev Proteomics* 4: 329–332.

Sudbo, J. 2004. Novel management of oral cancer: A paradigm of predictive oncology. *Clin Med Res* 2: 233–242.

Sudbo, J., W. Kildal, B. Risberg, et al. 2001. DNA content as a prognostic marker in patient with oral leukoplakia. *N Engl J Med* 344: 1270–1278.

Svirsky, J.A., J.C. Burns, D.G. Page, et al. 2001. Computer-assisted analysis of the oral brush biopsy. *Compend Contin Educ Dent* 22: 99–106.

Teresa, D.B., K.A. Neves, C.B. Neto, et al. 2007. Computer assisted analysis of cell proliferation markers in oral lesions. *Acta Histochem* 109: 377–387.

Thomas, F., W. Richard, P. Jens, et al. 2006. Cytokeratin 8/18 expression indicates a poor prognosis in squamous cell carcinomas of the oral cavity. *BMC Cancer* 6: 10.

Tumuluri, V., G.A. Thomas, and I.S. Fraser. 2002. Analysis of Ki-67 antigen at the invasive tumour front of the human oral squamous cell carcinoma. *J Oral Pathol Med* 31: 598–604.

Umudum, H., T. Rezanko, F. Dag, et al. 2005. Human papillomavirus genome detection by in situ hybridization in fine-needle aspirates of metastatic lesions from head and neck squamous cell carcinomas. *Cancer* 105: 171–777.

Uzawa, N., D. Akanuma, A. Negishi, et al. 2001. Homozygous deletions on the short arm of chromosome 3 in human oral squamous cell carcinomas. *Oral Oncol* 37: 897–877.

Vaidya, M.M., A.M. Borges, S.A. Pradhan, et al. 1989. Altered keratin expression in buccal mucosal squamous cell carcinoma. *J Oral Pathol Med* 18: 282–286.

Vaidya, M.M., A.M. Borges, S.A. Pradhan, et al. 1996. Cytokeratin expression in squamous cell carcinomas of the tongue and alveolar mucosa. *Eur J Cancer B Oral Oncol* 32B: 333–336.

Van Heerden, W.F., T.J. Swart, B. Robson, et al. 2001. FHIT RNA and protein expression in human oral squamous cell carcinomas. *Oral Oncol* 37: 498–504.

Vassilakos, P., D. Cossali, X. Albe, et al. 1996. Efficacy of monolayer preparations for cervical cytology—Emphasis on suboptimal specimens. *Acta Cytol* 40: 496–500.

Venkatakrishna, K., V.B. Kartha, M. Pai Keerthilatha, et al. 2003. HPLC-LIF for early detection of oral cancer. *Curr Sci* 84: 551–557.

Viet, C.T., and B.L. Schmidt. 2008. Methylation array analysis of preoperative and postoperative saliva DNA in oral cancer patients. *Cancer Epidemiol Biomarkers Prev* 17: 3603–3611.

Wakulich, C., L. Jackson-Boeters, T.D. Daley, et al. 2002. Immunohistochemical localization of growth factors fibroblast growth factor-1 and fibroblast growth factor-2 and receptors fibroblast growth factor receptor-2 and fibroblast growth factor-3 in normal oral epithelium, epithelial dysplasias, and squamous cell carcinoma. *Oral Surg Oral Med Oral Pathol Oral Radiol Endod* 93: 573–579.

Walling, D.M., C.M. Flaitz, K. Adler-Storthz, et al. 2003. A non-invasive technique for studying oral epithelial Epstein-Barr virus infection and disease. *Oral Oncol* 39: 436–444.

Walther, E.K., N. Dahlmann, and H.T. Gorgulla. 1990. Tumor markers in patients with head–neck carcinomas. *Laryngorhinootologie* 69: 271–274.

Warnakulasuriya, S., T. Soussi, R. Maher, et al. 2000. Expression of p53 in oral squamous cell carcinoma is associated with the presence of IgG and IgA p53 autoantibodies in sera and saliva of the patients. *J Pathol* 192: 52–57.

Weinberg, R.A. 1995. The retinoblastoma protein and cell cycle control. *Cell* 81: 323–330.

Werkmeister, R., B. Brandt, and V. Joos. 2000. Clinical relevance of erbB-1 and -2 oncogenes in oral carcinomas. *Oral Oncol* 194: 303–313.

Whyte, D.A., C.E. Bronton, and E.J. Shillitoe. 2002. The unexplained survival of cells in oral cancer: What is role of p53? *J Oral Pathol Med* 31: 25–33.

Wilder-Smith, P., K. Lee, S. Guo, et al. 2009. In vivo diagnosis of oral dysplasia and malignancy using optical coherence tomography: Preliminary studies in 50 patients. *Lasers Surg Med* 41: 353–357.

Wollenberg, B., N. Jan, W. Sautier, et al. 1997. Serum levels of intercellular adhesion molecule-1 in squamous cell carcinoma of the head and neck. *Tumour Biol* 18: 88–94.

Wong, D.T. 2006. Salivary diagnostics powered by nanotechnologies, proteomics and genomics. *J Am Dent Assoc* 137: 313–321.

Wright, W.E., O.M. Pereira-Smith, and J.W. Shay. 1989. Reversible cellular senescence: Implications for immortalization of normal human diploid fibroblasts. *Mol Cell Biol* 9: 3088–3092.

Wu, J.-Y., Y. Chen, C. Ho-Ren, et al. 2010. Potential biomarkers in saliva for oral squamous cell carcinoma. *Oral Oncol* 46: 226–231.

Wyllie, A.H., J.F. Kerr, and A.R. Currie. 1980. Cell death: The significance of apoptosis. *Int Rev Cytol* 68: 251–306.

Yanamoto, S., G. Kawasaki, I. Yoshitomi, et al. 2007. Clinicopathologic significance of EpCAM expression in squamous cell carcinoma of the tongue and its possibility as a potential target for tongue cancer gene therapy. *Oral Oncol* 43: 897–877.

Yoshimura, Y., M. Oka, and T. Harada. 1990. Squamous cell carcinoma antigen for detection of squamous cell and mucoepidermoid carcinoma after primary treatment: A preliminary report. *J Oral Maxillofac Surg* 48: 1288–1292.

Zhang, L., K.-J. Cheung, W.L. Lam, et al. 2001. Increased genetic damage in oral leukoplakia from high risk sites. *Cancer* 91: 2148–2155.

Zhong, L.P., C.P. Zhang, J.W. Zheng, et al. 2007a. Increased Cyfra 21-1 concentration in saliva from primary oral squamous cell carcinoma patients. *Arch Oral Biol* 52: 1079–1087.

Zhong, L.P., H.G. Zhu, C.P. Zhang, et al. 2007b. Detection of serum Cyfra 21-1 in patients with primary oral squamous cell carcinoma. *Int J Oral Maxillofac Surg* 36: 230–234.

Zhou, X., Z. Zhang, X. Yang, et al. 2009. Inhibition of cyclin D1 expression by cyclin D1 shRNAs in human oral squamous cell carcinoma cells is associated with increased cisplatin chemosensitivity. *Int J Cancer* 124: 483–489.

Zimmermann, B.G., N.J. Park, and D.T. Wong. 2007. Genomic targets in saliva. *Ann NY Acad Sci* 1098: 184–191.

Zimmermann, B.G., and D.T. Wong. 2008. Salivary mRNA targets for cancer diagnostics. *Oral Oncol* 44: 425–429.

Glossary

ABC drug transporters: A family of membrane proteins that mediate the efflux of drugs from cells in an energy (ATP)-dependent manner.

Acinus (plural *acini*): One of the small sac-like pouches composing some glands.

Acquired resistance: A term used to describe drug resistance in tumors that are initially responsive to drugs, but are resistant at relapse.

Adenoma: A benign or noncancerous tumor made up of glandular tissue.

Adjuvant therapy: Treatment given after the primary therapy, which is usually surgery.

Adverse effects: Problems that occur when treatment affects healthy cells.

Agent: In a cancer clinical trial, an agent is a substance that produces, or is capable of producing, an effect that fights cancer.

Allele: One of the alternate forms of a gene that may occupy a given locus (position on the chromosome).

Angiogenesis: The growth of blood vessels from surrounding tissue to a solid tumor.

Angiogenesis inhibitor: A substance that may prevent the formation of blood vessels.

Antiangiogenesis: Prevention of the growth of new blood vessels.

Antibody: An antigen-specific receptor, also called an immunoglobulin, made by B-lymphocytes.

Antibody therapy: Treatment with an antibody, a substance that can directly kill specific tumor cells or stimulate the immune system to kill tumor cells.

Antigens: Substances that cause the immune system to make a specific immune response.

Antioxidant: Any substance that delays or inhibits oxidative damage to a target molecule.

Apoptosis: A normal series of events in a cell that lead to its death.

Basement membrane: A specialized, sheet-like structure of the extracellular matrix that separates cells from the surrounding connective tissue, and thereby serves as a boundary of a tissue.

B-cells or B-lymphocytes: White blood cells that make antibodies and are an important part of the immune system.

Benign: A swelling or growth that is not cancerous and does not spread from one part of the body to another.

Biological therapy: Treatment to stimulate or restore the ability of the immune system to fight infection and disease.

Biomarker: A substance sometimes found in the blood, other body fluids, or tissues. Also called a tumor marker.

Biospecimens: Samples of tissue that can be used to study various aspects of cancer.

Bone marrow: The soft, sponge-like tissue in the center of bones that produces white blood cells, red blood cells, and platelets.

Bone marrow toxicity: The destruction of bone marrow using radiation or drugs.

Breast cancer: A malignant tumor that has developed from cells in the breast.

Cancer: A term for diseases in which abnormal cells divide without control. Also referred to as a malignant tumor or neoplasm.

Cancer in situ: The stage where the cancer is still confined to the tissue in which it started.

Cancer stem cells: A small population of cells inside tumors that have the ability to self-renew while giving rise to different types of cells.

Cancer vaccine: A form of biological therapy that may help a person's immune system to recognize cancer cells.

Carcinogen: A substance that increases the risk of developing cancer.

Carcinogenesis: The process by which normal cells are transformed into malignant cells.

Carcinoma: A type of cancer that arises in epithelial cells, such as those in the skin or lining of organs.

CCSS: Childhood cancer survivor study.

Cell: The fundamental living unit of animals and plants.

Cell cycle: The sequence of events by which a cell enlarges, duplicates its DNA. and divides.

Cell division: The process by which two daughter cells are produced from one parent cell.

Chemoprevention: The use of drugs, vitamins, or other agents to try to reduce the risk of or delay the development or recurrence of cancer.

Chemotherapy: Treatment with anticancer drugs.

Chimeric antibody: An antibody that is made of segments of both mouse and human antibodies.

Chondroma: A benign tumor of cartilage.

Chromosome: A long, tightly packaged DNA molecule that contains the genetic instructions essential for the life of a cell and allows the transmission of genetic information from generation to generation.

Clinical trial: A research study that tests how well new medical treatments or other interventions work in people.

Codon: A triplet of three bases or nucleotides that specify a single amino acid.

Colorectal cancer: A disease in which malignant (cancer) cells are found in the colon or rectum.

Combination therapy: Treatment using more than one anticancer drug.

Complete remission: The disappearance of all detectable signs of cancer. Also called complete response.

Concurrent therapy: A treatment that is given at the same time as another.

Consolidation therapy: Chemotherapy treatments given after induction chemotherapy to further reduce the number of cancer cells.

Control group: In a randomized clinical trial, the group that receives standard treatment.

Cytokines: Secreted proteins that are important for interactions between immune cells and help orchestrate an immune response. Also called interleukins or lymphokines.

Cytoplasm: The part of the cell between the cell membrane and the nucleus.

Cytotoxic: Having the ability to kill a cell.

Dendritic cell: A type of antigen-presenting cell named for its long arms or dendrites that is important for initiating and controlling the overall immune response.

Disease-free survival: The length of time after treatment during which no cancer is found.

Disease progression: Cancer that continues to grow or spread.

DNA (deoxyribonucleic acid): A nucleic acid made of deoxyribose and four different types of bases or nucleotides (A, T, G, C) that pair with each other to form a linear, double-stranded DNA helix.

DNA methylation: A type of epigenetic mark. Changes in methylation affect gene expression; the more methylated a stretch of DNA, the less likely it is to be transcribed to RNA.

DNA repair: The process through which mutations in DNA are repaired.

DNA replication: The copying or duplication of a DNA molecule.

Drug resistance: The ability of a tumor cell to survive in the presence of drugs that are normally toxic.

Duration of response: The length of time between anticancer treatments where a patient's cancer shrinks, disappears, or remains stable.

Dysplasia: Condition in which cells proliferate excessively and appear abnormal in shape and orientation.

EGFR (epidermal growth factor receptor): The protein found on the surface of some cells and to which epidermal growth factor binds, causing the cells to divide.

Epidemiology: The study of the incidence, distribution, causes, and control of disease in a population.

Epigenetic: Having to do with the chemical attachments to DNA or the histone proteins around which it coils.

Epithelium: The most common cell of origin for human cancer.

Etiology: The study of the cause or origin of a disease.

Extensive-stage small cell lung cancer: Cancer that has spread outside the lung to other tissues in the chest or to other parts of the body.

Extracellular: Outside of the cell.

Extracellular matrix: A complex network of fiber-forming proteins interwoven in a hydrated polysaccharide gel that helps hold cells and tissues together and provides a meshwork for cells to migrate on and interact with one another.

First-line therapy: The first therapy given in the treatment for cancer.

Food and Drug Administration (FDA): An agency of the U.S. Department of Health and Human Services.

Free radical: An atom or group of atoms that has at least one unpaired electron and is therefore unstable and highly reactive.

Gamete: A reproductive cell that contains a single set of unpaired chromosomes (egg or sperm).

Gene: The functional and physical unit of heredity passed from parent to offspring.

Genetic: Inherited, or having to do with information that is passed from parents to offspring through genes in sperm and egg cells.

Genetics research: Research that focuses on how someone's genetic makeup can assist in the early detection, diagnosis, or treatment of cancer.

Genetic testing: Analyzing DNA to look for a genetic alteration that may indicate an increased risk for developing a specific disease or disorder.

Genome: The total genetic information that is stored in the chromosomes and that governs how an organism develops.

Genotype: The genetic makeup of an organism, or the combination of alleles located on homologous chromosomes that determines a specific characteristic or trait.

Germ line: Genetic material that is passed down through the gametes (sperm and egg).

Glioma: Cancer of nonneuronal brain cells.

Growth factors: Substances made by the body that function to regulate cell division and cell survival.

Hazard ratio: A summary of the difference between two survival curves, representing the reduction in the risk of death on treatment compared to control.

Histones: Proteins that bind DNA and help wrap it into tightly packaged chromatin.

Humanized antibody: An antibody that contains over 90% human material.

Hyperplasia: An overgrowth of cells.

Imaging: Tests that produce pictures of areas inside the body.

Immune response: The activity of the immune system against foreign substances called antigens.

Immune system: The complex group of organs and cells that defend the body against infection or disease.

Immunotherapy: Treatment to stimulate or restore the ability of the immune system to fight infection and disease.

Induction therapy: Treatment designed to be used as a first step toward shrinking the cancer and evaluating the response to drugs and other agents.

Infection: Up to one-third of the world's cancers are associated with infectious diseases caused by pathogens such as *Helicobacter pylori* (stomach cancer) or viruses such as the human papillomavirus (HPV) (cervical cancer), hepatitis B virus (HBV) (liver), Epstein-Barr virus (EBV) (lymphoma), and HIV (Kaposi's sarcoma).

Inflammation: The cellular and vascular changes that occur after tissue injury.

Informed consent: The process of providing all relevant information about the trial's purpose, risks, benefits, alternatives, and procedures to a potential participant, who then, consistent with his or her own interests and circumstances, makes an informed decision about whether to participate.

In situ cancer: A tumor that has not broken through the basement membrane to invade surrounding tissue.

Institutional review board (IRB): A board designed to oversee the research process in order to protect participant safety.

Intracellular: Inside of the cell.

Intrinsic resistance: A term used to describe tumor types that are drug resistant at presentation.

Investigational group: In a clinical trial, the group that receives the new agent being tested.

Investigational new drug (IND): A drug that the Food and Drug Administration (FDA) allows.

Irreversible toxicity: Side effects that are caused by toxic substances or something harmful to the body and do not go away.

Karyotype: A photomicrograph of an individual's complete set of chromosomes arranged in homologous pairs and ordered by size.

Kegel excercises: Exercises to strengthen certain muscles in the bottom of the pelvis.

Kidney: An organ that filters the blood to remove excess water, salt, and waste in the form of urine.

Knockout mouse: A mouse that has had both alleles of a particular gene replaced with an inactive allele, usually through homologous recombination.

K-*ras*: A gene that can mutate into a cancer accelerator and allow the growth of colorectal cancer.

Leukemia: Cancer of white blood cells.

Leukocytes: White blood cells.

Li-Fraumeni syndrome: A rare, inherited familial cancer syndrome characterized by tumors at multiple sites.

Ligand: A linking or binding molecule that binds to a specific complementary site on (forms a complex with) another molecule.

LTFU: Long-term follow-up.

Lumen: The inner cavity or open space of a tube or tubular organ (e.g., blood vessel, intestine).

Lymphedema: A condition that manifests itself by swelling or pain in the tissue around any area on the body that has received surgery or radiation.

Lymphocyte: White blood cells that kill viruses and defend against the invasion of foreign material.

Lymphoma: Cancer of lymphocytes or the lymphoid system.

Metastasize: To spread from one part of the body to another.

MHC (major histocompatibility complex) molecules: Highly variant (or polymorphic) molecules expressed on the surface of most cells that flag the cell as "self" and present foreign antigens to T-cells.

Mitosis: The process of cell division, resulting in the formation of two daughter cells that are genetically identical to the parent cell.

Multidrug resistance: The ability of a tumor cell to develop resistance to a number of structurally and functionally unrelated drugs, following exposure to a single agent.

Murine (mouse) antibody: The antibody derived solely from mouse proteins and viewed as foreign by the host (human) body.

Mutation: A heritable change in the sequence of the DNA.

Myeloma: Cancer of the bone marrow, especially the antibody-producing cells in the bone marrow.

Natural killer (NK) cell: A cytotoxic cell of the immune system that can recognize and destroy a large variety of virally infected cells and tumor cells.

Neoadjuvant: Initial treatment that is not the primary therapy (for instance, chemotherapy or radiation, prior to surgery).

Neoplasm: An abnormal new growth of tissues or cells.

Non-Hodgkin's lymphoma: A group of cancers of the lymphoid system.

Non-small cell lung cancer: A group of lung cancers that includes squamous cell carcinoma, adenocarcinoma, and large cell carcinoma.

Nucleotides: The building blocks of nucleic acids (DNA and RNA).

Nucleus: The part of the cell bounded by its own nuclear membrane that contains the chromosomes.

Oncogene: A mutated proto-oncogene that is locked into an active state and continuously stimulates unregulated cell growth and proliferation that leads to tumor development.

Oncogenesis: The process by which normal cells turn cancerous.

Oncology: The branch of medicine that deals with tumors, including the study of their development, diagnosis, treatment, and prevention.

Overall survival: The percentage of subjects in a study who have survived for a defined period of time.

Palliative treatment: Treatment aimed at the relief of pain and symptoms of disease but not intended to cure the disease.

Partial response: A decrease in the size of a tumor, or the regression by more than 50% but less than 100% of the extent of cancer in the body.

Pathogenesis: The development of disease.

P-glycoprotein: The first identified and best studied of the ABC drug transporters.

Pharmacogenetics: The study of the genetic factors that influence each person's reaction to a drug.

Phase I trial: Small groups of people with cancer are treated with a certain dose of a new agent that has already been extensively studied in the laboratory.

Phase II trial: Phase II trials continue to test the safety of the new agent and begin to evaluate how well it works against a specific type of cancer.

Phase III trial: Phase III studies are designed to answer research questions across the disease continuum.

Phase IV trial: Phase IV trials are used to evaluate the long-term safety and effectiveness of a treatment.

Phenotype: The observable physical and biochemical characteristics or traits of an organism, such as hair color, blood type, or the presence or absence of a disease.

Pivotal trial: A controlled trial to evaluate the safety and efficacy of a drug in patients who have the disease or condition to be treated.

Placebo-controlled: Refers to a clinical study in which the control patients receive a placebo, or inert medication or procedure.

Preclinical testing: A process in which scientists test promising new anticancer agents in the laboratory and in animal models.

Prevention trials: Trials involving healthy people who are at high risk for developing cancer.

Progression-free survival: The length of time during and after treatment that the cancer does not grow.

Progressive disease: Defined in clinical trials as tumor growth of more than 20% since treatment began.

Proteome: All of the proteins encoded by the genome and made by a person's cells and tissues.

Proteomics: The study of the networks of proteins within cells and tissues.

Protocol: A written plan that acts as a "recipe" for conducting a clinical trial.

Proto-oncogenes: A gene that promotes normal cell growth and differentiation.

p-value: A measure of probability that a difference between groups during an experiment happened by chance.

Quality of life: The overall enjoyment of life.

Radiotherapy: The use of high-energy radiation from x-rays, neutrons, and other sources to kill cancer cells and shrink tumors.

Randomization: A method used to prevent bias in research.

Randomized clinical trial: A study in which the participants are assigned by chance to separate groups that compare different treatments.

Receptor: Proteins that bind to their specific ligands with high specificity.

RECIST criteria: The standard criteria used for performance measurement in solid tumor clinical trials.

Recurrence: The return of cancer, at the same site as the original (primary) tumor or in another location, after the tumor had disappeared.

Refractory: Cancer that has not responded to treatment.

Relapse: The return of signs and symptoms of cancer after a period of improvement.

Remission: A decrease in or disappearance of signs and symptoms of cancer.

Response rate: The percentage of patients whose cancer shrinks more than 50% following treatment.

RNA (ribonucleic acid): A single-stranded nucleic acid containing ribose instead of deoxyribose.

Sarcoma: Cancer of muscles or connective tissues, such as bone, cartilage, and fat.

Second-line treatment: Treatment that is given after the cancer has not responded to a first course of therapy or a patient ceases the first line of therapy.

Sequelae: A pathological condition resulting from a disease, or a secondary consequence or result.

Side effects: Problems that may occur when treatment affects healthy tissues or organs.

Signal transduction: Usually refers to the series of steps that occur in the cell cytoplasm after a receptor has bound its ligand to communicate/transduce the signal to the cell nucleus.

Small cell lung cancer: A type of lung cancer in which the cells appear small and round when viewed under the microscope.

Somatic cell: All cells in the body except the gametes.

Squamous cell carcinoma: Cancer that begins in squamous cells, which are thin, flat cells resembling fish scales.

Stage: The extent of a cancer, especially whether the disease has spread from the original site to other parts of the body, i.e., from localized to other organs.

Standard treatment: A currently accepted and widely used treatment for a certain type of cancer, based on the results of past research.

Stroma: The nonmalignant host cells and extracellular matrix in which a tumor grows.

Survivin: An inhibitor of apoptosis protein. Is highly expressed in most cancers and associated with chemotherapy resistance, increased tumor recurrence, and shorter patient survival, making antisurvivin therapy an attractive cancer treatment strategy.

Survivorship: Physical, psychosocial, and economic issues of cancer, from diagnosis until the end of life.

Symptom deterioration: A deterioration of health status requiring discontinuation of treatment without objective evidence of disease progression.

Targeted therapy: A type of treatment that uses drugs or other substances to identify and attack specific cancer cells while limiting the affect on normal cells.

T-cell or T-lymphocyte: White blood cells responsible for generating cell-mediated immune responses.

Telomerase: An enzyme that rebuilds telomeres.

Telomeres: Special DNA sequences at the ends of each chromosome that grow shorter each time a cell divides.

Toxicity: Harmful side effects from an agent being tested.

Transcription: The synthesis of a single stranded, complementary RNA molecule from a DNA template in the cell nucleus.

Transcription factors: Proteins that bind to DNA and help initiate gene transcription.

Transformation: The process by which a normal cell undergoes a series of changes that cause it to become cancerous.

Transgenic mouse: A mouse that has had DNA artificially introduced into one or more of its cells.

Translation: The synthesis of a polypeptide chain (protein) from its mRNA template.

Translational research: Research that bridges the gap between laboratory research and its application to the clinic.

Translocation: An abnormal rearrangement of the DNA in a chromosome.

Tumor: An abnormal mass of tissue that results from excessive cell division.

Tumorigenesis: The process by which normal cells turn into a tumor.

Tumor suppressor gene: A gene that normally inhibits excessive cell proliferation.

Tyrosine kinases: A large group of enzymes important in cell growth, differentiation, and development.

Vaccine: A substance or group of substances meant to cause the immune system to respond to a tumor or to microorganisms, such as bacteria or viruses.

VEGF (vascular endothelial growth factor): A protein that is secreted by oxygen-deprived cells, such as cancerous cells.

Virus: A very small infectious agent that carries proteins and nucleic acids (DNA or RNA) in a protective coat.

White blood cells: A variety of cells that fight invading germs, infection, and allergy-causing agents. They are also called leukocytes.

Index

Printed and bound by CPI Group (UK) Ltd, Croydon, CR0 4YY

18/10/2024

01776261-0005